RSMeans

Square Foot Estimating Methods

Second Edition

Billy J. Cox
F. William Horsley

RS Means

Square Foot Estimating Methods

Second Edition

Billy J. Cox
F. William Horsley

Copyright 1996

R.S. MEANS COMPANY, INC.
CONSTRUCTION PUBLISHERS & CONSULTANTS

100 Construction Plaza
P.O. Box 800
Kingston, MA 02364-0800
(617) 585-7880

Southam
Construction
Information
Network

The editors for this book were John Ferguson and Suzanne Morris; the managing editor was Mary Greene. The production manager was Helen Marcella; the production coordinator was Marion Schofield. Composition was supervised by Karen O'Brien. The book and cover were designed by Norman Forgit.

Printed in the United States of America

10 9 8 7 6 5 4 3 2 1

Library of Congress Cataloging in Publication Data
ISBN 0-87629-418-2

Table of Contents

Foreword

Means Square Foot Estimating Methods is a practical guide to a widely used method of estimating the cost of building construction. Square foot estimating is normally done when only preliminary information on the project is available. At this stage there are usually no plans, and only basic requirements are known. The object of this book is to explain how the square foot method of estimating is used correctly and advantageously under these circumstances.

In this new edition, all references to *Means Square Foot Costs* and *Means Assemblies Cost Data* have been updated, along with all costs in the sample project estimate. In addition, a new chapter addresses the use of the UniFormat system of organizing costs, including the latest available information on the new, revised UniFormat II, and how it applies to better design and planning. This chapter was written by Robert P. Charette, P.E., CVS, a building economics consultant and by Anik Shooner, an architect with Menkes, Shooner, and Dagenais in Montréal. Mr. Charette is a member of both the CSI Formats Committee and the ASTM Building Economics Committee. Ms. Shooner is Vice President of Construction Specifications Canada, Montréal chapter.

For over 50 years, R.S. Means Company has been developing detailed cost information for projects, from small residential homes to complex buildings costing millions of dollars. Square foot estimates utilize the systems or assemblies format, based on the concept that the various trades and small items involved in each portion of a building can be grouped together into constructed elements called *assemblies.* Means draws upon its extensive detailed cost database to create these assemblies. Assemblies provide a way of obtaining a reliable cost estimate when complete contract drawings are not available, and in far less time than would be required to generate a detailed unit price estimate.

There is more than one source of square foot cost information available to the construction industry. For consistency in this book, the 1996 edition of *Means Assemblies Cost Data** is referenced throughout the examples. The line numbers, tables, and reference numbers mentioned in this book can be found in *Means Assemblies Cost Data.*

Care must be taken when using cost data from more than one source in preparing an estimate. While cost data taken from multiple sources can be used to estimate the cost of a single project when using the square foot method, those source projects must be identical in all important respects to the project being estimated.

* Note: The prices presented in Means Assemblies Cost Data *are averaged for 30 major cities in the United States, although in some cases they are obtained from a definite locality. This data has been accumulated from actual job costs in 1995; and material dealers' 1996 quotations have been combined with negotiated 1996 labor rates. Hundreds of contractors, subcontractors, and manufacturers have furnished cost information on their products.*

Variations in wage rates, labor efficiency, union restrictions, and material prices will result in local fluctuations. Local, regional, or national shortages of construction materials can severely influence material costs as well as cause considerable job delays with a corresponding increase in indirect job costs.

Sales tax is not included in material prices. Prices include the installing contractor's overhead and profit (O&P). The prices are those that would be quoted to either the general or prime contractor. Thus an allowance for the general or prime contractor's mark-up, overhead, supervision and management should be added to the prices in this book. The usual range for this item is 5% to 15%. A figure of 10% is the most typical allowance.

Part 1

Design and
Development

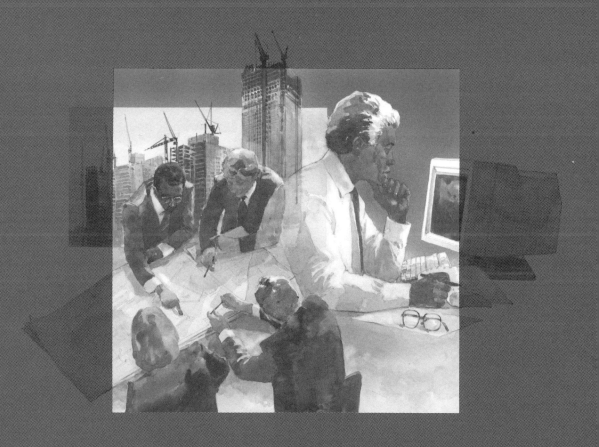

Introduction

A reliable cost estimate in the early stages of a construction project can be difficult to attain, yet that's when an estimate has its greatest value — during the design and development of a building. This part of the book shows how to obtain reliable square foot costs and arrive at a conceptual estimate. An assemblies square foot estimate is a tool that allows the estimator to analyze all the elements involved in a project and to assign specific systems to each area of construction. This type of estimate can be done when only basic building parameters are available.

Chapters 1 through 3 show how to create conceptual estimates for space planning, construction with cost limitations, and projects with size and site limitations. Chapters 4 through 10 explain the step-by-step procedures for assembling the loads and other basic elements, with each item explained in detail. Each element requires a specialized approach when initial decisions about costs are being made.

Square foot costs are not just numbers derived from a simple analysis of completed projects; they are resources that must be defined in relation to specific building components. To make these costs work, all tables and systems must be viewed with the specific loads and applications. This fast, logical approach to each estimate makes much more sense in the final analysis than does guesswork or poorly matched cost information.

After completing this section of the book, you will be ready to go on to the examples in Part 2 for a closer look at how the final cost figures are put together.

Chapter 1

Preliminary Estimating

What does it cost to construct an office building today? About $65 per square foot, give or take a few dollars depending on the location. Right?

In a nutshell, that type of guess is what square foot cost estimating is all about. It is common in the construction industry to think of costs in those terms, and a few general figures for basic building types tend to be the accepted standard at any given time. Most builders have in their minds these kinds of numbers based on their experience over the years in their type of construction. Other industry professionals, such as architects and appraisers, have their own basic feeling for what a house or office building is worth, depending on its size. Square foot cost estimating develops the specific techniques for determining the estimated construction cost of a building based on its size and selected other criteria.

Experience teaches that general square foot cost figures come from a number of sources:

- Past experience or cost figures taken from jobs of similar construction.
- Published costs for similar building types.
- Historical costs with appropriate mark-ups.
- Unit price estimates.
- Assemblies (or systems) estimates.
- Order of Magnitude estimates.

The next section defines four major types of estimates in order to put square foot estimating into context.

Four Types of Estimates

All final square foot costs are derived from the thousands of components that go into a building. The figures include the cost of the actual material used to build the structure, as well as the cost of installation. Other expenses such as delivery costs, and even miscellaneous items such as the cost of the contractor's overhead and profit requirements, must be included. Most buildings use similar materials or combinations of materials to create the whole. Installation is often accomplished using similar methods, with contractors using a similar type of labor force, either union or non-union. That is the basic idea behind estimating the cost of a construction project. Almost any two structures can be compared in relation to the basic elements they have in common. This idea is the cornerstone of estimating.

There are several different methods that an estimator can use to arrive at the estimated cost of a particular project. To better understand what an estimate is, take a look at the four different types of estimates and evaluate where each type of estimate fits into the overall estimating process. The four basic types are: *Unit Price; Assemblies or Systems; Square Foot (S.F.) and Cubic Foot (C.F.);* and *Order of Magnitude.* Each type is described in the following sections.

Unit Price Estimates

Unit price estimates involve a careful breakdown of all the elements that go into the building process. This type of estimate is assembled division by division. Each unit price is matched to a specific quantity. To complete a unit price estimate, it is necessary to know the specified or required prices for all material, equipment, and labor. In general, this type of estimate is made from working drawings and specifications, and can be done at various stages of completion of drawings. The accuracy of unit price estimates is generally + or −5%.

Because of the extreme level of detail involved, unit price estimating is a lengthy and costly process. The basic purpose of this type of estimate is to develop prices for a designed structure and to arrive at an estimated cost for construction. It is usually not economically feasible or even possible to use unit price estimating in the conceptual stages of a building construction project. Typically, unit price estimates are performed on a project that is being negotiated or competitively bid.

CSI Format

Cost manuals used for unit price estimates such as *Means Building Construction Cost Data* divide all data into 16 CSI divisions. These divisions are patterned after the MasterFormat adopted by the Construction Specifications Institute, which is widely used by the building construction industry. Unit price estimates follow the 16-division format and are assembled one division at a time. The main focus of the estimating process is the quantity takeoff, which is a detailed analysis of the material and equipment required to construct a project.

Extended Costs

To start a unit price estimate, the estimator examines the plans for the project and develops a quantity takeoff. The quantities are then matched to the appropriate unit prices. For instance, if a wall in a building requires 50 cubic yards of concrete, that quantity is multiplied by the appropriate unit price for concrete to produce the cost for the concrete in the wall. If the unit price for 3,000 psi concrete is $60.00 per cubic yard, the estimated cost of the concrete itself (the cost of the material only) would be $3,000. This price is called an "extended cost." Such costs can be calculated for any part of a construction project: material, equipment, labor, and eventually total project cost including overhead and profit. Totals of all these costs, organized along the lines of the CSI format, constitute the total unit price estimate.

Assemblies (or Systems) Estimates

Everyone from the designer to the ultimate user of a building finds it increasingly important to consider the cost relationship between major components in the structure. The assemblies approach to preliminary cost estimating involves the grouping of several units into components. It is based on the size of the structure and other parameters as they pertain to a particular site and a particular owner's requirements. The degree of accuracy of an assemblies estimate is generally + or −10%. (Note: Throughout the remainder of this text, the term "assemblies" will be used for this type of estimate.)

Breaking a Project Down into Components

One logical approach to developing an assemblies estimate is to break down the structure into several convenient components. Perhaps the most conventional approach is to take each component of the project in the order it is constructed. The list below provides an outline of how this approach works:

1. Foundations, including excavation and slab on grade.
2. Substructures, including excavation and slab on grade.
3. The superstructure: suspended floors, roof, and stairs (concrete, steel, wood, or a combination of the three).
4. Exterior closure: building envelope including walls, windows, and doors.
5. Roofing: related moisture protection and insulation.
6. Interior construction: partitions (fixed and movable), floors, ceilings, and decorating.
7. Conveying: elevators, escalators, dumbwaiters, etc.
8. Mechanical: plumbing, roof drains, heating, ventilation, air conditioning, and controls.
9. Electrical: service, power, and lighting.
10. General conditions.
11. Special construction: architectural equipment, furnishings, and special owner's requirements which, when added together, raise the square foot cost considerably.
12. Site work: excavation, trenching, roadways, and parking areas.

In many cases, it is possible to combine estimate types within the same project estimate. For example, a detailed unit price estimate may be necessary for the structural framing and exterior closure systems, while an assemblies estimate may be sufficient for the balance of the project estimate.

Using the Assemblies Approach

The conceptual stage of a building is no time for the restraints of individual trades and materials, which can impair good design. One must be free to think in terms of broad systems or whole elements of the structure. A designer can spend many weeks designing a building only to find, when the estimate is done, that the design and the budget are incompatible. This problem provided the impetus for the development of assemblies estimating.

It is not necessary for the designer to assign specific materials or detailed dimensions to every element in a design before arriving at a cost. Instead, it is possible to decide on systems in various combinations within basic, predetermined limitations, then determine the direction to be taken in each system to accommodate the project's budget, building codes, and the owner's special requirements.

Understanding the Assemblies Format

A close examination of a building system reveals that each item normally included in a unit price estimate finds a new slot in the assemblies estimate.

Unit price estimates conform to the 16 CSI MasterFormat divisions. These divisions have been developed to help maintain consistency in the classification of construction materials. They are particularly useful when comparisons must be made between construction projects that have little else in common. The unit price approach assigns costs for each piece of material, each equipment item, and all installation costs to a specific CSI division. An assemblies estimate essentially reshuffles the deck, reorganizing certain items that were formerly grouped into a single trade breakdown or division.

In the "UniFormat" division organization of the assemblies estimate, the various construction materials in an estimate end up being included in many different systems. Concrete is a good example. In a unit price estimate, all the concrete items are priced in the Concrete division of the estimate. In the assemblies estimate, concrete is found in the Foundation division, the Substructures division, the Site Work division, and possibly in the Superstructures division, depending on the structural system used. Conversely, other items formerly listed in separate trade breakdowns are now combined into one system.

This approach consolidates the many different elements that go into a complex system. A system is a concept that is easy to use, since it reflects the way builders think about the actual construction of a building. This makes it easier to get an overview of exactly what each part of the building costs. Interior partitions are a good case in point. In a unit price estimate, interior partitions are comprised of Division 6 (Wood & Plastics) and Division 9 (Finishes). In the assemblies estimate, these are combined in Division 6 (Interior Construction). This relocation of familiar items may seem confusing at first, but once the concept is understood, the resultant increase in estimating speed is well worth the initial familiarization required.

Making Trade-offs During Initial Design

Assemblies estimating is usually not a substitute for unit price estimating. It is normally done during the earlier conceptual stage before detailed plans and specifications have even been developed. Assemblies estimating enables the designer to bring the project within the owner's budget. During the initial design process, the designer is often forced to make important decisions and comparisons of each of the various systems. Some of the criteria may include:

- Price of each system
- Appearance and quality
- Story height
- Clear span
- Site complications and restrictions
- Thermal characteristics
- Life cycle costs
- Maintenance costs
- Acoustical characteristics
- Fireproofing characteristics
- Owner's special requirements in excess of code requirements

Organizing the Assemblies Estimate

When organizing an assemblies estimate, it is a good idea to gather all available information on the project. A typical preliminary estimate form that can be used to assemble project information is shown in Figure 1.1. At the start of the estimate, all blanks should be filled in as much as possible. Information can be gathered from:

- Code requirements
- Owner's requirements
- Preliminary assumptions
- Site inspection and investigation

The information as it is gathered will be analyzed based on many factors including:

- Budget restraints
- Architectural treatment
- Fireproofing requirements
- Energy considerations
- Zoning limitations
- Site characteristics
- Acoustical requirements
- Life cycle cost considerations
- Type of occupancy
- Environmental controls
- Thermal requirements
- Building height restriction

Assemblies Estimating Forms

The purpose of good organization is to save time, to make the estimating task as easy as possible, to provide a checklist, and to increase overall confidence in the estimate. A standard preprinted estimating form is the key to proper organization. Again, see Figure 1.1. The Preliminary Estimate form is flexible enough to be used for an entire building cost estimate, or it can be used to estimate the cost of only one or two divisions. It is not unusual to find preliminary estimates that have been priced out using a combination of square foot costs, assemblies costs, and unit prices.

In addition to the Preliminary Estimate form, special cost tables in books such as *Means Assemblies Cost Data* can be used as follows:

- To determine costs of individual footings, columns, etc.
- To develop cost comparisons of floor systems, roof systems, etc.
- To select materials from a worksheet and combine them to develop the system price (for an item like a wall or ceiling) from those materials chosen.
- To calculate the loads for structural, mechanical, and electrical components used to determine total foundation and superstructure costs.

Square Foot (S.F.) and Cubic Foot (C.F.) Estimates

Very early in the process of deciding whether or not to build, a rough idea of the cost of the project must be made. At this stage there are usually no plans; only basic requirements are known. At this point, the planning of a square foot or cubic foot estimate can be completed in less than an hour using square foot and cubic foot tables such as those found in *Means Assemblies Cost Data.* Estimates are usually within + or −15% of the final bid price, providing allowances are made for the following:

- Classification of the job (apartment, hotel, school, etc.).
- Type of owner (government, industrial, etc.).
- Location of project (using City Cost Indexes).

PRELIMINARY ESTIMATE

PROJECT	TOTAL SITE AREA
BUILDING TYPE	OWNER
LOCATION	ARCHITECT
DATE OF CONSTRUCTION	ESTIMATED CONSTRUCTION PERIOD
BRIEF DESCRIPTION	

TYPE OF PLAN	TYPE OF CONSTRUCTION
QUALITY	BUILDING CAPACITY

Floor			**Wall Areas**				
Below Grade Levels			Foundation Walls	L.F.		Ht.	S.F.
Area		S.F.	Frost Walls	L.F.		Ht.	S.F.
Area		S.F.	Exterior Closure		Total		S.F.
Total Area		S.F.	Comment				
Ground Floor			Fenestration			%	S.F.
Area		S.F.				%	S.F.
Area		S.F.	Exterior Wall			%	S.F.
Total Area		S.F.				%	S.F.
Supported Levels			**Site Work**				
Area		S.F.	Parking		S.F. (For		Cars)
Area		S.F.	Access Roads		L.F. (X		Ft. Wide)
Area		S.F.	Sidewalk		L.F. (X		Ft. Wide)
Area		S.F.	Landscaping		S.F. (		% Unbuilt Site)
Area		S.F.	**Building Codes**				
Total Area		S.F.	City		County		
Miscellaneous			National		Other		
Area		S.F.	**Loading**				
Area		S.F.	Roof	psf	Ground Floor		psf
Area		S.F.	Supported Floors	psf	Corridor		psf
Area		S.F.	Balcony	psf	Partition, allow		psf
Total Area		S.F.	Miscellaneous				psf
Net Finished Area		S.F.	Live Load Reduction				
Net Floor Area		S.F.	Wind				
Gross Floor Area		S.F.	Earthquake			Zone	
Roof			Comment				
Total Area		S.F.	Soil Type				
Comments			Bearing Capacity				K.S.F.
			Frost Depth				Ft.
Volume			**Frame**				
Depth of Floor System			Type		Bay Spacing		
Minimum		In.	Foundation, Standard				
Maximum		In.	Special				
Foundation Wall Height		Ft.	Substructure				
Floor to Floor Height		Ft.	Comment				
Floor to Ceiling Height		Ft.	Superstructure, Vertical			Horizontal	
Subgrade Volume		C.F.	Fireproofing		☐ Columns		Hrs.
Above Grade Volume		C.F.	☐ Girders	Hrs.	☐ Beams		Hrs.
Total Building Volume		C.F.	☐ Floor	Hrs.	☐ None		

Figure 1.1

- Relative size of the project (using size modifier).
- Construction period (accounting for future escalation of costs).

Square Foot Cost Estimates Table

Figure 1.2 shows a typical square foot (S.F.) and cubic foot (C.F.) cost table based on thousands of buildings constructed within the ten years prior to 1996 and adjusted to January 1, 1996 prices.

The table in Figure 1.2 is in ranges of 1/4, Median, and 3/4. For the 1/4 figures, three-quarters of the projects studied had higher cost requirements and 1/4 had lower cost requirements. For the median figures, one-half of the projects studied had higher cost requirements and one-half had lower. For the 3/4 figures, one-quarter had higher cost requirements and three-quarters had lower cost requirements. If nothing else about a project is known at this stage, it is advisable to use the median column to approximate the building size.

Figure 1.3 is a Space Planning Guide that can be used to approximate the square foot requirements of specified building types to determine the final building size.

Once the project size has been determined, the numbers in the tables are used to arrive at a representative cost per square foot for the appropriate type of building. Annually-updated tables are published in *Means Assemblies Cost Data.*

The Area Conversion Scale

The next decision to make is whether to price the building at the 1/4, Median, or 3/4 range, or at some other point. For this decision, analyze:

- The quality the owner requires.
- Unusual design or construction requirements.
- Extent of site work, installed equipment and furnishings.
- Extent of air conditioning, heating, and electrical requirements.

One factor that affects the square foot cost of buildings is the size. In general, for buildings with similar design and specifications in the same locality, the larger building has the lower square foot cost. This is mainly because of the decreasing contribution of the exterior walls plus the economy of scale that is usually achieved in larger buildings.

The Area Conversion Scale in Figure 1.4 provides factors to convert median costs for the typical size building to an adjusted cost for the particular size project. The Square Foot Base Size table in Figure 1.4 lists the median costs, the typical project size for the accumulated data, and the size range of the projects.

The size factor for the project is found by dividing the actual area by the typical project size for the building type. Enter the Area Conversion Scale in Figure 1.4 at the point of the appropriate size factor, and determine the appropriate cost multiplier on the cost modifier curve.

$$\text{Project Size Factor} = \frac{\text{Actual Project Area (S.F.)}}{\text{Typical Project Size (S.F.)}}$$

There are two stages in an estimate when square foot costs are useful. The first is at the conceptual stage when no details are available, where square foot costs make a useful starting point. The second stage is after the bids are in and the costs can be worked back into their appropriate units for information purposes. As soon as details become available in the project design, the square foot approach should be discontinued and the project priced by individual components.

14.1 000	S.F. & C.F. Costs			UNIT COSTS			% OF TOTAL		
		UNIT	1/4	MEDIAN	3/4	1/4	MEDIAN	3/4	
010	0010	APTMENTS Low Rise (1 to 3 story) R14.1 -010	S.F.	40.20	50.65	67.35				010
	0020	Total project cost	C.F.	3.61	4.74	5.95				
	0100	Site work	S.F.	3.34	4.81	7.60	6.30%	10.50%	13.90%	
	0500	Masonry		.73	1.87	3.18	1.50%	3.90%	6.50%	
	1500	Finishes		4.22	5.40	7.15	8.90%	10.70%	12.90%	
	1800	Equipment		1.31	1.99	2.96	2.70%	4.10%	6.30%	
	2720	Plumbing		3.13	4.03	5.06	6.70%	8.90%	10.10%	
	2770	Heating, ventilating, air conditioning		1.99	2.46	3.56	4.20%	5.60%	7.60%	
	2900	Electrical		2.32	3.08	4.19	5.20%	6.70%	8.40%	
	3100	Total: Mechanical & Electrical		6.95	8.50	10.90	15.90%	18.20%	22%	
	9000	Per apartment unit, total cost	Apt.	31,200	46,700	68,900				
	9500	Total: Mechanical & Electrical	"	5,700	8,400	12,100				
020	0010	APTMENTS Mid Rise (4 to 7 story) R14.1 -010	S.F.	52.80	63.90	77.90				020
	0020	Total project costs	C.F.	4.14	5.65	7.85				
	0100	Site work	S.F.	2.06	4.03	7.50	5.20%	6.70%	9.10%	
	0500	Masonry		3.23	4.52	6.70	5.20%	7.30%	10.50%	
	1500	Finishes		6.55	8.30	10.85	10.40%	11.90%	16.90%	
	1800	Equipment		1.66	2.38	3.12	2.80%	3.50%	4.40%	
	2500	Conveying equipment		1.20	1.48	1.77	2%	2.20%	2.60%	
	2720	Plumbing		3.12	4.93	5.40	6.20%	7.40%	8.90%	
	2900	Electrical		3.59	4.77	5.85	6.60%	7.20%	8.90%	
	3100	Total: Mechanical & Electrical		9.80	12.35	15.90	17.90%	20.10%	22.30%	
	9000	Per apartment unit, total cost	Apt.	38,400	58,300	67,400				
	9500	Total: Mechanical & Electrical	"	12,200	13,700	21,500				
030	0010	APTMENTS High Rise (8 to 24 story) R14.1 -010	S.F.	60.60	73	89.20				030
	0020	Total project costs	C.F.	4.98	6.90	8.45				
	0100	Site work	S.F.	1.85	3.55	4.96	2.50%	4.80%	6.10%	
	0500	Masonry		3.43	6.20	7.85	4.70%	9.60%	10.70%	
	1500	Finishes		6.55	8.40	9.60	9.30%	11.70%	13.50%	
	1800	Equipment		1.92	2.37	3.16	2.70%	3.30%	4.30%	
	2500	Conveying equipment		1.22	2.02	2.88	2.20%	2.70%	3.30%	
	2720	Plumbing		4.54	5.25	6.61	6.90%	9.10%	10.60%	
	2900	Electrical		4.15	5.20	7.10	6.40%	7.60%	8.80%	
	3100	Total: Mechanical & Electrical		12.25	14.85	18.35	18.20%	21.80%	23.90%	
	9000	Per apartment unit, total cost	Apt.	55,700	66,500	73,400				
	9500	Total: Mechanical & Electrical	"	13,500	15,500	16,800				
040	0010	AUDITORIUMS R14.1 -010	S.F.	61	85.30	110				040
	0020	Total project costs	C.F.	4.01	5.60	8				
	2720	Plumbing	S.F.	3.89	5.15	6.70	5.80%	7%	8.60%	
	2900	Electrical		4.96	7.05	9	6.70%	8.80%	10.90%	
	3100	Total: Mechanical & Electrical		10.10	13.65	23.70	14.40%	18.50%	23.60%	
050	0010	AUTOMOTIVE SALES R14.1 -010	S.F.	42.45	50.80	76.60				050
	0020	Total project costs	C.F.	3.12	3.56	4.63				
	2720	Plumbing	S.F.	2.26	3.68	4.09	4.70%	6.40%	7.80%	
	2770	Heating, ventilating, air conditioning		3.26	4.99	5.40	6.30%	10%	10.30%	
	2900	Electrical		3.74	5.60	6.45	7.40%	9.90%	12.30%	
	3100	Total: Mechanical & Electrical		7.90	11.55	15.10	16.60%	19.10%	27%	
060	0010	BANKS R14.1 -010	S.F.	91.05	113	143				060
	0020	Total project costs	C.F.	6.60	8.85	11.65				
	0100	Site work	S.F.	9.05	16.60	24.65	7%	13.80%	17.50%	
	0500	Masonry		4.60	7.65	16.85	2.90%	5.80%	11.30%	
	1500	Finishes		7.75	10.55	13.55	5.50%	7.60%	9.90%	
	1800	Equipment		3.63	7.55	16.85	3.20%	8.20%	12.50%	
	2720	Plumbing		2.89	4.10	6	2.80%	3.90%	4.90%	
	2770	Heating, ventilating, air conditioning		5.65	7.30	9.80	4.90%	7.10%	8.50%	

Figure 1.2

Order of Magnitude Estimates

The Order of Magnitude estimate can be completed with minimal information. It could be loosely described as an educated guess for advanced planning of projects. The accuracy of Order of Magnitude estimates is + or −20%.

Order of Magnitude costs are defined in relation to the usable units that have been designed for a facility. If, for example, a hospital administrator is planning to enlarge a hospital, he or she needs to know the projected cost per bed. Tables such as the one in Figure 1.5 indicate usable unit costs for several types of structures. This table also includes square foot and cubic foot costs. If an estimator knows the quantity of beds in a proposed hospital (or the number of apartments in an apartment building, or parking stalls in a garage) the cost of the structure can be estimated.

Order of Magnitude estimating is discussed in more detail in Chapter 2.

The figures in the table below indicate typical ranges in square feet as a function of the "occupant" unit. This table is best used in the preliminary design stages to help determine the probable size requirement for the total project.

Unit Gross Area Requirements

Building Type	Unit	Gross Area in S.F.		
		1/4	Median	3/4
Apartments	Unit	660	860	1,100
Auditorium & Play Theaters	Seat	18	25	38
Bowling Alleys	Lane		940	
Churches & Synagogues	Seat	20	28	39
Dormitories	Bed	200	230	275
Fraternity & Sorority Houses	Bed	220	315	370
Garages, Parking	Car	325	355	385
Hospitals	Bed	685	850	1,075
Hotels	Rental Unit	475	600	710
Housing for the elderly	Unit	515	635	755
Housing, Public	Unit	700	875	1,030
Ice Skating Rinks	Total	27,000	30,000	36,000
Motels	Rental Unit	360	465	620
Nursing Homes	Bed	290	350	450
Restaurants	Seat	23	29	39
Schools, Elementary	Pupil	65	77	90
Junior High & Middle		85	110	129
Senior High		102	130	145
Vocational		110	135	195
Shooting Ranges	Point		450	
Theaters & Movies	Seat		15	

Figure 1.3

Square Foot Project Size Modifier

One factor that affects the S.F. cost of a particular building is the size. In general, for buildings built to the same specifications in the same locality, the larger building will have the lower S.F. cost. This is due mainly to the decreasing contribution of the exterior walls plus the economy of scale usually achievable in larger buildings. The Area Conversion Scale shown below will give a factor to convert costs for the typical size building to an adjusted cost for the particular project.

The Square Foot Base Size lists the median costs, most typical project size in our accumulated data and the range in size of the projects.

The Size Factor for your project is determined by dividing your project area in S.F. by the typical project size for the particular Building Type. With this factor, enter the Area Conversion Scale at the appropriate Size Factor and determine the appropriate cost multiplier for your building size.

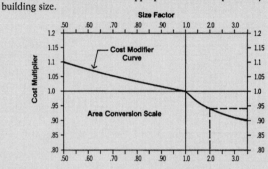

Example: Determine the cost per S.F. for a 100,000 S.F. Mid-rise apartment building.

$$\frac{\text{Proposed building area} = 100,000 \text{ S.F.}}{\text{Typical size from below} = 50,000 \text{ S.F.}} = 2.00$$

Enter Area Conversion scale at 2.0, intersect curve, read horizontally the appropriate cost multiplier of .94. Size adjusted cost becomes .94 x $63.90 = $60.05 based on national average costs.

Note: For Size Factors less than .50, the Cost Multiplier is 1.1
For Size Factors greater than 3.5, the Cost Multiplier is .90

Building Type	Median Cost per S.F.	Typical Size Gross S.F.	Typical Range Gross S.F.	Building Type	Median Cost per S.F.	Typical Size Gross S.F.	Typical Range Gross S.F.
Apartments, Low Rise	$ 50.65	21,000	9,700.00 - 37,200	Jails	$148.00	13,700	7,500 - 28,000
Apartments, Mid Rise	63.90	50,000	32,000 - 100,000	Libraries	90.40	12,000	7,000 - 31,000
Apartments, High Rise	73.00	310,000	100,000 - 650,000	Medical Clinics	86.80	7,200	4,200 - 15,700
Auditoriums	85.30	25,000	7,600 - 39,000	Medical Offices	80.90	6,000	4,000 - 15,000
Auto Sales	50.80	20,000	10,800 - 28,600	Motels	62.10	27,000	15,800 - 51,000
Banks	113.00	4,200	2,500 - 7,500	Nursing Homes	83.10	23,000	15,000 - 37,000
Churches	76.25	9,000	5,300 - 13,200	Offices, Low Rise	68.20	8,600	4,700 - 19,000
Clubs, Country	75.90	6,500	4,500 - 15,000	Offices, Mid Rise	72.10	52,000	31,300 - 83,100
Clubs, Social	73.80	10,000	6,000 - 13,500	Offices, High Rise	89.95	260,000	151,000 - 468,000
Clubs, YMCA	77.65	28,300	12,800 - 39,400	Police Stations	114.00	10,500	4,000 - 19,000
Colleges (Class)	99.05	50,000	23,500 - 98,500	Post Offices	84.05	12,400	6,800 - 30,000
Colleges (Science Lab)	140.00	45,600	16,600 - 80,000	Power Plants	642.00	7,500	1,000 - 20,000
College (Student Union)	110.00	33,400	16,000 - 85,000	Religious Education	70.60	9,000	6,000 - 12,000
Community Center	79.15	9,400	5,300 - 16,700	Research	117.00	19,000	6,300 - 45,000
Court Houses	106.00	32,400	17,800 - 106,000	Restaurants	102.00	4,400	2,800 - 6,000
Dept. Stores	47.10	90,000	44,000 - 122,000	Retail Stores	50.00	7,200	4,000 - 17,600
Dormitories, Low Rise	77.35	24,500	13,400 - 40,000	Schools, Elementary	72.35	41,000	24,500 - 55,000
Dormitories, Mid Rise	98.00	55,600	36,100 - 90,000	Schools, Jr. High	74.90	92,000	52,000 - 119,000
Factories	45.45	26,400	12,900 - 50,000	Schools, Sr. High	75.00	101,000	50,500 - 175,000
Fire Stations	81.20	5,800	4,000 - 8,700	Schools, Vocational	75.20	37,000	20,500 - 82,000
Fraternity Houses	69.60	12,500	8,200 - 14,800	Sports Arenas	58.35	15,000	5,000 - 40,000
Funeral Homes	87.65	7,800	4,500 - 11,000	Supermarkets	50.10	20,000	12,000 - 30,000
Garages, Commercial	55.55	9,300	5,000 - 13,600	Swimming Pools	92.55	13,000	7,800 - 22,000
Garages, Municipal	70.65	8,300	4,500 - 12,600	Telephone Exchange	135.00	4,500	1,200 - 10,600
Garages, Parking	27.80	163,000	76,400 - 225,300	Theaters	75.20	10,500	8,800 - 17,500
Gymnasiums	72.95	19,200	11,600 - 41,000	Town Halls	80.35	10,800	4,800 - 23,400
Hospitals	139.00	55,000	27,200 - 125,000	Warehouses	33.60	25,000	8,000 - 72,000
House (Elderly)	69.10	37,000	21,000 - 66,000	Warehouse & Office	38.75	25,000	8,000 - 72,000
Housing (Public)	63.25	36,000	14,400 - 74,400				
Ice Rinks	72.20	29,000	27,200 - 33,600				

Figure 1.4

141 | S.F., C.F. and % of Total Costs

			UNIT	UNIT COSTS			% OF TOTAL			
14.1 000 | S.F. & C.F. Costs				1/4	MEDIAN	3/4	1/4	MEDIAN	3/4	
390	2730	Heating & ventilating	R14.1 -010	S.F.	3.30	4.54	5.50	5.20%	7.20%	9.50%	390
	2900	Electrical			3.26	5.10	7.15	7.10%	9%	11.10%	
	3100	Total: Mechanical & Electrical		↓	6.45	12.80	18.50	15.40%	20.20%	26%	
400	0010	GARAGES, MUNICIPAL (Repair)	R14.1 -010	S.F.	53.45	70.65	99.50				400
	0020	Total project costs		C.F.	3.33	4.17	6.90				
	0500	Masonry		S.F.	5	9.70	14.95	5.90%	10%	15.50%	
	2720	Plumbing			2.39	4.59	8.65	3.50%	6.60%	7.90%	
	2730	Heating & ventilating			2.85	4.68	7.90	4.90%	7.20%	11.60%	
	2900	Electrical			3.93	5.85	8.90	6.30%	8%	11.70%	
	3100	Total: Mechanical & Electrical		↓	10.35	18.45	30.50	15.20%	25.50%	32.70%	
410	0010	GARAGES, PARKING	R14.1 -010	S.F.	20.05	27.80	45.65				410
	0020	Total project costs		C.F.	1.95	2.51	3.88				
	2720	Plumbing		S.F.	.53	.91	1.40	2.20%	3.30%	3.80%	
	2900	Electrical			.85	1.23	1.89	4.20%	5.40%	6.30%	
	3100	Total: Mechanical & Electrical		↓	1.47	1.98	3.29	6.80%	8.70%	9.50%	
	3200										
	9000	Per car, total cost		Car	8,600	9,900	13,800				
430	0010	GYMNASIUMS	R14.1 -010	S.F.	57.45	72.95	93.10				430
	0020	Total project costs		C.F.	2.88	3.67	4.76				
	1800	Equipment		S.F.	1.21	2.47	5.04	2%	3.30%	6.70%	
	2720	Plumbing			3.65	4.51	5.60	5.40%	7.20%	8.30%	
	2770	Heating, ventilating, air conditioning			3.92	6	12.05	9%	11%	22.60%	
	2900	Electrical			4.44	5.50	7.45	6.70%	8.20%	10.70%	
	3100	Total: Mechanical & Electrical		↓	8.30	13.15	19.25	15.70%	19.60%	26.30%	
	3500	See also division 114-801 & 114-805									
460	0010	HOSPITALS	R14.1 -010	S.F.	118	139	203				460
	0020	Total project costs		C.F.	8.60	10.60	14.65				
	1800	Equipment		S.F.	2.87	5.30	9.10	1.60%	3.80%	5.90%	
	2720	Plumbing			9.75	12.90	17.75	7.50%	9.10%	10.90%	
	2770	Heating, ventilating, air conditioning			14	18.55	25.60	8.40%	14.40%	17%	
	2900	Electrical			11.90	15.75	24.40	9.80%	11.70%	14.30%	
	3100	Total: Mechanical & Electrical		↓	33	44.30	69.20	26.10%	33.50%	39.80%	
	9000	Per bed or person, total cost		Bed	32,700	63,300	81,200				
	9900	See also division 117-001									
480	0010	HOUSING For the Elderly	R14.1 -010	S.F.	54.75	69.10	85.55				480
	0020	Total project costs		C.F.	3.95	5.50	6.95				
	0100	Site work		S.F.	3.79	6.10	8.65	5%	8.20%	12%	
	0500	Masonry			1.30	5.95	9.10	2.10%	7.40%	12.20%	
	1800	Equipment			1.32	1.81	2.87	1.80%	3.20%	4.30%	
	2510	Conveying systems			1.33	1.78	2.42	1.80%	2.20%	2.80%	
	2720	Plumbing			4	5.40	7.10	8.10%	9.70%	10.70%	
	2730	Heating, ventilating, air conditioning			1.92	2.88	4.18	3.20%	5.60%	7.10%	
	2900	Electrical			4.07	5.45	7.30	7.40%	8.70%	10.50%	
	3100	Total: Mechanical & Electrical		↓	9.90	13.80	17.25	18.20%	21.30%	24.70%	
	9000	Per rental unit, total cost		Unit	49,900	58,700	65,500				
	9500	Total: Mechanical & Electrical		"	9,550	11,700	14,200				
500	0010	HOUSING Public (Low Rise)	R14.1 -010	S.F.	46.90	63.25	82.80				500
	0020	Total project costs		C.F.	3.64	5.05	6.35				
	0100	Site work		S.F.	5.45	8	13.35	8%	11.60%	16.10%	
	1800	Equipment			1.23	2.01	3.31	2.10%	2.90%	4.60%	
	2720	Plumbing			3.11	4.37	5.50	6.80%	9%	11.50%	
	2730	Heating, ventilating, air conditioning		↓	1.66	3.22	3.44	4.20%	6%	6.40%	

Figure 1.5

Time and Cost of Estimating

Figure 1.6 indicates the relative time and cost requirements for various types of estimates. The final accuracy required must be weighed against the cost of creating the estimate. The "best" estimate depends on the purpose of the estimate and is a function of how much specific design detail is available when the estimate is done.

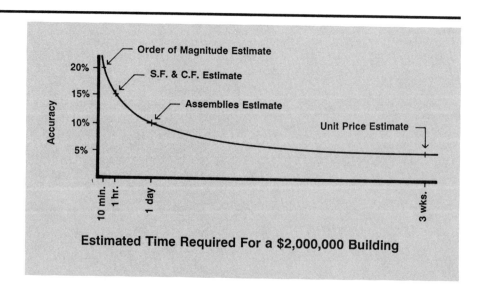

Estimated Time Required For a $2,000,000 Building

Figure 1.6

Chapter 2

Order of Magnitude Estimates

Because Order of Magnitude estimates are used for advance planning of projects, they are considered first in this part of the book. The Order of Magnitude estimate can be completed with a minimum of information. The proposed use and size of the planned structure should be known and may be the only requirement. The "units" can be very general and need not be well defined.

Using Historical Costs

As mentioned in Chapter 1, an Order of Magnitude estimate lists historical costs per usable unit of facility. For example, assume that a developer would like an approximate cost for financial planning on a proposed 20-story apartment building with 200 apartments. Figure 1.2 shows the historical cost per apartment.

A quick look at the table reveals a median historical cost per apartment of $46,700.

> 200 apartments × $46,700/apartment = $9,340,000

Space Planning

To carry the estimate one step further, use the Space Planning Guide shown in Figure 1.3, which shows the typical gross square footage for a high-rise apartment unit to be 925 square feet.

> 200 apartments × 925 S.F./apartment = 185,000 S.F.

Project Size Modifiers

The Square Foot Base Size chart in Figure 1.4 shows the typical high-rise apartment building to be 310,000 square feet.

Dividing the proposed building area of 185,000 S.F. by 310,000 S.F. yields a size factor of 0.6.

Enter the Area Conversion Scale in Figure 1.4 with 0.6. Project vertically to intersect the line on the graph. Now, project horizontally. The figure for the cost multiplier is 1.08.

> $9,340,000 × 1.08 = $10,087,200

Costs may also be modified by a City Cost Index, as published in Means' annual cost data publications. Any cost, when totaled and multiplied by the appropriate City Cost Index, will result in a total cost for an entire building that will be close to the actual cost for construction at a specific location in the United States and Canada. The relative index for material, labor, and total construction costs is listed for the 305 largest U.S. and Canadian cities.

Conclusion

Order of Magnitude and square foot estimates may be used to develop preliminary costs only if certain limitations are taken into account. Always consider the following facts when using Order of Magnitude and square foot estimates:

- Neither type uses cost figures that reflect floor and roof loading, structural systems, clear spans, or column spacing. All of these items can be important cost factors.
- When these estimates are assembled, the materials used for the exterior closure system, as well as the extent and type of sash and glazing, are not defined. System selection can affect cost here as well.
- Neither type makes allowances for the type of heating and cooling system, the extent of plumbing requirements, and the need for fire protection sprinklers. Once again, these are important cost considerations.
- Some building square foot costs include site work, and others do not. If possible, determine whether or not this cost is part of the figures being used for the estimate.

Always carefully evaluate the parameters of an Order of Magnitude or square foot estimate to ensure its validity.

The next chapter explains how square foot estimates are put together and includes examples of how they are used in the design stage of various projects.

Chapter 3

Square Foot and Cubic Foot Estimates

As shown in Figure 1.6, "Estimated Time To Complete an Estimate," a square foot estimate takes more time to complete than an Order of Magnitude estimate. This is because in many instances the required square footage of the building must be developed or determined.

The examples in this chapter illustrate four possible cases in which a square foot estimate is used to advantage during the development stage of the building design. All related square foot and cubic foot costs are included, along with the necessary tables.

Example One: Three-Story Office Building

A developer has a site and potential rental clients. The developer would like a price for a three-story office building with approximately 22,000 square feet of area to lease. A 40,000-square-foot level site with 160' of frontage in an office park is the probable location. A suitable number of parking spaces and a certain amount of landscaping are desirable.

What is a reasonable size in square feet for this structure, given all the factors? What would the cost of such a structure be, based on the total square feet and cubic feet?

Step One: Develop the Building's First Floor, or "Footprint"

Use the table in Figure 3.1 to determine the Gross to Net Ratio for an office building: 135%. The additional 35% is the non-rentable area required for hallways, stairwells, toilet rooms, exterior walls, etc.

22,000 S.F. of Rental Space × 1.35 = 29,700 S.F. Round to: 30,000 S.F.

$$\frac{30,000 \text{ S.F.}}{3 \text{ stories}} = 10,000 \text{ S.F./floor}$$

Base the estimate on a square building.

Find the square root of 10,000: 100' × 100' = 10,000

Step Two: Sketch the Site

See Figure 3.2.

Step Three: Develop Square Foot and Cubic Foot Costs

Square Foot Cost: Use the table in Figure 1.5: $68.20/S.F.

30,000 S.F. × $68.20/S.F. = $3,046,000

Note: This cost includes site work.

Use Size Modifier (see Figure 1.4).

$$\frac{30,000}{8,600} = 3.5 \text{ Size Factor}$$

Cost Multiplier = 0.90

Modified Cost

$2,046,000 × .9   = $1,841,400

Add the architect's fee.

Use Figure 3.3: 7.3%
.073 × 1,841,000   = + 134,420
Total Square Foot Cost: $1,975,820

Cubic Foot Cost

Assume: 9'-0" Ceilings
           12'-0" Floor to Floor

Use Figure 2.1: $5.40/C.F.

3 Floors × 10,000 S.F. × 12' × $5.40/C.F. = $1,944,000

Add the architect's fee.

Use Figure 3.3: 7.3%   + 141,910
Total Cubic Foot Cost $2,085,910

Floor Area Ratios

Table below lists commonly used gross to net area and net to gross area ratios expressed in % for various building types.

Building Type	Gross to Net Ratio	Net to Gross Ratio	Building Type	Gross to Net Ratio	Net to Gross Ratio
Apartment	156	64	School Buildings (campus type)		
Bank	140	72	Administrative	150	67
Church	142	70	Auditorium	142	70
Courthouse	162	61	Biology	161	62
Department Store	123	81	Chemistry	170	59
Garage	118	85	Classroom	152	66
Hospital	183	55	Dining Hall	138	72
Hotel	158	63	Dormitory	154	65
Laboratory	171	58	Engineering	164	61
Library	132	76	Fraternity	160	63
Office	135	75	Gymnasium	142	70
Restaurant	141	70	Science	167	60
Warehouse	108	93	Service	120	83
			Student Union	172	59

The gross area of a building is the total floor area based on outside dimensions.

The net area of a building is the usable floor area for the function intended and excludes such items as stairways, corridors and mechanical rooms. In the case of a commercial building, it might be considered as the "leasable area."

Figure 3.1

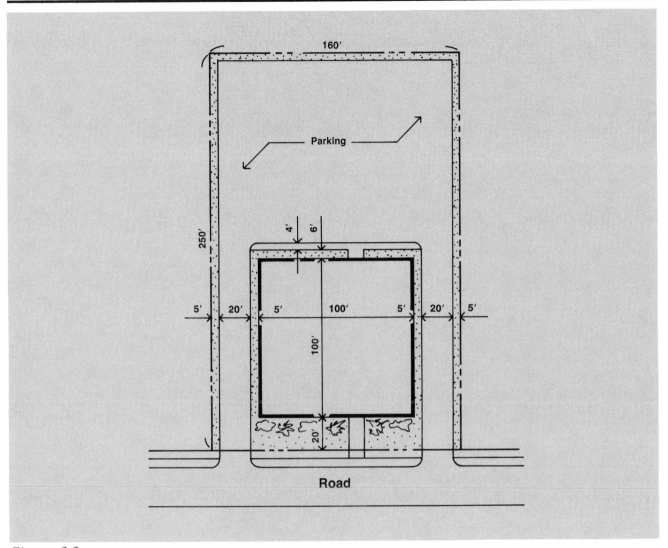

Figure 3.2

Architectural Fees

Tabulated below are typical percentage fees by project size, for good professional architectural service. Fees may vary from those listed depending upon degree of design difficulty and economic conditions in any particular area.

Rates can be interpolated horizontally and vertically. Various portions of the same project requiring different rates should be adjusted proportionately. For alterations, add 50% to the fee for the first $500,000 of project cost and add 25% to the fee for project cost over $500,000.

Architectural fees tabulated below include Engineering Fees.

Building Types	Total Project Size in Thousands of Dollars						
	100	250	500	1,000	5,000	10,000	50,000
Factories, garages, warehouses, repetitive housing	9.0%	8.0%	7.0%	6.2%	5.3%	4.9%	4.5%
Apartments, banks, schools, libraries, offices, municipal buildings	11.7	10.8	8.5	7.3	6.4	6.0	5.6
Churches, hospitals, homes, laboratories, museums, research	14.0	12.8	11.9	10.9	8.5	7.8	7.2
Memorials, monumental work, decorative furnishings	–	16.0	14.5	13.1	10.0	9.0	8.3

Figure 3.3

Example Two: Two-Story Office Building, Limited Cost

A client wants to construct a two-story office building, and has limited the project cost to $1,000,000. How big a facility can be built within this budget?

Step One: Establish Square Foot Cost

Consult Figure 1.2:	Office Building (Low Rise)
Median Cost	$68.20
Architect's Fee (Figure 3.3): 7.3%	+4.98
	$73.18

$1,000,000/$73.18 = 13,660 S.F.

Step Two: Modify for Size

Use the information in Figure 1.4.

$$\frac{18,790 \text{ S.F.}}{8,600} = 2.18$$

Cost Multiplier = 0.95

$$\frac{\$1,000,000}{73.18 \times .95} = 14,380 \text{ S.F.}$$

Figure 3.4 is a sketch of a two-story office building with 14,380 S.F. of floor area.

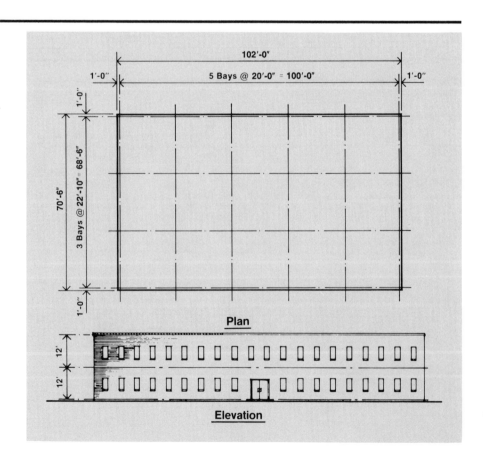

Figure 3.4

Example Three: Three-Story Office Building, Limited Site

A developer requires that as much building as possible be constructed on a limited site. What is the ideal dimension for the building footprint?

Also, what would be the total project cost of the building based on that ideal dimension, taking into account certain criteria as outlined below?

- Restrictions: Local building codes
- Height: 3 floors above grade (or 45')
- Parking: .66 stalls per occupant
- Landscaping: 20% of site

Step One: Develop Total Area Available for Building and Parking

Site Area:	40,000 S.F.
Required Landscaping (20%)	− 8,000 S.F.
	32,000 S.F.
Walks, Entrances, etc. (Assume 2%)	− 800 S.F.
Available for Building and Parking:	31,200 S.F.

Consult Figure 3.5. Minimum occupancy for an office building is 100 S.F. per person.

Consult Figure 3.1. Floor area ratio for an office building: 135%

100 S.F. × 1.35 = 135 S.F. of building per person

Parking Stalls: Use Figure 3.6: 390 S.F. each including room for parking and maneuvering.

.66 stalls per occupant × 390 S.F. = 260 S.F. per occupant

Occupancy Determinations

Description		S.F. Required per Person		
		BOCA	SBC	UBC
Assembly Areas	Fixed Seats	**	6	7
	Movable Seats		15	15
	Concentrated	7		
	Unconcentrated	15		
	Standing Space	3		
Educational	Unclassified			
	Classrooms	20	40	20
	Shop Areas	50	100	50
Institutional	Unclassified		125	
	In-Patient Areas	240		
	Sleeping Areas	120		
Mercantile	Basement	30	30	20
	Ground Floor	30	30	30
	Upper Floors	60	60	50
Office		100	100	100

BOCA = Building Officials & Code Administrators
SBC = Southern Building Code
UBC = Uniform Building Code

** The occupancy load for assembly area with fixed seats shall be determined by the number of fixed seats installed.

Figure 3.5

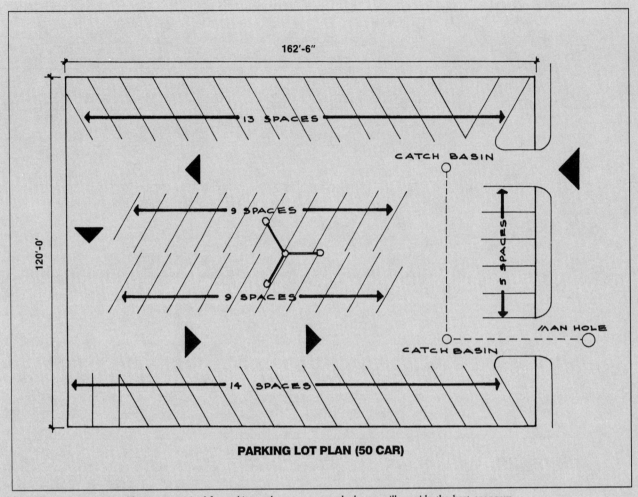

PARKING LOT PLAN (50 CAR)

Preliminary Design Data: The space required for parking and maneuvering is between 300 and 400 S.F. per car, depending upon engineering layout and design.

Ninety degree (90°) parking, with a central driveway and two rows of parked cars, will provide the best economy.

Diagonal parking is easier than 90° for the driver and reduces the necessary driveway width, but requires more total space.

Figure 3.6

Step Two: Develop the Area Required for Parking and the Building Footprint

Let "U" equal the number of persons per floor

Parking area		Footprint area
(3 floors × 260 S.F. × U)	+	(135 × U) = 31,200 S.F.
Solving for U		
915 U = 31,200		U = 34 persons per floor

Parking Area: 3 Floors × 34 persons × 260 S.F. = 26,520 S.F.
Building Footprint: 34 persons × 135 S.F. = 4,590 S.F.

Assume a building 3 bays long x 2 bays wide with the entrance at the center of the building and approximately 25'-0" column spacing.

Front: 3 × 25'-0" = 75'-0" + 2 × 8" = 76'-4"
Side: 2 × 30'-0" = 60'-0" + 2 × 8" = 61'-4"
Footprint area: 76.33 × 61.33 = 4,681 S.F.
Building area: 3 floors = 4,681 S.F. × 3 = 14,040 S.F.

Sketch the building floor plan and elevation as shown on Figure 3.7.

Summary:

Building Footprint	4,681 S.F.
Parking	26,520 S.F.
Landscape	8,000 S.F.
Walks and Miscellaneous	800 S.F.
Total	40,001 S.F.

Square Foot Cost:

14,040 S.F. × $68.20/S.F.	= $ 957,528
Use Size Modifier (.96)	− 919,227
Architect's fee (7.3%)	+ 67,103
Total	$ 986,330

Cubic Foot Cost:

12' Floor to Floor	
14,040 S.F. × 12' × $5.40/C.F.	= $ 909,792
Architect's fee (7.3%)	+ 66,415
Total	$ 976,207

Example Four: Two-Story, 100-Unit Motel

A developer needs to know how much it would cost to build a 100-unit motel. What would be a reasonable estimate of the cost based on three types of estimates (square foot, cubic foot, and Order of Magnitude)?

Assume the following requirements:

- 100-Unit Motel
- 2 Stories
- Includes: restaurant, lobby, and shops

Step One: Determine the Building Area Required for the Project

Use Figure 1.3 to determine the appropriate number of square feet per unit:

100 units × 450 S.F. per unit = 45,000 S.F.

Use the information in Figure 3.1 to develop a Gross to Net Ratio: 158%.

Determine an average room size:

450 / 1.58 = 284 S.F.

Approximate Room Size: 12'-0" x 24'-0".

Step Two: Sketch the Building

See Figure 3.8. It is assumed that 13'-0" center-to-center partition spacing is used to separate rooms. A 6'-0" corridor is also included. To service the rooms, separate the wings with an area for a pool and landscaping.

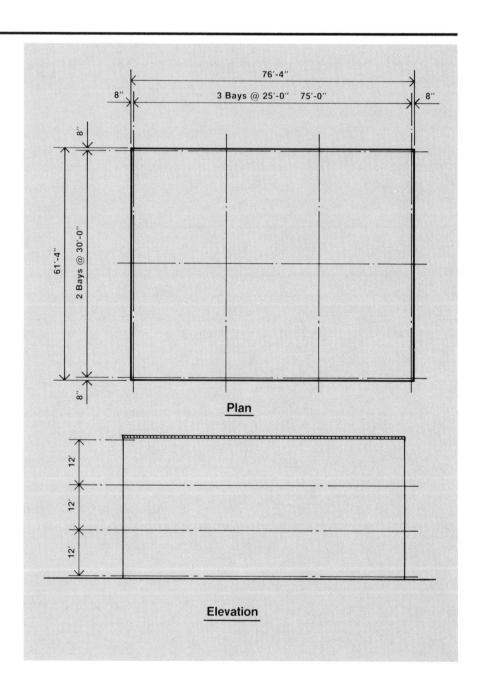

Plan

Elevation

Figure 3.7

26

Step Three: Determine the Various Areas Involved

Gross Area:	45,000 S.F.
Motel Area from the Sketch:	
2 × 169 × 54 × 2 floors	− 36,504 S.F.
Service Area:	8,496 S.F.

From the sketch, the service area length can be determined by adding the room plus corridor width on each wing, plus the distance between the wings.

$$(2 \times 30') + 50' = 110'$$

$$\frac{8{,}496/\text{S.F.}}{110'} = 77'$$

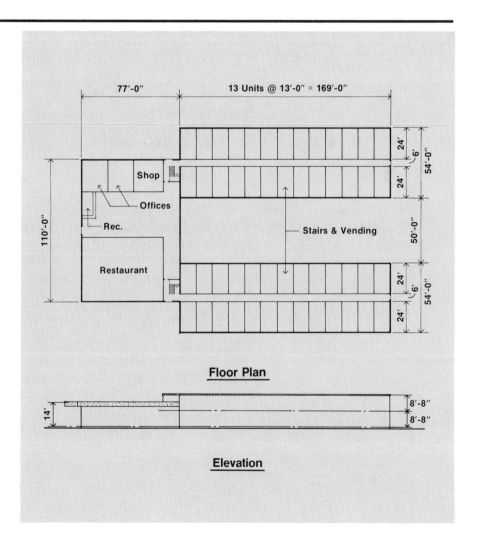

Floor Plan

Elevation

Figure 3.8

Step Four: Develop a Square Foot Cost
Use Figure 3.9 to develop a square foot cost: $62.10/S.F.

44,974 S.F. × $62.10/S.F.	= $2,792,885
Cost Modifier: (.96 × 2,792,885)	− 111,715
	$2,681,170
Architect's fee: 5.6%	+ 150,145
Total	$2,831,315

Step Five: Develop a Cubic Foot Cost

Motel	Res. & Office	Stairs
[4 × 169' × 54' × 8.67'] +	[110 × 77 × 14] +	[2 × 20 × 10 × 3.33] = 436,400 C.F.

436,400 C.F. × $5.20/C.F.	= $2,269,280
Architect's Fee: 5.6%	+ 127,080
Total (No Site Work Included):	$2,396,360

Step Six: Develop an Order of Magnitude Estimate
Use Figure 3.9 to develop an Order of Magnitude estimate.

100 Units × $31,500 = $3,150,000

Step Seven: Compare the Square Foot and Order of Magnitude Estimates (No Architect's Fee)

Square Foot Estimate:	$2,831,315
Order of Magnitude:	3,150,000
Difference between the two:	$318,685 (10%)

Conclusion

Order of Magnitude, square foot, and cubic foot estimates can be developed for buildings with site, occupancy, building code, and construction cost restrictions. However, the following limitations must be recognized:

- If extensive site work or costly foundation systems are required, then the costs must be modified. This is also true for complex wall systems, mechanical and electrical systems, and structural systems with large bay sizes.
- Always make a sketch of the proposed building and its site to ensure that the specified requirements assumed in the estimating process are fulfilled.
- When projects have square foot requirements significantly different from the typical (shown in Figure 1.4), use the size modifier to properly adjust the calculated cost.
- If the source of cost information being used is a national average, as found in *Means Assemblies Cost Data,* the estimate should be adjusted using a City Cost Index to more realistically represent the cost at the location of the construction project.

If these modifications are realistically applied, the final cost of the project should be within 20% of the Order of Magnitude estimate, and 15% of the square foot or cubic foot estimate. Keeping these limitations in mind, both square foot and Order of Magnitude estimating methods can be used with confidence.

Chapters 4 through 10—the remaining chapters in Part 1—explain in detail how to analyze the loads and other important elements that go into a square foot estimate using the assemblies approach to estimating.

141 | S.F., C.F. and % of Total Costs

		14.1 000	S.F. & C.F. Costs			UNIT COSTS			% OF TOTAL			
				UNIT	1/4	MEDIAN	3/4	1/4	MEDIAN	3/4	
500	2900	Electrical	R14.1 -010	S.F.	2.77	3.97	5.70	5%	6.50%	8.10%	500
	3100	Total: Mechanical & Electrical		↓	8.35	12	16.55	15.60%	19.20%	23.50%	
	9000	Per apartment, total cost		Apt.	49,500	54,300	68,800				
	9500	Total: Mechanical & Electrical		"	8,000	11,000	13,700				
510	0010	ICE SKATING RINKS	R14.1 -010	S.F.	41.35	72.20	100				510
	0020	Total project costs		C.F.	2.88	2.95	3.40				
	2720	Plumbing		S.F.	1.47	2.33	2.75	3.10%	3.20%	5.60%	
	2900	Electrical		↓	4.20	4.63	6.46	6.70%	10.10%	15%	
	3100	Total: Mechanical & Electrical		↓	7.43	10.70	13.40	9.90%	25.90%	29.80%	
520	0010	JAILS	R14.1 -010	S.F.	118	148	196				520
	0020	Total project costs		C.F.	11	13	17.60				
	1800	Equipment		S.F.	5	13.25	23.05	4%	8.90%	15.10%	
	2720	Plumbing			11.75	15.45	18.80	7%	8.90%	13.30%	
	2770	Heating, ventilating, air conditioning			10.75	14.65	27.15	7.40%	9.40%	17.70%	
	2900	Electrical			13.15	17	20.80	8.10%	11.30%	12.40%	
	3100	Total: Mechanical & Electrical		↓	30.75	44.40	58.55	24.30%	30.10%	32.60%	
530	0010	LIBRARIES	R14.1 -010	S.F.	72.40	90.40	116				530
	0020	Total project costs		C.F.	5.10	6.20	7.95				
	0500	Masonry		S.F.	4.05	8.05	14.75	5.50%	7.40%	11.70%	
	1800	Equipment			1.05	2.70	4.42	1.30%	2.80%	4.80%	
	2720	Plumbing			2.86	4.03	5.50	3.40%	4.60%	5.60%	
	2770	Heating, ventilating, air conditioning			6.20	10.45	13.55	7.90%	11%	14.60%	
	2900	Electrical			7.30	9.55	12.20	8.10%	10.80%	11.90%	
	3100	Total: Mechanical & Electrical		↓	15.20	20	28.75	17.10%	21.30%	27.20%	
550	0010	MEDICAL CLINICS	R14.1 -010	S.F.	70.60	86.80	109				550
	0020	Total project costs		C.F.	5.30	6.80	9.10				
	1800	Equipment		S.F.	1.89	3.98	6.45	1.80%	4.80%	7.80%	
	2720	Plumbing			4.79	6.65	8.90	6.10%	8.40%	10%	
	2770	Heating, ventilating, air conditioning			5.86	7.40	10.85	6.70%	9%	11.30%	
	2900	Electrical			6.10	8.55	11.25	8.10%	10%	12.10%	
	3100	Total: Mechanical & Electrical		↓	14.85	19.23	26.45	18.80%	23.80%	29.60%	
	3500	See also division 117-001		↓							
570	0010	MEDICAL OFFICES	R14.1 -010	S.F.	65.60	80.90	101				570
	0020	Total project costs		C.F.	4.98	6.85	9.20				
	1800	Equipment		S.F.	2.20	4.36	6.20	2.80%	5.80%	7.10%	
	2720	Plumbing			3.69	5.70	7.75	5.60%	6.80%	8.50%	
	2770	Heating, ventilating, air conditioning			4.39	6.55	8.50	6.10%	8%	9.60%	
	2900	Electrical			5.25	7.70	10.60	7.60%	9.70%	11.60%	
	3100	Total: Mechanical & Electrical		↓	12.25	16.80	22.85	16.60%	21.20%	25.90%	
590	0010	MOTELS	R14.1 -010	S.F.	42.85	62.10	78.05				590
	0020	Total project costs		C.F.	3.67	5.20	8.50				
	2720	Plumbing		S.F.	4.27	5.35	6.50	9.40%	10.50%	12.50%	
	2770	Heating, ventilating, air conditioning			2.25	3.88	5.65	4.90%	5.60%	8.20%	
	2900	Electrical			3.98	5	6.50	7.10%	8.20%	10.40%	
	3100	Total: Mechanical & Electrical		↓	9.60	12.35	17	18.50%	23.10%	26.10%	
	5000										
	9000	Per rental unit, total cost		Unit	21,500	31,500	41,700				
	9500	Total: Mechanical & Electrical		"	4,025	5,700	6,200				
600	0010	NURSING HOMES	R14.1 -010	S.F.	63.55	83.10	103				600
	0020	Total project costs		C.F.	5.20	6.60	9.05				
	1800	Equipment		S.F.	2.08	2.77	4.23	2%	3.60%	5.90%	
	2720	Plumbing		↓	5.80	7.05	10.35	9.30%	10.30%	13.30%	

Figure 3.9

29

Chapter 4

Superstructures: Assemblies Estimate

This chapter explains how to develop the basic structural loads and other design criteria required for a construction estimate. The procedures shown throughout Part 1 of this book are intended for assemblies cost and square foot cost development only and are not intended to replace competent design practice.

Developing the assemblies square foot estimate begins by calculating the loads that are applied to the foundation. This chapter outlines how these loads are analyzed and assembled. The rest of Part 1 deals with other important elements that must be examined in the development stage.

Part 2 of this book demonstrates how information about loads and system selection can be applied to an actual cost estimate. The cost figures in Part 2 are based on the loads listed for the Three-Story Office Building example analyzed in this part of the book.

Several other examples are used in the following chapters of this section to show how loads are accumulated when various floor systems are used, such as steel, concrete, or precast concrete. Other examples explain cases where support systems such as columns, bearing walls, and pile foundations are employed.

Developing Loads

When structural loads are developed during the first stage of a complete square foot estimate, it is necessary to rely on assemblies costs for basic information. In addition, the limitations of the overall project must be known before selecting the correct systems. There are four pieces of general information that are required in this "development" stage:

- Cost limitations.
- Building size, site, and height limitations.
- Occupancy requirements.
- An acceptable sketch of the building, or drawings of an existing building that is similar to the proposed structure. This set of plans can be updated for current prices, and cost comparisons can be made.

Assembly Versatility

An assemblies estimate of any superstructure must be extremely versatile to be accurate. Data must be available for comparison and substitution of assemblies and combinations of assemblies to truly analyze the merits of each particular combination. Basically, designs have to be determined and costs provided for the following:

- Columns
- Floor assemblies
- Roof assemblies
- Stairs

Each of the components of these subsystems has characteristics that vary by:

- Use
- Cost
- Code requirements
- Weight to foundation
- Life cycle costs
- Thermal and fireproofing qualities
- Appearance

If soil-bearing capacity is limited, particular attention must be paid to the weight of the systems and the bay spacing in the superstructure. It is far more economical to build a structure with many short spans and no piles or caissons than a long-span structure that requires piles or caissons. For that reason, it is a good idea to examine the soil bearing capacity of the site before proceeding.

Soil Bearing Capacity

Before any wall and column loads are developed, determine the soil bearing capacity to be certain that the proper foundation is selected for the structure. Without soil bearing capacity information, it is difficult to develop accurate foundation costs.

The term "kip" has been used by engineers so much over the years that it has become a standard abbreviation. "Kip" is a short form of the term "kilopound." One kilopound is equivalent to 1,000 pounds (454 kilograms). A kip is a convenient unit of force for use in structure calculations.

Soil bearing capacity is the amount of pressure in "kips per square foot" that the soil will support without excessive settlement. For the purposes of the assemblies square foot estimates that follow in this book, assume that the soil bearing capacity used is in the range of 3 to 6 kips per square foot (KSF). This is a range between stiff clay at the lower end of the scale, and compact sand at the upper end of the scale.

> NOTE: To further clarify the concept of soil pressure, simply think of a 200-pound man standing on the ball of his foot on sand and assume that the contact area of the ball of his foot is 6 sq. in. (3″ x 2″).
>
> 200 lbs./6 sq. in. = 33.33 lbs./sq. in.
> 33.33 lbs./sq. in. × 144 sq. in./S.F. = 4800 lbs./S.F. or 4.8 kips/S.F.

For purposes of a square foot estimate, approximate soil bearing capacities may be established as follows:

- Call the local building inspector.
- Contact or talk to local contractors, engineers, or architects.
- Check borings or boring logs, usually available in building inspectors' offices, for other jobs in close proximity.
- See the table in Figure 4.1.

Load Requirements

There are many different types of loads involved in a structure. To develop floor and roof assemblies for any building, an accurate accounting must be made of three groups of information:

- **Live loads:** Minimum requirements determined by planned occupancy and minimum requirements by code (see Figure 4.2).
- **Miscellaneous loads:** Partitions, ceilings, or mechanical systems supported by the structure (see Figure 4.3).
- **Bay sizes:** Clear spans or column spacing required by planned building occupancy.

Floor and roof assemblies can be selected by cost comparison. The procedure for doing so is described in Part 2. At the start of an assemblies square foot estimate, it is more important to consider the loads themselves and carefully examine each possible assembly in terms of its specific application in the structure.

Total loads in pounds per square foot (psf) of a floor system are a function of the following:

- Roof and floor live loads required by code or building occupancy.
- The miscellaneous loads supported by the structure, such as partitions, ceilings, and mechanical loads.
- The dead load of the structure.

The superimposed loads applied to a floor assembly are a combination of the live and miscellaneous loads. The total load of a floor or roof assembly is the superimposed load plus the weight of the floor system.

COST DETERMINATION

1. Determine Soil Bearing Capacity by a known value or using this table as a guide.

Soil Bearing Capacity in Kips per S.F.

Bearing Material	Typical Allowable Bearing Capacity
Hard sound rock	120 KSF
Medium hard rock	80
Hardpan overlaying rock	24
Compact gravel and boulder-gravel; very compact sandy gravel	20
Soft rock	16
Loose gravel; sandy gravel; compact sand; very compact sand-inorganic silt	12
Hard dry consolidated clay	10
Loose coarse to medium sand; medium compact fine sand	8
Compact sand-clay	6
Loose fine sand; medium compact sand-inorganic silts	4
Firm or stiff clay	3
Loose saturated sand-clay; medium soft clay	2

Figure 4.1

Minimum Design Live Loads in Pounds per S.F. for Various Building Codes

Occupancy	Description	Minimum Live Loads, Pounds per S.F.		
		BOCA	ANSI	UBC
Armories		150	150	
Assembly	Fixed seats	60	60	50
	Movable seats	100	100	100
	Platforms or stage floors	100	100	125
Commercial & Industrial	Light manufacturing	125	125	75
	Heavy manufacturing	250	250	125
	Light storage	125	125	75
	Heavy storage	250	250	100
	Stores, retail, first floor	100	100	75
	Stores, retail, upper floors	75	75	
	Stores, wholesale	125	125	100
Court rooms		100		
Dance halls	Ballrooms	100	100	
Dining rooms	Restaurants	100	100	
Fire escapes	Other than below	100	100	
	Multi or single family residential	40		
Garages	Passenger cars only	50	50	50
Gymnasiums	Main floors and balconies	100	100	
Hospitals	Operating rooms, laboratories	60	60	
	Private room	40	40	
	Wards	40	40	
	Corridors, above first floor	80	80	
Libraries	Reading rooms	60	60	
	Stack rooms	150	150	125
	Corridors, above first floor		80	
Marquees		75	75	
Office Buildings	Offices	50	50	50
	Lobbies	100	100	
	Corridors, above first floor	80	*	100
Residential	Multi family private apartments	40	40	40
	Multi family, public rooms	100	100	
	Multi family, corridors	80	*	100
	Dwellings, first floor	40	40	40
	Dwellings, second floor & habitable attics	30	30	
	Dwellings, uninhabitable attics	20	20	
	Hotels, guest rooms	40	40	40
	Hotels, public rooms	100	100	
	Hotels, corridors serving public rooms	100	100	
	Hotels, corridors	80	80	
Roofs	Flat	12-20		20
	Pitched	12-16		
Schools	Classrooms	40	40	40
	Corridors	80	80	100
Sidewalks	Driveways, etc. subject to trucking	250	250	250
Stairs	Exits	100		100
Theaters	Aisles, corridors and lobbies	100	100	
	Orchestra floors	60	100	
	Balconies	60	100	
	Stage floors	150	150	
Yards	Terrace, pedestrian	100	100	

BOCA = Building Officials & Code Administration International, Inc. National Building Code
ANSI = Standard A58.1
UBC = Uniform Building Code, International Conference of Building Officials
* Corridor loading equal to occupancy loading.

Figure 4.2

The following examples show the actual "bookkeeping" involved in the process of assembling the systems that transmit actual loads to the foundation of a specific building.

We begin with a straightforward concrete structure where only the superimposed floor loads are analyzed. The first two examples show a step-by-step procedure for developing the total load to the foundation.

Example One: Load

Start by developing the superimposed loads.

- Building: 30,000 S.F.
- Structure: Concrete Flat Plate
- Stories: 3
- Dimensions: 10,000 S.F./floor; 12'-0" Floor to Floor
- Building Occupancy: Office
- Bay Size: 25'-0" x 25'-0"
- Code: BOCA

Step One: Analyze the Floor Loads

Minimum Live Load for Office Building	50 psf	BOCA Code (Figure 4.2)
Partitions	15 psf	(Figure 4.3)
Ceiling	5 psf	(Figure 4.3)
Mechanical	5 psf	(Figure 4.3)
Superimposed Floor Load	75 psf	

Step Two: Develop the Superimposed Roof Load

Snow Load	20 psf	(Figure 4.4)
Roofing and Insulation	10 psf	(Figure 4.5)
Ceiling	5 psf	(Figure 4.3)
Mechanical	5 psf	(Figure 4.3)
Superimposed Roof Load	40 psf	

Step Three: Sketch the Floor Plan and Elevation

At this point, it is a good idea to make a sketch of the floor plan to show column spacing and bay sizes. Also sketch an elevation to show the number of suspended floors, the roof, and the floor-to-floor relationship. See Figure 4.6.

Superimposed Dead Load Ranges

Component	Load Range (PSF)
Ceiling	5-10
Partitions	10-20
Mechanical	4-8

Figure 4.3

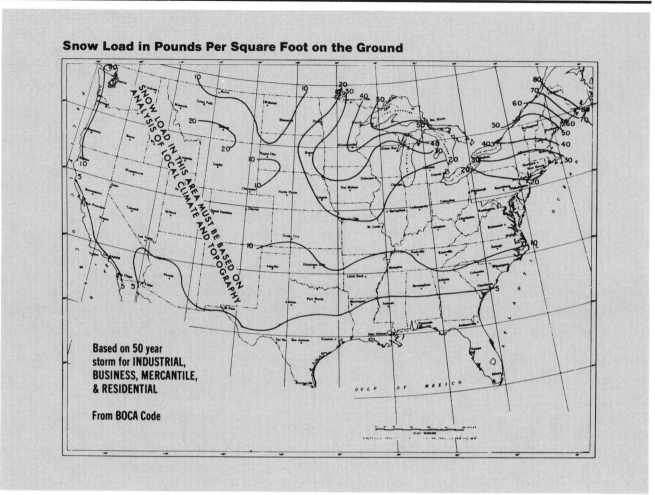

Snow Load in Pounds Per Square Foot on the Ground

Based on 50 year storm for INDUSTRIAL, BUSINESS, MERCANTILE, & RESIDENTIAL

From BOCA Code

Figure 4.4

Design Weight per S.F. for Roof Coverings

Type		Description	Weight Per S.F.	Type	Wall Thickness	Description	Weight Per S.F.
Sheathing	Gypsum	1" thick	4	Metal	Aluminum	Corr. & ribbed, .024" to .040"	.4-.8
	Wood	¾" thick	3		Copper	or tin	1.0.
Insulation	per 1"	Loose	.5		Steel	Corrugated, 29 ga. to 12 ga.	.6-5.0
		Poured in place	2	Shingles	Asphalt	Strip shingles	3
		Rigid	1.5		Clay	Tile	9-20
Built-up	Tar & gravel	3 ply felt	5.5		Slate	¼" thick	10
		5 ply felt			Wood		2

Figure 4.5

Example Two: Total Load Derivation

Continue by following the steps below. Remember, the superimposed loads in psf, bay sizes, and a building layout have already been developed. The next step is to begin the selection of various building systems.

Step One: Determine Maximum Column Size

- Use the information in Figure 4.7, Cast in Place Flat Plate Floor System.
- Enter the table, knowing the bay size is 25' x 25'.
- Go to the line in the bay size listing for a 75 psf superimposed floor load.
- Total Load (psf): 194 psf (from Figure 4.7)
- Minimum Column Size (in.): 24" (from Figure 4.7)

Step Two: Determine the Total Load

- Use Figure 4.7, knowing the bay size is 25' x 25'.
- The superimposed roof load from the Load Example is 40 psf.
- Total Load (psf): 152 psf.

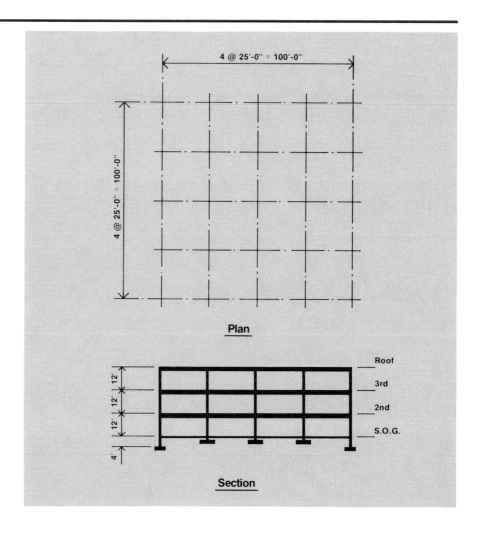

Figure 4.6

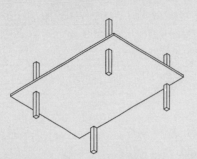

General: Flat Plates: Solid uniform depth concrete two-way slab without drops or interior beams. Primary design limit is shear at columns.

Design and Pricing Assumptions:
Concrete f'c to 4 KSI, placed by concrete pump.
Reinforcement, fy = 60 KSI.
Forms, four use.
Finish, steel trowel.
Curing, spray on membrane.
Based on 4 bay x 4 bay structure.

System Components	QUANTITY	UNIT	COST PER S.F.		
			MAT.	INST.	TOTAL
SYSTEM 3.5-150-2000					
15'X15' BAY 40 PSF S. LOAD, 12" MIN. COL.					
Forms in place, flat plate to 15' high, 4 uses	.992	S.F.	.83	3.41	4.24
Edge forms to 6" high on elevated slab, 4 uses	.065	L.F.	.03	.16	.19
Reinforcing in place, elevated slabs #4 to #7	1.706	Lb.	.51	.46	.97
Concrete ready mix, regular weight, 3000 psi	.459	C.F.	.98		.98
Place and vibrate concrete, elevated slab less than 6", pump	.459	C.F.		.44	.44
Finish floor, monolithic steel trowel finish for finish floor	1.000	S.F.		.61	.61
Cure with sprayed membrane curing compound	.010	C.S.F.	.02	.05	.07
TOTAL			2.37	5.13	7.50

3.5-150		Cast in Place Flat Plate						
	BAY SIZE (FT.)	SUPERIMPOSED LOAD (P.S.F.)	MINIMUM COL. SIZE (IN.)	SLAB THICKNESS (IN.)	TOTAL LOAD (P.S.F.)	COST PER S.F.		
						MAT.	INST.	TOTAL
2000	15 x 15	40	12	5-1/2	109	2.37	5.15	7.52
2200	R3.5 -010	75	14	5-1/2	144	2.39	5.15	7.54
2400		125	20	5-1/2	194	2.49	5.20	7.69
2600		175	22	5-1/2	244	2.54	5.20	7.74
3000	15 x 20	40	14	7	127	2.71	5.20	7.91
3400	R3.5 -100	75	16	7-1/2	169	2.88	5.30	8.18
3600		125	22	8-1/2	231	3.18	5.45	8.63
3800		175	24	8-1/2	281	3.20	5.45	8.65
4200	20 x 20	40	16	7	127	2.71	5.20	7.91
4400		75	20	7-1/2	175	2.92	5.35	8.27
4600		125	24	8-1/2	231	3.19	5.45	8.64
5000		175	24	8-1/2	281	3.21	5.50	8.71
5600	20 x 25	40	18	8-1/2	146	3.16	5.45	8.61
6000		75	20	9	188	3.27	5.50	8.77
6400		125	26	9-1/2	244	3.53	5.65	9.18
6600		175	30	10	300	3.68	5.70	9.38
7000	25 x 25	40	20	9	152	3.27	5.50	8.77
7400		75	24	9-1/2	194	3.46	5.60	9.06
7600		125	30	10	250	3.68	5.75	9.43
8000								

Figure 4.7

When using concrete floor and column systems, it is usually more economical to use the same size concrete column for each floor to minimize formwork changes. In each step to this point, the minimum column size for the floor system has been used exclusively, since the larger size governs the overall selection process.

The total load as recorded in Steps One and Two is actually the sum of the superimposed load plus the dead load of the structure:

25'-0" Square Bay Superimposed Load		75 psf
Floor Dead Load:		
Slab Thickness (from Figure 4.7: 9'-1/2")	.79" × 150 pcf =	119 psf
Total Load (Floor)		194 psf
25'-0" Square Bay Superimposed Load		40 psf
Roof Dead Load:		
Slab Thickness (from Figure 4.7: 9")	.75" × 150 pcf =	112 psf
Total Load (Roof)		152 psf

The total load for floor assemblies using steel beams and concrete decks includes the beams, steel deck (if used in the system), concrete slab, and sprayed-on fireproofing required for the system. Floor assemblies using joists do not include fireproofing, since the assemblies are usually fireproofed by applying a rated ceiling below the joists.

The total load for roof systems of steel joists and beams or steel joists and joist girders with steel roof decks includes an allowance for insulation, roofing, and miscellaneous loads, but does not include fireproofing.

It is important to carefully examine each assembly for the makeup of the total load, since roof assemblies have an allowance for insulation, and floor systems used as roofs do not.

Examine each assembly used to determine the makeup of the total load.

Step Three: Develop the Area Supported by an Interior Column

Sketch a cross section of an interior column, allowing adequate space for calculations and load accumulation as shown in Figure 4.8.

Develop the area supported by an interior column as shown in Figure 4.9.

Step Four: Determine the Load Imposed by the Roof or Floor

To determine the load imposed by the roof or floor, multiply the bay size by the total load in psf. The column load for concrete columns is the column size in feet multiplied by the height of the column, all multiplied by 150 pcf (the weight of reinforced concrete in pounds per cubic foot).

To ensure that all roof, floor, and column loads are accumulated, the method shown in Figure 4.10 is recommended. With the calculations shown, these loads have been determined:

The minimum column load (95 kips + 7 kips)	=	102 kips
The maximum column load	=	358 kips
The load to the foundation	=	358 kips

Step Five: Determine the Required Concrete Column

As previously determined, the minimum allowable column size required for a 25'-0" square bay with 75 psf superimposed load using a cast in place flat plate is 24" x 24".

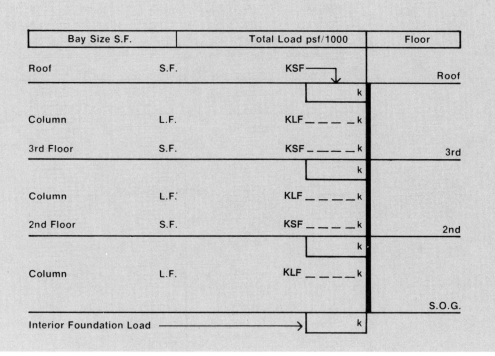

Bay Size S.F.		Total Load psf/1000		Floor
Roof	S.F.	KSF		Roof
			k	
Column	L.F.	KLF _ _ _ _ k		
3rd Floor	S.F.	KSF _ _ _ _ k		3rd
			k	
Column	L.F.	KLF _ _ _ _ k		
2nd Floor	S.F.	KSF _ _ _ _ k		2nd
			k	
Column	L.F.	KLF _ _ _ _ k		
				S.O.G.
Interior Foundation Load			k	

Figure 4.8

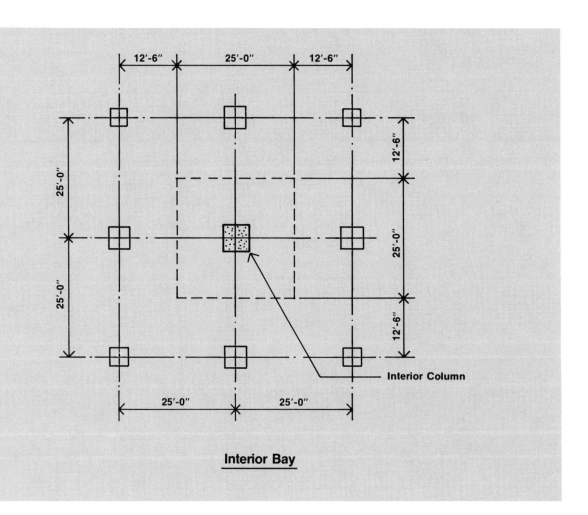

Interior Bay

Figure 4.9

The perimeter of the column is $4 \times 24'' = 96''$. If round columns are chosen, it is necessary to supply a column with a perimeter of 96".

$$96''/3.14 = 30.57''$$

Round this figure down to allow for a 30" diameter column. In this example, use square columns.

Enter the second portion of Figure 4.11 (Columns—Square Tied Minimum Reinforcing). Enter the table with the column size (minimum column required 24") and note the allowable load in kips (700 kips).

700 kips Allowable > 358 kips Required: Column O.K.

If the column load is in excess of 700 kips allowable, enter the first portion of Figure 4.11 under Column Size (24") and proceed down the row to the first 24" column with 12'-0" story height. The allowable load is: 900 kips.

The cost of the column is determined in this way:

$$\frac{\text{Cost of 900 kips Column} + \text{Cost of Minimum Reinforced Column} \times \text{Total Length}}{2}$$

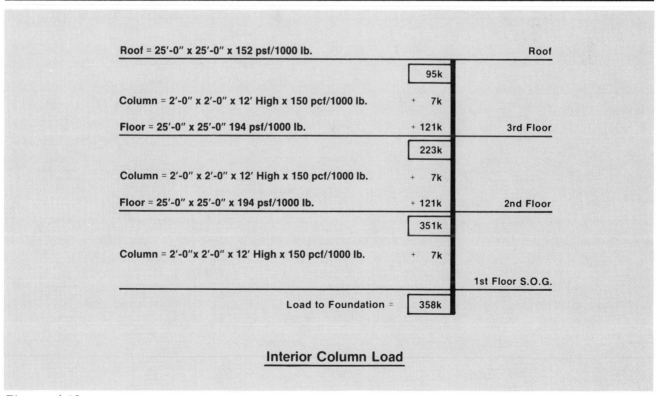

Interior Column Load

Figure 4.10

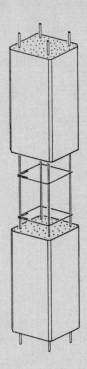

CONCRETE COLUMNS

General: It is desirable for purposes of consistency and simplicity to maintain constant column sizes throughout the building height. To do this, concrete strength may be varied (higher strength concrete at lower stories and lower strength concrete at upper stories), as well as varying the amount of reinforcing.

The first portion of the table provides probable minimum column sizes with related costs and weights per lineal foot of story height for bottom level columns.

The second portion of the table provides costs by column size for top level columns with minimum code reinforcement. Probable maximum loads for these columns are also given.

How to Use Table:

1. Enter the second portion (minimum reinforcing) of the table with the minimum allowable column size from the selected cast in place floor system.

   If the total load on the column does not exceed the allowable working load shown, use the cost per L.F. multiplied by the length of columns required to obtain the column cost.

2. If the total load on the column exceeds the allowable working load shown in the second portion of the table, enter the first portion of the table with the total load on the column and the minimum allowable column size from the selected cast in place floor system.

   Select a cost per L.F. for bottom level columns by total load or minimum allowable column size.

   Select a cost per L.F. for top level columns using the column size required for bottom level columns from the second portion of the table.

$$\frac{\text{Btm.} + \text{Top Col. Costs/L.F.}}{2} = \text{Avg. Col. Cost/L.F}$$

Column Cost = Average Col. Cost/L.F. x Length of Cols. Required.

See reference section in back of book to determine total loads.

Design and Pricing Assumptions:
Normal wt. concrete, f'c = 4 or 6 KSI, placed by pump.
Steel, fy = 60 KSI, spliced every other level.
Minimum design eccentricity of 0.1t.
Assumed load level depth is 8" (weights prorated to full story basis).
Gravity loads only (no frame or lateral loads included).

Please see the reference section for further design and cost information.

System Components	QUANTITY	UNIT	COST PER V.L.F.		
			MAT.	INST.	TOTAL
SYSTEM 3.1-114-0640					
SQUARE COLUMNS, 100K LOAD,10' STORY, 10" SQUARE					
Forms in place, columns, plywood, 10" x 10", 4 uses	3.323	SFCA	3.13	18.47	21.60
Chamfer strip,wood, 3/4" wide	4.000	L.F.	2.72	2.44	5.16
Reinforcing in place, columns, #3 to #7	3.653	Lb.	1.50	2.60	4.10
Reinforcing in place, column ties	1.405	Lb.	.30	.52	.82
Concrete ready mix, regular weight, 4000 psi	.026	C.Y.	1.57		1.57
Placing concrete, incl. vibrating, 12" sq./round columns, pumped	.026	C.Y.		1.66	1.66
Finish, break ties, patch voids, burlap rub w/grout	3.323	S.F.	.27	2.17	2.44
TOTAL			9.49	27.86	37.35

3.1-114				C.I.P. Column, Square Tied				
	LOAD (KIPS)	STORY HEIGHT (FT.)	COLUMN SIZE (IN.)	COLUMN WEIGHT (P.L.F.)	CONCRETE STRENGTH (PSI)	COST PER V.L.F.		
						MAT.	INST.	TOTAL
0640	100	10	10	96	4000	9.50	28	37.50
0680	R3.1 -112	12	10	97	4000	9.35	27.50	36.85
0700		14	12	142	4000	11.50	33.50	45
0710								
0740	150	10	10	96	4000	9.50	28	37.50
0780		12	12	142	4000	11.55	33.50	45.05
0800		14	12	143	4000	11.50	33.50	45

Figure 4.11

| 3.1-114 | C.I.P. Column, Square Tied |

	LOAD (KIPS)	STORY HEIGHT (FT.)	COLUMN SIZE (IN.)	COLUMN WEIGHT (P.L.F.)	CONCRETE STRENGTH (PSI)	COST PER V.L.F.		
						MAT.	INST.	TOTAL
0840	200	10	12	140	4000	11.60	34	45.60
0860		12	12	142	4000	11.55	33.50	45.05
0900		14	14	196	4000	13.55	39.50	53.05
0920	300	10	14	192	4000	13.85	40	53.85
0960		12	14	194	4000	13.75	39.50	53.25
0980		14	16	253	4000	15.25	43.50	58.75
1020	400	10	16	248	4000	16.35	45.50	61.85
1060		12	16	251	4000	16.25	45	61.25
1080		14	16	253	4000	16.15	45	61.15
1200	500	10	18	315	4000	20.50	54	74.50
1250		12	20	394	4000	21	56	77
1300		14	20	397	4000	21	56	77
1350	600	10	20	388	4000	24	61.50	85.50
1400		12	20	394	4000	24	61	85
1600		14	20	397	4000	23.50	61	84.50
1900	700	10	20	388	4000	24.50	62	86.50
2100		12	22	474	4000	24.50	61.50	86
2300		14	22	478	4000	24	61.50	85.50
2600	800	10	22	388	4000	33.50	77.50	111
2900		12	22	474	4000	33.50	77.50	111
3200		14	22	478	4000	33	77	110
3400	900	10	24	560	4000	33	78	111
3800		12	24	567	4000	32.50	77	109.50
4000		14	24	571	4000	32.50	76.50	109
4250	1000	10	24	560	4000	39.50	89	128.50
4500		12	26	667	4000	36.50	80	116.50
4750		14	26	673	4000	36	79.50	115.50
5600	100	10	10	96	6000	9.60	28	37.60
5800		12	10	97	6000	9.45	28	37.45
6000		14	12	142	6000	11.55	33.50	45.05
6200	150	10	10	96	6000	9.50	28	37.50
6400		12	12	98	6000	11.55	33.50	45.05
6600		14	12	143	6000	11.55	33.50	45.05
6800	200	10	12	140	6000	11.95	34.50	46.45
7000		12	12	142	6000	11.55	33.50	45.05
7100		14	14	196	6000	13.55	39.50	53.05
7300	300	10	14	192	6000	13.70	39.50	53.20
7500		12	14	194	6000	13.65	39.50	53.15
7600		14	14	196	6000	13.55	39.50	53.05
7700	400	10	14	192	6000	13.70	39.50	53.20
7800		12	14	194	6000	13.65	39.50	53.15
7900		14	16	253	6000	16.10	44.50	60.60
8000	500	10	16	248	6000	16.35	45.50	61.85
8050		12	16	251	6000	16.25	45	61.25
8100		14	16	253	6000	16.15	45	61.15
8200	600	10	18	315	6000	19.20	52.50	71.70
8300		12	18	319	6000	19.05	52	71.05
8400		14	18	321	6000	18.95	52	70.95
8500	700	10	18	315	6000	20.50	55	75.50
8600		12	18	319	6000	20.50	54.50	75
8700		14	18	321	6000	20	54	74
8800	800	10	20	388	6000	21	56.50	77.50
8900		12	20	394	6000	21	56.50	77.50
9000		14	20	397	6000	21	56	77
9100	900	10	20	388	6000	29	70	99
9300		12	20	394	6000	28.50	69.50	98
9600		14	20	397	6000	28	69	97

Figure 4.11 (cont.)

If the load to be supported is in excess of the load allowable on a 24″ square column, it may be necessary to use a larger size column or a higher strength concrete column to satisfy conditions. Follow the same procedure to obtain costs.

Conclusion   It should be clear from the examples in this chapter that it is important to know and understand the loads in a building if an accurate final building cost estimate is to be made. To ensure sound decisions when developing loads in the conceptual stages of a project, always consider the following four points when developing an assemblies estimate:

- Know the basic characteristics of the local soil. With this information available, it is sometimes possible to consider alternatives, such as a lighter building design that will save money on costly foundation design.
- Always keep in mind the structural assemblies being used. A careful assemblies analysis ensures an accurate inventory of all loads, eliminating the possibility of redesign at an advanced stage.
- Use simple sketches and drawings to answer the many "what if" questions of owners and contractors. The same sketches can serve as a basic outline of the project during conceptual stages.
- Check local building codes for unusual load requirements. Weather conditions in some locations may dictate certain structural and exterior wall materials. Also, consider the contractor's ability to construct the selected type of structure in the time allowed for the project.

SUPERSTRUCTURES		A3.1-114	C.I.P. Column, Square Tied					
3.1-114		C.I.P. Column, Square Tied						
	LOAD (KIPS)	STORY HEIGHT (FT.)	COLUMN SIZE (IN.)	COLUMN WEIGHT (P.L.F.)	CONCRETE STRENGTH (PSI)	COST PER V.L.F.		
						MAT.	INST.	TOTAL
9800	1000	10	22	469	6000	26	64.50	90.50
9840		12	22	474	6000	26	64	90
9900		14	22	478	6000	26	65	91
3.1-114		C.I.P. Column, Square Tied-Minimum Reinforcing						
	LOAD (KIPS)	STORY HEIGHT (FT.)	COLUMN SIZE (IN.)	COLUMN WEIGHT (P.L.F.)	CONCRETE STRENGTH (PSI)	COST PER V.L.F.		
						MAT.	INST.	TOTAL
9912	150	10-14	12	135	4000	11.35	33.50	44.85
9918	300	10-14	16	240	4000	15.15	43	58.15
9924	500	10-14	20	375	4000	20.50	56	76.50
9930	700	10-14	24	540	4000	28.50	69.50	98
9936	1000	10-14	28	740	4000	35	83	118
9942	1400	10-14	32	965	4000	47	96	143
9948	1800	10-14	36	1220	4000	56	110	166
9954	2300	10-14	40	1505	4000	63	122	185

Figure 4.11 (cont.)

Chapter 5

Foundation: Assemblies Estimate

Selection of the proper foundation for the imposed loads is a function of the allowable soil bearing capacity at a given depth below the slab on grade, or at a given depth below grade.

Spread Footings

Spread footings are used to convert a column or grade beam load into an allowable area load on the supporting soil. Spread footings are the most widely used type of footing for concentrated loads. They tend to minimize excavation, and soil conditions at the bearing area can be easily examined. The depth of interior footings should be below topsoil or soil materials containing loose fill or vegetation. Exterior footings must be lower than frost penetration. See Figures 5.1a and 5.1b.

Exterior or 1/2 bay footings and corner or 1/4 bay footings assume less load from the superstructure than full bay footings. In Figure 5.1a there are:

- 9 Interior Footings
- 12 Exterior Footings
- 4 Corner Footings

It would appear that the load reduction for exterior or corner footings should be 1/2 or 1/4 respectively, but usually exterior walls, slab overhangs, and other elements of the structure contribute to the column loads on the exterior. Use the following guide to obtain foundation loads for exterior and corner footings:

- To determine the load on exterior footings, multiply the interior load to foundation by 0.6.
- To determine the load on corner footings, multiply the interior load to foundation by 0.45.

These factors apply to buildings with normal fascia and floor-to-floor height. For special cases, calculate wall loads or use the interior column load for the load on exterior or corner footings.

Spread Footing Selection

Assume a 6 KSF soil bearing capacity. Using the 358k interior column load that was derived in the Total Load example in Chapter 4, the foundation loads and sizes for interior, exterior, and corner columns can be determined as follows:

Interior Column Load 358k

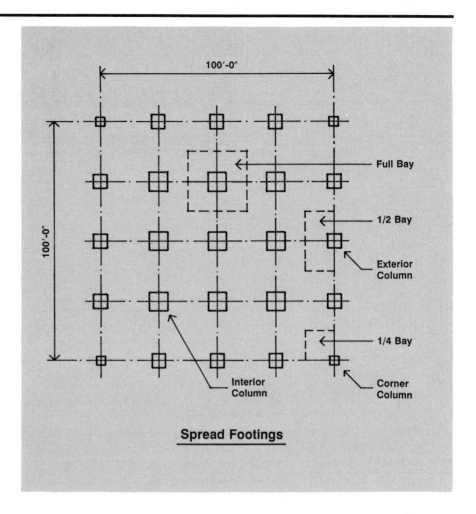

Figure 5.1a

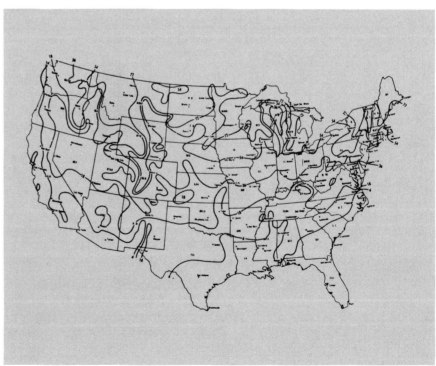

Figure 5.1b

Exterior Column Load	358k × 0.6 = 215k
Corner Column Load	358k × 0.45 = 161k

Enter the Spread Footings table in Figure 5.2.

358k (400k)	8'-6" square x 27" deep
215k (300k)	7'-6" square x 25" deep
161k (200k)	6'-0" square x 20" deep

Concrete Pier

In many instances a concrete pier is required to extend the superstructure column at the ground or basement floor to the spread footing because of frost penetration, basement walls, or poor soil conditions.

Use the information in Figure 4.11 for piers. For concrete columns, assume the pier is the same size as the concrete column. For steel or wood columns assume the pier is approximately 10" larger than the column size to allow for base plates, anchor bolts, and grouting. Exterior piers usually protrude inside concrete walls because column lines are set back to allow the building fascia to cover the column. Protruding exterior piers interrupt wall formwork and are both time-consuming and costly to erect.

Figure 5.3 shows pier sizes for a 24" square concrete column, a 10" steel column, and an 8" square wood column.

Strip Footings

Strip footings are used to convert a lineal wall load to an allowable soil bearing capacity. They may also be employed as a leveling pad to facilitate the placement of formwork for concrete walls.

For low-rise, normal-span buildings, the soil bearing capacity is of little consequence. Choose a foundation width twice the bearing wall thickness, with a minimum thickness of the strip footing equal to the bearing wall.

Example One: Loads to Strip Footings

A two-story building for elderly housing is to be constructed. This example shows how to develop the load to the footings in units of one thousand pounds per linear foot (KLF). It then shows how to select the appropriate system, given the load information. Remember, calculate loads and then select the appropriate system.

This is the only information known about the proposed structure:

Floor to Ceiling Height: 8'
Span: 25'
Structure: 8" Concrete Plank on 8" Block Wall
Building Occupancy: Elderly Housing
Code: Requires Reinforced Footings

Step One: Develop Loads

Using the same basic steps outlined in the previous examples, develop the following superimposed loads:

Floor Loads:		Roof Loads:	
Live Load Floor	40 psf	Live Load Roof	20 psf
Partitions	20 psf	Roofing & Insul.	10 psf
Flooring	5 psf	Ceiling	5 psf
Ceiling	5 psf	Mechanical	5 psf
	70 psf		40 psf

Step Two: Sketch a Cross Section of the Building

See Figure 5.4.

FOUNDATIONS	A1.1-120	Spread Footings

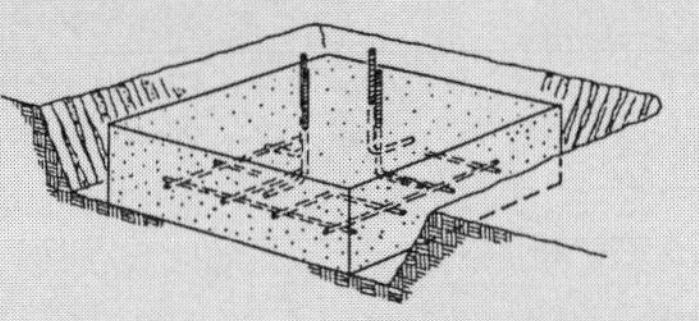

The Spread Footing System includes: excavation; backfill; forms (four uses); all reinforcement; 3,000 p.s.i. concrete (chute placed); and screed finish.

Footing systems are priced per individual unit. The Expanded System Listing at the bottom shows footings that range from 3′ square x 12″ deep, to 18′ square x 52″ deep. It is assumed that excavation is done by a truck mounted hydraulic excavator with an operator and oiler.

Backfill is with a dozer, and compaction by air tamp. The excavation and backfill equipment is assumed to operate at 30 C.Y. per hour.

Please see the reference section for further design and cost information.

System Components	QUANTITY	UNIT	COST EACH MAT.	COST EACH INST.	COST EACH TOTAL
SYSTEM 1.1-120-7100					
SPREAD FOOTINGS, LOAD 25K, SOIL CAPACITY 3 KSF, 3′ SQ X 12″ DEEP					
Bulk excavation	.590	C.Y.		3.18	3.18
Hand trim	9.000	S.F.		4.77	4.77
Compacted backfill	.260	C.Y.		.55	.55
Formwork, 4 uses	12.000	S.F.	7.20	36.12	43.32
Reinforcing, fy = 60,000 psi	.006	Ton	3.45	4.50	7.95
Dowel or anchor bolt templates	6.000	L.F.	4.02	15	19.02
Concrete, f′c = 3,000 psi	.330	C.Y.	18.98		18.98
Place concrete, direct chute	.330	C.Y.		5	5
Screed finish	9.000	S.F.		2.97	2.97
TOTAL			33.65	72.09	105.74

1.1-120	Spread Footings	COST EACH MAT.	COST EACH INST.	COST EACH TOTAL
7090	Spread footings, 3000 psi concrete, chute delivered			
7100	Load 25K, soil capacity 3 KSF, 3′-0″ sq. x 12″ deep	33.50	72	105.50
7150	Load 50K, soil capacity 3 KSF, 4′-6″ sq. x 12″ deep	69.50	125	194.50
7200	Load 50K, soil capacity 6 KSF, 3′-0″ sq. x 12″ deep	33.50	72	105.50
7250	Load 75K, soil capacity 3 KSF, 5′-6″ sq. x 13″ deep	109	177	286
7300	Load 75K, soil capacity 6 KSF, 4′-0″ sq. x 12″ deep	57	107	164
7350	Load 100K, soil capacity 3 KSF, 6′-0″ sq. x 14″ deep	138	212	350
7410	Load 100K, soil capacity 6 KSF, 4′-6″ sq. x 15″ deep	85.50	147	232.50
7450	Load 125K, soil capacity 3 KSF, 7′-0″ sq. x 17″ deep	217	305	522
7500	Load 125K, soil capacity 6 KSF, 5′-0″ sq. x 16″ deep	110	177	287
7550	Load 150K, soil capacity 3 KSF 7′-6″ sq. x 18″ deep	261	360	621
7610	Load 150K, soil capacity 6 KSF, 5′-6″ sq. x 18″ deep	145	223	368
7650	Load 200K, soil capacity 3 KSF, 8′-6″ sq. x 20″ deep	370	475	845
7700	Load 200K, soil capacity 6 KSF, 6′-0″ sq. x 20″ deep	189	275	464
7750	Load 300K, soil capacity 3 KSF, 10′-6″ sq. x 25″ deep	675	770	1,445
7810	Load 300K, soil capacity 6 KSF, 7′-6″ sq. x 25″ deep	355	465	820
7850	Load 400K, soil capacity 3 KSF, 12′-6″ sq. x 28″ deep	1,075	1,150	2,225
7900	Load 400K, soil capacity 6 KSF, 8′-6″ sq. x 27″ deep	490	605	1,095
7950	Load 500K, soil capacity 3 KSF, 14′-0″ sq. x 31″ deep	1,475	1,525	3,000
8010	Load 500K, soil capacity 6 KSF, 9′-6″ sq. x 30″ deep	675	785	1,460
8050	Load 600K, soil capacity 3 KSF, 16′-0″ sq. x 35″ deep	2,150	2,075	4,225
8100	Load 600K, soil capacity 6 KSF, 10′-6″ sq. x 33″ deep	905	1,025	1,930
8150	Load 700K, soil capacity 3 KSF, 17′-0″ sq. x 37″ deep	2,525	2,375	4,900
8200	Load 700K, soil capacity 6 KSF, 11′-6″ sq. x 36″ deep	1,150	1,250	2,400

R1.1 -120

Figure 5.2

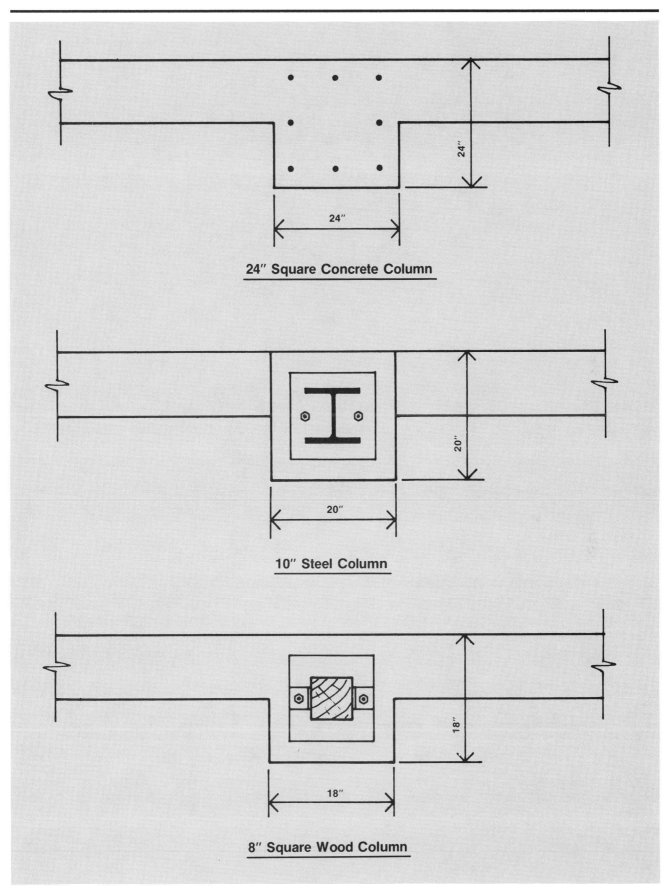

24" Square Concrete Column

10" Steel Column

8" Square Wood Column

Figure 5.3

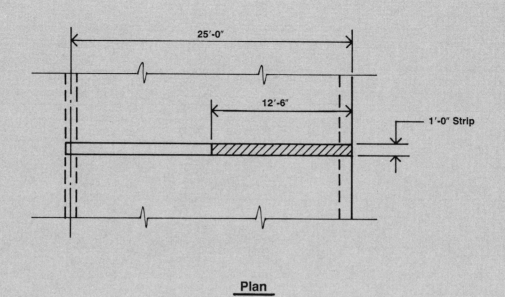

Plan

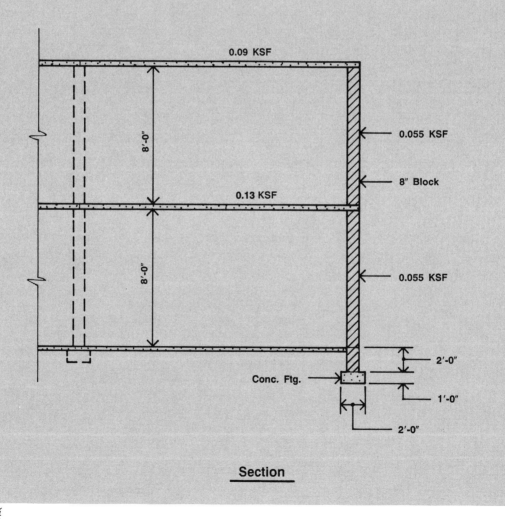

Section

Figure 5.4

Enter the table in Figure 5.5 for a 25'-0" span.

For a 75 psf superimposed load, Total Load = 130 psf.
For a 40 psf superimposed load, Total Load = 90 psf.

Show the appropriate loads on the building cross section.

Step Three: Select an Appropriate Footing System

Develop total loads in KLF, and use the cost tables to select an appropriate footing system.

Load to interior strip footing:

Roof	25.0'	×	0.09 KSF	=	2.25k
Wall	8.0'	×	0.055 KSF	=	0.44k
Floor	25.0'	×	0.13 KSF	=	3.25k
Wall	8.33'	×	0.055 KSF	=	0.46k
					6.40 KLF

Load to exterior bearing wall footing:

Roof	12.5'	×	0.09 KSF	=	1.13k
Wall	8.0'	×	0.055 KSF	=	0.44k
Floor	12.5'	×	0.13 KSF	=	1.63k
Wall	10.0'	×	0.055 KSF	=	0.55k
					3.75 KLF

Assume 3 KSF soil bearing capacity. Also, assume that a reinforced footing is required by code. Enter the Strip Footings table in Figure 5.6 at the listing for a strip footing with a load of 6.4 KSF. A 12" deep x 32" wide reinforced footing is required.

The exterior bearing wall footing load is 3.75k. Use a 12" deep x 24" reinforced footing (required by code).

Grade Beams

Grade beams are used primarily to support walls or slabs where footings are at a greater depth (because of poor soil conditions) than the depth required by code or frost requirements.

Grade beams are stiff, self-supporting structural members that are usually designed as simply supported beams. This minimizes the effects of unequal footing settlement.

Example Two: Grade Beams

The following example shows how to develop the load (in KLF) applied to a grade beam. It also develops the load to the pier and total load to the spread footing. See Figure 5.7.

Construction: Steel Joists on Walls. Use Figure 5.8 for load figures.
Superimposed Load: 40 psf
Frost Requirements: 36"
Roof Total Load: 64 psf (from Figure 5.8)

Roof Load:	40 × .064 KSF	=	2.56 KLF
Wall 12" Block:	16.0' × .085 KSF	=	1.36 KLF
Total Load to Grade Beam:			3.92 KLF

Grade Beam for Bearing Walls:

Enter the Grade Beams table in Figure 5.9, at listing for a 36" minimum depth and a 30'-0" span.

Choose 40" deep x 12" wide grade beam 4 KLF > 3.92 required.

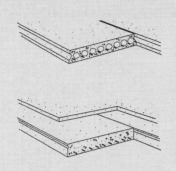

General: Units priced here are for plant produced prestressed members, transported to site and erected.

Normal weight concrete is most frequently used. Lightweight concrete may be used to reduce dead weight.

Structural topping is sometimes used on floors: insulating concrete or rigid insulation on roofs.

Camber and deflection may limit use by depth considerations.

Prices are based upon 10,000 S.F. to 20,000 S.F. projects, and 50 mile to 100 mile transport.

Concrete is f'c = 5 KSI and Steel is fy = 250 or 300 KSI

Note: Deduct from prices 20% for Southern states. Add to prices 10% for Western states.

Description of Table: Enter table at span and load. Most economical sections will generally consist of normal weight concrete without topping. If acceptable, note this price, depth and weight. For topping and/or lightweight concrete, note appropriate data.

Generally used on masonry and concrete bearing or reinforced concrete and steel framed structures.

The solid 4" slabs are used for light loads and short spans. The 6" to 12" thick hollow core units are used for longer spans and heavier loads. Cores may carry utilities.

Topping is used structurally for loads or rigidity and architecturally to level or slope surface.

Camber and deflection and change in direction of spans must be considered (door openings, etc.), especially untopped.

System Components		QUANTITY	UNIT	COST PER S.F.		
				MAT.	INST.	TOTAL
SYSTEM 3.5-210-2000						
10' SPAN, 40 LBS S.F. WORKING LOAD, 2" TOPPING						
Precast prestressed concrete roof/floor slabs 4" thick, grouted		1.000	S.F.	3.90	1.08	4.98
Edge forms to 6" high on elevated slab, 4 uses		.100	L.F.	.04	.25	.29
Welded wire fabric 6 x 6 - W1.4 x W1.4 (10 x 10), 21 lb/csf, 10% lap		.010	C.S.F.	.08	.23	.31
Concrete ready mix, regular weight, 3000 psi		.170	C.F.	.36		.36
Place and vibrate concrete, elevated slab less than 6", pumped		.170	C.F.		.16	.16
Finishing floor, monolithic steel trowel finish for resilient tile		1.000	S.F.		.54	.54
Curing with sprayed membrane curing compound		.010	C.S.F.	.02	.05	.07
	TOTAL			4.40	2.31	6.71

3.5-210		Precast Plank with No Topping							
	SPAN (FT.)	SUPERIMPOSED LOAD (P.S.F.)	TOTAL DEPTH (IN.)	DEAD LOAD (P.S.F.)	TOTAL LOAD (P.S.F.)	COST PER S.F.			
						MAT.	INST.	TOTAL	
0720	10	40	4	50	90	3.90	1.08	4.98	
0750	R3.5 -010	75	6	50	125	4.18	.90	5.08	
0770		100	6	50	150	4.18	.90	5.08	
0800	15	40	6	50	90	4.18	.90	5.08	
0820	R3.5 -100	75	6	50	125	4.18	.90	5.08	
0850		100	6	50	150	4.18	.90	5.08	
0875	20	40	6	50	90	4.18	.90	5.08	
0900		75	6	50	125	4.18	.90	5.08	
0920		100	6	50	150	4.18	.90	5.08	
0950	25	40	6	50	90	4.18	.90	5.08	
0970		75	8	55	130	4.47	.73	5.20	
1000		100	8	55	155	4.47	.73	5.20	
1200	30	40	8	55	95	4.47	.73	5.20	
1300		75	8	55	130	4.47	.73	5.20	
1400		100	10	70	170	4.90	.59	5.49	
1500	40	40	10	70	110	4.90	.59	5.49	
1600		75	12	70	145	5.15	.55	5.70	
1700	45	40	12	70	110	5.15	.55	5.70	

Figure 5.5

FOUNDATIONS	A1.1-140	Strip Footings

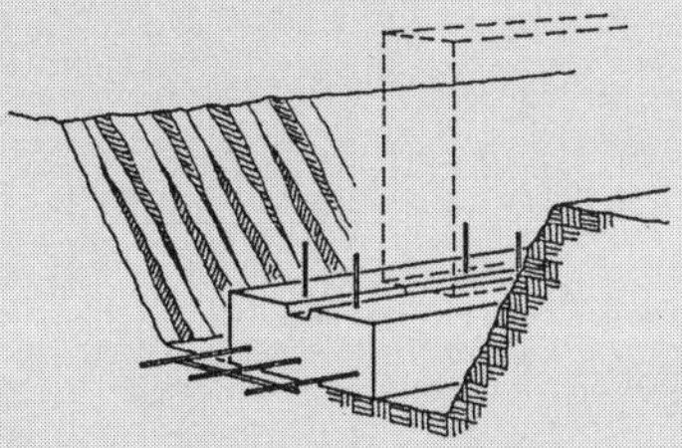

The Strip Footing System includes: excavation; hand trim; all forms needed for footing placement; forms for 2″ x 6″ keyway (four uses); dowels; and 3,000 p.s.i. concrete.

The footing size required varies for different soils. Soil bearing capacities are listed for 3 KSF and 6 KSF. Depths of the system range from 8″ to 24″. Widths range from 16″ to 96″. Smaller strip footings may not require reinforcement.

Please see the reference section for further design and cost information.

System Components	QUANTITY	UNIT	COST PER L.F. MAT.	INST.	TOTAL
SYSTEM 1.1-140-2500					
STRIP FOOTING, LOAD 5.1KLF, SOIL CAP. 3 KSF, 24″WIDE X12″DEEP, REINF.					
Trench excavation	.148	C.Y.		.78	.78
Hand trim	2.000	S.F.		1.06	1.06
Compacted backfill	.074	C.Y.		.16	.16
Formwork, 4 uses	2.000	S.F.	.94	5.14	6.08
Keyway form, 4 uses	1.000	L.F.	.28	.64	.92
Reinforcing, fy = 60000 psi	3.000	Lb.	.90	1.11	2.01
Dowels	2.000	Ea.	2.32	3.94	6.26
Concrete, f'c = 3000 psi	.074	C.Y.	4.26		4.26
Place concrete, direct chute	.074	C.Y.		1.03	1.03
Screed finish	2.000	S.F.		.92	.92
TOTAL			8.70	14.78	23.48

1.1-140	Strip Footings	COST PER L.F. MAT.	INST.	TOTAL
2100	Strip footing, load 2.6KLF,soil capacity 3KSF, 16″wide x 8″deep plain	4.27	8.80	13.07
2300	Load 3.9 KLF soil capacity, 3 KSF, 24″wide x 8″deep, plain	5.25	9.75	15
2500	Load 5.1KLF, soil capacity 3 KSF, 24″wide x 12″deep, reinf. R1.1 -140	8.70	14.80	23.50
2700	Load 11.1KLF, soil capacity 6 KSF, 24″wide x 12″deep, reinf.	8.70	14.80	23.50
2900	Load 6.8 KLF, soil capacity 3 KSF, 32″wide x 12″deep, reinf.	10.45	16.15	26.60
3100	Load 14.8 KLF, soil capacity 6 KSF, 32″wide x 12″deep, reinf.	10.45	16.15	26.60
3300	Load 9.3 KLF, soil capacity 3 KSF, 40″wide x 12″deep, reinf.	12.20	17.50	29.70
3500	Load 18.4 KLF, soil capacity 6 KSF, 40″wide x 12″deep, reinf.	12.30	17.65	29.95
3700	Load 10.1KLF, soil capacity 3 KSF, 48″wide x 12″deep, reinf.	13.55	18.90	32.45
3900	Load 22.1KLF, soil capacity 6 KSF, 48″wide x 12″deep, reinf.	14.45	20	34.45
4100	Load 11.8KLF, soil capacity 3 KSF, 56″wide x 12″deep, reinf.	15.80	21	36.80
4300	Load 25.8KLF, soil capacity 6 KSF, 56″wide x 12″deep, reinf.	17.05	22.50	39.55
4500	Load 10KLF, soil capacity 3 KSF, 48″wide x 16″deep, reinf.	16.70	21.50	38.20
4700	Load 22KLF, soil capacity 6 KSF, 48″wide, 16″deep, reinf.	17.05	22	39.05
4900	Load 11.6KLF, soil capacity 3 KSF, 56″wide x 16″deep, reinf.	18.90	30.50	49.40
5100	Load 25.6KLF, soil capacity 6 KSF, 56″wide x 16″deep, reinf.	19.95	31.50	51.45
5300	Load 13.3KLF, soil capacity 3 KSF, 64″wide x 16″deep, reinf.	21.50	26	47.50
5500	Load 29.3KLF, soil capacity 6 KSF, 64″wide x 16″deep, reinf.	23	27.50	50.50
5700	Load 15KLF, soil capacity 3 KSF, 72″wide x 20″deep, reinf.	28.50	31	59.50
5900	Load 33KLF, soil capacity 6 KSF, 72″wide x 20″deep, reinf.	30	33	63
6100	Load 18.3KLF, soil capacity 3 KSF, 88″wide x 24″deep, reinf.	40	38.50	78.50
6300	Load 40.3KLF, soil capacity 6 KSF, 88″wide x 24″deep, reinf.	43.50	43	86.50
6500	Load 20KLF, soil capacity 3 KSF, 96″wide x 24″deep, reinf.	43.50	41.50	85
6700	Load 44 KLF, soil capacity 6 KSF, 96″ wide x 24″ deep, reinf.	46	44.50	90.50

Figure 5.6

Load to Pier:

Wall and Roof Load: 30′ × 3.92 KLF		=	117.6k
Grade Beam Load: 30′ × 3.33′ × .15 KCF		=	15.0k
Load to Pier:			132.6k

Enter the table in Figure 4.11, C.I.P Columns — Square Tied, Minimum Reinforcing: 16 sq. pier maximum load 300k > 132.6k required.

Load to Spread Footing: 3 KSF soil bearing capacity

Load to Pier:		132.6k
Pier 5′ × 1.33′ × 1.33′ × .15 KCF	=	1.3k
Load to Spread Footing:		133.9k

The Spread Footing System selection, Figure 5.2:

133.9k (150k), 7′-6″ square x 18″ deep.

Piles, Pile Caps, and Caissons

The recommendations of the Geotechnical Report (which will specify the foundation system best suited to the project) should be followed.

Piles

Piles are column-like shafts that receive superstructure loads, overturning forces, or uplift forces. They receive these loads from isolated column or pier foundations (pile caps), foundation walls, grade beams, or foundation mats. The piles then transfer these loads through shallower poor soil layers to deeper soil with adequate support strength and acceptable settlement.

Pile foundations are costly. Piles usually are associated with difficult foundation problems and substructure conditions. Ground conditions determine the type of pile to use, and different pile types have been developed to suit ground conditions. A full investigation of ground conditions, early in the selection process, is essential to provide maximum information for professional foundation engineering and an acceptable structure. The Geotechnical Report is the best source of information for the specific system to be used.

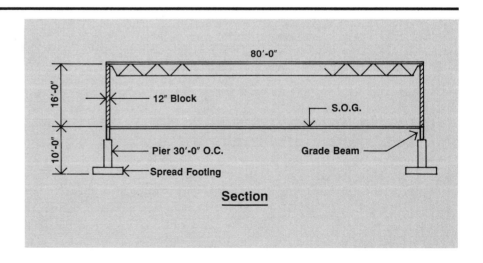

Figure 5.7

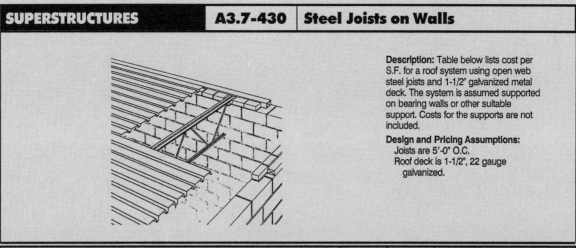

Description: Table below lists cost per S.F. for a roof system using open web steel joists and 1-1/2″ galvanized metal deck. The system is assumed supported on bearing walls or other suitable support. Costs for the supports are not included.

Design and Pricing Assumptions:
  Joists are 5′-0″ O.C.
  Roof deck is 1-1/2″, 22 gauge galvanized.

System Components	QUANTITY	UNIT	COST PER S.F.		
			MAT.	INST.	TOTAL
SYSTEM 3.7-430-1100					
METAL DECK AND JOISTS, 20′ SPAN, 20 PSF S. LOAD					
Open web joists, horiz. bridging T.L. lot, to 30′ span	1.114	Lb.	.48	.25	.73
Metal decking, open type, galv 1-1/2″ deep	1.050	S.F.	.76	.37	1.13
TOTAL			1.24	.62	1.86

3.7-430		Steel Joists & Deck on Bearing Walls					
	BAY SIZE (FT.)	SUPERIMPOSED LOAD (P.S.F.)	DEPTH (IN.)	TOTAL LOAD (P.S.F.)	COST PER S.F.		
					MAT.	INST.	TOTAL
1100	20	20	13-1/2	40	1.24	.62	1.86
1200		30	15-1/2	50	1.27	.63	1.90
1300		40	15-1/2	60	1.36	.68	2.04
1400	25	20	17-1/2	40	1.36	.68	2.04
1500		30	17-1/2	50	1.47	.73	2.20
1600		40	19-1/2	60	1.49	.75	2.24
1700	30	20	19-1/2	40	1.49	.74	2.23
1800		30	21-1/2	50	1.52	.76	2.28
1900		40	23-1/2	60	1.64	.82	2.46
2000	35	20	23-1/2	40	1.64	.69	2.33
2100		30	25-1/2	50	1.69	.72	2.41
2200		40	25-1/2	60	1.80	.76	2.56
2300	40	20	25-1/2	41	1.82	.77	2.59
2400		30	25-1/2	51	1.94	.81	2.75
2500		40	25-1/2	61	2.01	.84	2.85
2600	45	20	27-1/2	41	2.05	1.10	3.15
2700		30	31-1/2	51	2.16	1.16	3.32
2800		40	31-1/2	61	2.28	1.23	3.51
2900	50	20	29-1/2	42	2.29	1.23	3.52
3000		30	31-1/2	52	2.50	1.36	3.86
3100		40	31-1/2	62	2.65	1.44	4.09
3200	60	20	37-1/2	42	2.55	1.26	3.81
3300		30	37-1/2	52	2.82	1.40	4.22
3400		40	37-1/2	62	2.82	1.40	4.22
3500	70	20	41-1/2	42	2.81	1.40	4.21
3600		30	41-1/2	52	2.99	1.49	4.48
3700		40	41-1/2	64	3.62	1.80	5.42
3800	80	20	45-1/2	44	3.93	2.02	5.95
3900		30	45-1/2	54	3.93	2.02	5.95
4000		40	45-1/2	64	4.35	2.24	6.59

Figure 5.8

The Grade Beam System includes: excavation with a truck mounted backhoe; hand trim; backfill; forms (four uses); reinforcing steel; and 3,000 p.s.i. concrete placed from chute.

Superimposed loads vary in the listing from 8 Kips per linear foot (KLF) to 50 KLF. In the Expanded System Listing, the span of the beams varies from 15' to 40'. Depth varies from 28" to 52". Width varies from 12" to 28".

Please see the reference section for further design and cost information.

System Components	QUANTITY	UNIT	COST PER L.F.		
			MAT.	INST.	TOTAL
SYSTEM 1.1-230-2220					
GRADE BEAM, 15' SPAN, 28" DEEP, 12" WIDE, 8 KLF LOAD					
Excavation, trench, hydraulic backhoe, 3/8 CY bucket	.260	C.Y.		1.39	1.39
Trim sides and bottom of trench, regular soil	2.000	S.F.		1.06	1.06
Backfill, by hand, compaction in 6" layers, using vibrating plate	.170	C.Y.		.91	.91
Forms in place, grade beam, 4 uses	4.700	SFCA	3.10	14.99	18.09
Reinforcing in place, beams & girders, #8 to #14	.019	Ton	10.64	11.12	21.76
Concrete ready mix, regular weight, 3000 psi	.090	C.Y.	5.18		5.18
Place and vibrate conc. for grade beam, direct chute	.090	C.Y.		1	1
TOTAL			18.92	30.47	49.39

1.1-230	Grade Beams		COST PER L.F.		
			MAT.	INST.	TOTAL
2220	Grade beam, 15' span, 28" deep, 12" wide, 8 KLF load		18.90	30.50	49.40
2240	14" wide, 12 KLF load		19.50	30.50	50
2260	40" deep, 12" wide, 16 KLF load	R1.1 -230	19.70	37.50	57.20
2280	20 KLF load		22.50	40.50	63
2300	52" deep, 12" wide, 30 KLF load		26	51	77
2320	40 KLF load		31.50	57	88.50
2340	50 KLF load		37.50	62.50	100
3360	20' span, 28" deep, 12" wide, 2 KLF load		12.20	23.50	35.70
3380	16" wide, 4 KLF load		15.85	26	41.85
3400	40" deep, 12" wide, 8 KLF load		19.70	37.50	57.20
3420	12 KLF load		25	42.50	67.50
3440	14" wide, 16 KLF load		30	47.50	77.50
3460	52" deep, 12" wide, 20 KLF load		32.50	57.50	90
3480	14" wide, 30 KLF load		44	68.50	112.50
3500	20" wide, 40 KLF load		51.50	72	123.50
3520	24" wide, 50 KLF load		63.50	82.50	146
4540	30' span, 28" deep, 12" wide, 1 KLF load		12.75	24	36.75
4560	14" wide, 2 KLF load		20	33.50	53.50
4580	40" deep, 12" wide, 4 KLF load		23.50	40	63.50
4600	18" wide, 8 KLF load		33.50	48.50	82
4620	52" deep, 14" wide, 12 KLF load		42.50	66.50	109
4640	20" wide, 16 KLF load		51.50	73	124.50
4660	24" wide, 20 KLF load		63.50	83	146.50
4680	36" wide, 30 KLF load		90.50	104	194.50
4700	48" wide, 40 KLF load		119	127	246
5720	40' span, 40" deep, 12" wide, 1 KLF load		17.50	35.50	53
5740	2 KLF load		21.50	39.50	61
5760	52" deep, 12" wide, 4 KLF load		31.50	57	88.50

Figure 5.9

Piles support loads by end bearing and friction. Both of these are generally present, but piles are designated by their principle method of load transfer to soil.

End bearing piles have shafts that pass through soft strata or thin hard strata, and they have tips that bear on bedrock, or they penetrate some distance into a dense, adequate soil (sand or gravel).

Friction piles have shafts that may be entirely embedded in cohesive soil (moist clay), and provide support mainly by adhesion or "skin-friction" between soil and shaft area.

Seldom are piles installed singly; they are generally installed in clusters (or groups). Some codes require a minimum of three piles per major column load or two per foundation wall or grade beam. Single pile capacity is limited by pile structural strength or support strength of soil. Support capacity of a pile cluster is almost always less than the sum of its individual pile capacities because of overlapping of bearing and friction stresses.

Pile Caps

A *pile cap* is a structural member placed on, and fastened to, the top of all piles in a group.

> NOTE: The function of a pile cap is to transfer a column or pier load equally to each pile in its supporting cluster.

Caissons

Caissons, as covered in this section, are drilled or dug-out cylindrical foundation shafts that function primarily as short column-like compression members. They transfer superstructure loads through inadequate soils to bedrock or hard strata. They may be either reinforced or unreinforced concrete, and either straight or belled out at the bearing level.

Shaft diameters range in size from 20″ to 84″, with the most typical sizes beginning at 34″. If inspection of the bottom is required, the minimum practical diameter is 30″. If handwork is required (in addition to mechanical belling) the minimum diameter is 32″. The most frequently used shaft diameter is probably 36″ with a 5- or 6-foot bell diameter. The maximum practical bell diameter is 3 times the shaft diameter.

Plain concrete is commonly used, poured directly against the excavated face of soil. Permanent casings add to the cost and should be avoided for economic reasons. Wet or loose strata are undesirable. The associated installation sometimes involves a mudding operation with bentonite clay slurry to keep the walls of excavation stable (this cost is not included in the tables).

Reinforcement is sometimes used, especially for heavy loads. It is required if uplift, bending moment, or lateral loads are present. A small amount of reinforcement is desirable at the top portion of each caisson, even if the above conditions theoretically are not present. This is a provision for construction eccentricities and other possibilities. Reinforcement, if present, should extend below the soft strata. Horizontal reinforcement is not required for belled bottoms.

There are three basic types of caisson bearing details:
- **Belled** — generally recommended to provide reduced bearing pressure on soil. These are not for shallow depths or poor soils.
  - Good soils for belling include most clays, hardpan, soft shale, and decomposed rock.
  - Soils requiring handwork include hard shale, limestone, and sandstone.

- Soils not recommended include sand, gravel, silt, and igneous rock. Compact sand and gravel above water table may stand. Water in the bearing strata is undesirable.
  - **Straight Shafted** — have no bell but the entire length is enlarged to permit safe bearing pressures. They are most economical for light loads on high bearing capacity soil.
  - **Socketed or Keyed**

Advantages of using caissons include:
- Shafts can pass through soils that piles cannot penetrate.
- There is no soil heaving or displacement during installation.
- There is no vibration during installation.
- They create less noise than pile driving.
- Bearing strata can be visually inspected and tested.

Uses of caissons include:
- Situations where unsuitable soil exists to moderate depth
- Tall structures
- Heavy structures
- Underpinning (extensive use)

Selecting Pilings, Pile Caps, and Caissons

Selecting the type of foundation assembly to use is based on engineering judgment and relative costs. System selection and cost analysis are dealt with in this example.

Select an appropriate piling foundation system, given the following information.

Construction: 8-story building

> 30'-0" x 35'-0" bays composite beam and deck and lightweight concrete slab
> 12'-0" floor to floor

Superimposed Loads:

> Floor: 75 psf
> Roof: 40 psf

Total Load (Figure 5.10):

> Roof: 117 psf
> Floor: 82 psf

Interior Column Load:

Roof	30' × 35'	× .087 KSF	=	91k
Floors	7 (30' × 35'	× .117 KSF)	=	860k
Column (estimated)	8 floors	× 12' × .15 KLF	=	14k
Interior Column Load:				965k

See the boring log in Figure 5.11. Firm fine sand and gravel start at 20' depth and continue to refusal at 45'-6" depth below grade.

Good bearing strata at 25'-0" allows the use of pressure injected piles. End bearing piles or caissons would be driven or placed to the rock strata or 50'-0" + below grade to allow for cutoffs and variations in the elevation of the rock strata.

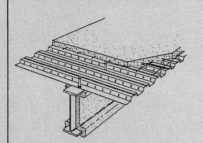

| SUPERSTRUCTURES | A3.5-540 | Composite Beam, Deck & Slab |

Description: Table below lists costs ($/S.F.) for a floor system using composite steel beams with welded shear studs, composite steel deck, and light weight concrete slab reinforced with W.W.F. Price includes sprayed fiber fireproofing on steel beams.

Design and Pricing Assumptions:
Structural steel is A36, high strength bolted.
Composite steel deck varies from 22 gauge to 16 gauge, galvanized.

Shear Studs are 3/4".
W.W.F., 6 x 6 – W1.4 x W1.4 (10 x 10)
Concrete f'c = 3 KSI, lightweight.
Steel trowel finish and cure.
Fireproofing is sprayed fiber (non-asbestos).

Spandrels are assumed the same as interior beams and girders to allow for exterior wall loads and bracing or moment connections.

System Components	QUANTITY	UNIT	COST PER S.F.		
			MAT.	INST.	TOTAL
SYSTEM 3.5-540-2400					
20X25 BAY, 40 PSF S. LOAD, 5-1/2" SLAB, 17-1/2" TOTAL THICKNESS					
Structural steel	4.320	Lb.	2.64	1.12	3.76
Welded shear connectors 3/4" diameter 4-7/8" long	.163	Ea.	.09	.22	.31
Metal decking, non-cellular composite, galv. 3" deep, 22 gauge	1.050	S.F.	.98	.59	1.57
Sheet metal edge closure form, 12", w/2 bends, 18 ga	.045	L.F.	.10	.07	.17
Welded wire fabric rolls, 6 x 6 - W1.4 x W1.4 (10 x 10), 21 lb/csf	1.000	S.F.	.08	.23	.31
Concrete ready mix, light weight, 3,000 PSI	.333	C.F.	1.19		1.19
Place and vibrate concrete, elevated slab less than 6", pumped	.333	C.F.		.32	.32
Finishing floor, monolithic steel trowel finish for finish floor	1.000	S.F.		.61	.61
Curing with sprayed membrane curing compound	.010	C.S.F.	.02	.05	.07
Shores, erect and strip vertical to 10' high	.020	Ea.		.25	.25
Sprayed mineral fiber/cement for fireproof, 1" thick on beams	.483	S.F.	.20	.35	.55
TOTAL			5.30	3.81	9.11

| 3.5-540 | Composite Beams, Deck & Slab | | | | | | | |

	BAY SIZE (FT.)	SUPERIMPOSED LOAD (P.S.F.)	SLAB THICKNESS (IN.)	TOTAL DEPTH (FT. - IN.)	TOTAL LOAD (P.S.F.)	COST PER S.F.		
						MAT.	INST.	TOTAL
2400	20x25	40	5-1/2	1 - 5-1/2	80	5.30	3.81	9.11
2500	R3.5 -100	75	5-1/2	1 - 9-1/2	115	5.50	3.82	9.32
2750		125	5-1/2	1 - 9-1/2	167	6.80	4.50	11.30
2900		200	6-1/4	1 - 11-1/2	251	7.65	4.84	12.49
3000	25x25	40	5-1/2	1 - 9-1/2	82	5.55	3.65	9.20
3100		75	5-1/2	1 - 11-1/2	118	6.10	3.69	9.79
3200		125	5-1/2	2 - 2-1/2	169	6.40	4.01	10.41
3300		200	6-1/4	2 - 6-1/4	252	8.30	4.65	12.95
3400	25x30	40	5-1/2	1 - 11-1/2	83	5.65	3.61	9.26
3600		75	5-1/2	1 - 11-1/2	119	6.05	3.66	9.71
3900		125	5-1/2	1 - 11-1/2	170	6.95	4.12	11.07
4000		200	6-1/4	2 - 6-1/4	252	8.35	4.66	13.01
4200	30x30	40	5-1/2	1 - 11-1/2	81	5.35	3.74	9.09
4400		75	5-1/2	2 - 2-1/2	116	5.85	3.89	9.74
4500		125	5-1/2	2 - 5-1/2	168	7.05	4.37	11.42
4700		200	6-1/4	2 - 9-1/4	252	8.45	5.05	13.50
4900	30x35	40	5-1/2	2 - 2-1/2	82	5.60	3.84	9.44
5100		75	5-1/2	2 - 5-1/2	117	6.10	3.95	10.05
5300		125	5-1/2	2 - 5-1/2	169	7.25	4.46	11.71
5500		200	6-1/4	2 - 9-1/4	254	8.45	5.05	13.50
5750	35x35	40	5-1/2	2 - 5-1/2	84	6.25	3.87	10.12
6000		75	5-1/2	2 - 5-1/2	121	7.10	4.14	11.24
7000		125	5-1/2	2 - 8-1/2	170	8.30	4.71	13.01
7200		200	5-1/2	2 - 11-1/2	254	9.25	5.20	14.45

Figure 5.10

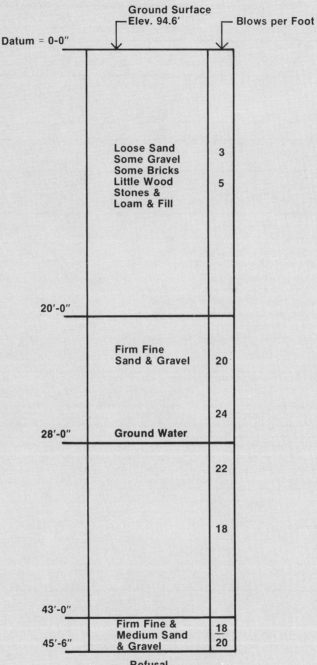

Boring #3

Figures in right hand column indicate number of blows required to drive a **Two Inch Split Sampler**, 12 inches, using a 140 lb. weight falling 30 inches ±.

Unless otherwise specified, water levels noted were observed at completion of borings, and do not necessarily represent permanent ground water levels.

Figure 5.11

Option One

End bearing, try steel H piles, 8 piles required.
Pile cost per interior column (Figure 5.12) $ 9,575
Pile cap, Figure 5.13 (1243k, 8 piles, 3'-0" spacing)   1,465

$11,040 per column

Option Two

Pressure injected piles (require 4'-6" pile spacing),
   6 piles required.
Pile cost per interior column (Figure 5.14) $5,900
Pile cap, Figure 5.13 (1413k, 6 piles, 4'-6" spacing)   1,185

$7,085

Option Three

Caissons (Figure 5.15).
Caisson 4' × 100' @ 1,200k = $13,375/2 (50' required) = $6,688

The choice appears to be caissons or pressure-injected piles, depending on availability in the area. Add mobilization, demobilization, and pile testing costs to obtain complete prices.

To satisfy the column loading of 965k the following pile clusters were selected from *Means Assemblies Cost Data*.

C.I.P. concrete piles, 50' long 1200k	15 piles	$25,475
Precast concrete piles, 50' long 1200k	19 piles	29,400
Steel pipe piles, 50' long 1200k	16 piles	25,200
Steel H piles, 50' long 1200k	8 piles	9,575
Step tapered steel piles, 50' long 1200k	16 piles	11,350
Pressure injected footings, 25' long 1200k	6 piles	5,900

Steel H piles appear to be the economical choice for end bearing piles. Check steel H piles 50'-0" long against pressure injected footings 25'-0" long and caissons 50'-0" long to support a 965k column load.

Both steel H pile clusters and pressure-injected footing clusters require a pile cap to transfer the column load to each pile in the cluster.

Pile caps are shown in the table in Figure 5.13. Enter the table with a column or pier load, number of piles in the cluster, and the required pile spacing. Pressure-injected footings require 4'-6" pile spacing while 3'-0" pile spacing is adequate for other pile types.

Conclusion

Spread and strip footings are the most economical foundation system to use if there is adequate soil bearing capacity at a reasonable depth. Also, these systems can be used with piers, columns, walls, and grade beams.

Pile or caisson assemblies are more costly to install, and more difficult to price since there are so many unknown factors involved in their installation.

An assemblies square foot estimate for a foundation system that involves piles or caissons requires additional research. Always collect as much information as possible about subsoil types and the specific systems being used by contacting the following:

- Area building inspector or building department
- Local pile or caisson subcontractors
- Local contractors or engineers

Often, a phone call to the right source provides sufficient information to determine approximate depths to good bearing strata in an area.

At best, assembling an assemblies square foot estimate for pile foundation systems is useful only for general budget figures.

H

A Steel "H" Pile System includes: steel H sections; heavy duty driving point; splices where applicable and allowance for cutoffs.

The Expanded System Listing shows costs per cluster of piles. Clusters range from one pile to seventeen piles. Loads vary from 50 Kips to 2,000 Kips. All loads for Steel H Pile systems are given in terms of end bearing capacity.

Steel sections range from 10" x 10" to 14" x 14" in the Expanded System Listing. The 14" x 14" steel section is used for all H piles used in applications requiring a working load over 800 Kips.

Please see the reference section for cost of mobilization of the pile driving equipment and other design and cost information.

System Components	QUANTITY	UNIT	COST EACH		
			MAT.	INST.	TOTAL
SYSTEM 1.4-143-2220					
STEEL H PILES, 50' LONG, 100K LOAD, END BEARING, 1 PILE					
Steel H piles 10" x 10", 42 #/L.F.	53.000	V.L.F.	612.15	372.59	984.74
Heavy duty point, 10"	1.000	Ea.	88	104	192
Pile cut off, steel pipe or H piles	1.000	Ea.		18.80	18.80
TOTAL			700.15	495.39	1,195.54

1.4-143	Steel H Piles		COST EACH		
			MAT.	INST.	TOTAL
2220	Steel H piles, 50' long, 100K load, end bearing, 1 pile		700	495	1,195
2260	2 pile cluster		1,400	995	2,395
2280	200K load, end bearing, 2 pile cluster	R1.4 -100	1,400	995	2,395
2300	3 pile cluster		2,100	1,475	3,575
2320	400K load, end bearing, 3 pile cluster		2,100	1,475	3,575
2340	4 pile cluster		2,800	1,975	4,775
2360	6 pile cluster		4,200	2,975	7,175
2380	800K load, end bearing, 5 pile cluster		3,500	2,475	5,975
2400	7 pile cluster		4,900	3,475	8,375
2420	12 pile cluster		8,400	5,925	14,325
2440	1200K load, end bearing, 8 pile cluster		5,600	3,975	9,575
2460	11 pile cluster		7,700	5,450	13,150
2480	17 pile cluster		11,900	8,425	20,325
2500	1600K load, end bearing, 10 pile cluster		8,675	5,200	13,875
2520	14 pile cluster		12,200	7,275	19,475
2540	2000K load, end bearing, 12 pile cluster		10,400	6,225	16,625
2560	18 pile cluster		15,600	9,325	24,925
2580					
3580	100' long, 50K load, end bearing, 1 pile		2,350	1,075	3,425
3600	100K load, end bearing, 1 pile		2,350	1,075	3,425
3620	2 pile cluster		4,725	2,150	6,875
3640	200K load, end bearing, 2 pile cluster		4,725	2,150	6,875
3660	3 pile cluster		7,075	3,225	10,300
3680	400K load, end bearing, 3 pile cluster		7,075	3,225	10,300
3700	4 pile cluster		9,425	4,300	13,725
3720	6 pile cluster		14,100	6,475	20,575
3740	800K load, end bearing, 5 pile cluster		11,800	5,375	17,175
3760	7 pile cluster		16,500	7,550	24,050

Figure 5.12

A1.1-330 | **Pile Caps**

These pile cap systems include excavation with a truck mounted hydraulic excavator, hand trimming, compacted backfill, forms for concrete, templates for dowels or anchor bolts, reinforcing steel and concrete placed and screeded.

Pile embedment is assumed as 6". Design is consistent with the Concrete Reinforcing Steel Institute Handbook f'c = 3000 psi, fy = 60,000.

Please see the reference section for further design and cost information.

System Components

System Components	QUANTITY	UNIT	COST EACH		
			MAT.	INST.	TOTAL
SYSTEM 1.1-330-5100					
CAP FOR 2 PILES, 6'-6"X3'-6"X20", 15 TON PILE, 8" MIN. COL., 45K COL. LOAD					
Excavation, bulk, hyd excavator, truck mtd. 30" bucket 1/2 CY	2.890	C.Y.		15.55	15.55
Trim sides and bottom of trench, regular soil	23.000	S.F.		12.19	12.19
Dozer backfill & roller compaction	1.500	C.Y.		3.20	3.20
Forms in place pile cap, square or rectangular, 4 uses	33.000	SFCA	24.42	107.58	132
Templates for dowels or anchor bolts	8.000	Ea.	5.36	20	25.36
Reinforcing in place footings, #8 to #14	.025	Ton	14	10.88	24.88
Concrete ready mix, regular weight, 3000 psi	1.400	C.Y.	80.50		80.50
Place and vibrate concrete for pile caps, under 5CY, direct chute	1.400	C.Y.		25.96	25.96
Monolithic screed finish	23.000	S.F.		7.59	7.59
TOTAL			124.28	202.95	327.23

1.1-330 Pile Caps

	NO. PILES	SIZE FT-IN X FT-IN X IN	PILE CAPACITY (TON)	COLUMN SIZE (IN)	COLUMN LOAD (KIPS)	COST EACH		
						MAT.	INST.	TOTAL
5100	2	6-6x3-6x20	15	8	45	124	203	327
5150	R1.1 -330	26	40	8	155	152	247	399
5200		34	80	11	314	206	320	526
5250		37	120	14	473	221	340	561
5300	3	5-6x5-1x23	15	8	75	148	235	383
5350		28	40	10	232	165	265	430
5400		32	80	14	471	191	300	491
5450		38	120	17	709	220	345	565
5500	4	5-6x5-6x18	15	10	103	166	234	400
5550		30	40	11	308	241	335	576
5600		36	80	16	626	284	395	679
5650		38	120	19	945	300	410	710
5700	6	8-6x5-6x18	15	12	156	274	340	614
5750		37	40	14	458	445	500	945
5800		40	80	19	936	505	560	1,065
5850		45	120	24	1413	565	620	1,185
5900	8	8-6x7-9x19	15	12	205	410	465	875
5950		36	40	16	610	590	605	1,195
6000		44	80	22	1243	730	735	1,465
6050		47	120	27	1881	795	790	1,585
6100	10	11-6x7-9x21	15	14	250	580	570	1,150
6150		39	40	17	756	880	815	1,695
6200		47	80	25	1547	1,075	965	2,040
6250		49	120	31	2345	1,125	1,025	2,150

Figure 5.13

Pressure Injected Piles are usually uncased up to 25' and cased over 25' depending on soil conditions.

These costs include excavation and hauling of excess materials; steel casing over 25'; reinforcement; 3,000 p.s.i. concrete; plus mobilization and demobilization of equipment for a distance of up to fifty miles to and from the job site.

The Expanded System lists cost per cluster of piles. Clusters range from one pile to eight piles. End-bearing loads range from 50 Kips to 1,600 Kips.

Please see the reference section for further design and cost information.

System Components	QUANTITY	UNIT	COST EACH		
			MAT.	INST.	TOTAL
SYSTEM 1.4-820-4200					
PRESSURE INJECTED FOOTING, END BEARING, 50' LONG, 50K LOAD, 1 PILE					
Pressure injected footings, cased, 30-60 ton cap., 12" diameter	50.000	V.L.F.	467.50	781	1,248.50
Pile cutoff, concrete pile with thin steel shell	1.000	Ea.		9.40	9.40
TOTAL			467.50	790.40	1,257.90

1.4-820	Pressure Injected Footings		COST EACH		
			MAT.	INST.	TOTAL
2200	Pressure injected footing, end bearing, 25' long, 50K load, 1 pile		205	555	760
2400	100K load, 1 pile		205	555	760
2600	2 pile cluster	R1.4 -100	410	1,100	1,510
2800	200K load, 2 pile cluster		410	1,100	1,510
3200	400K load, 4 pile cluster		820	2,225	3,045
3400	7 pile cluster		1,425	3,875	5,300
3800	1200K load, 6 pile cluster		1,700	4,200	5,900
4000	1600K load, 7 pile cluster		1,975	4,925	6,900
4200	50' long, 50K load, 1 pile		470	790	1,260
4400	100K load, 1 pile		680	790	1,470
4600	2 pile cluster		1,350	1,575	2,925
4800	200K load, 2 pile cluster		1,350	1,575	2,925
5000	4 pile cluster		2,700	3,150	5,850
5200	400K load, 4 pile cluster		2,700	3,150	5,850
5400	8 pile cluster		5,425	6,325	11,750
5600	800K load, 7 pile cluster		4,750	5,550	10,300
5800	1200K load, 6 pile cluster		4,450	4,725	9,175
6000	1600K load, 7 pile cluster		5,175	5,550	10,725

Figure 5.14

65

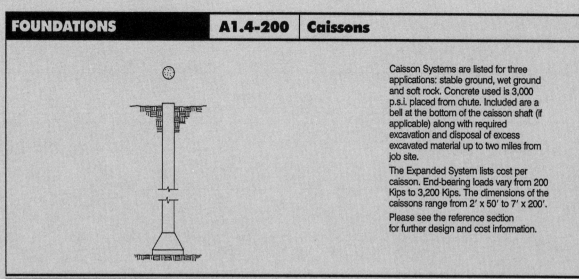

FOUNDATIONS · A1.4-200 · Caissons

Caisson Systems are listed for three applications: stable ground, wet ground and soft rock. Concrete used is 3,000 p.s.i. placed from chute. Included are a bell at the bottom of the caisson shaft (if applicable) along with required excavation and disposal of excess excavated material up to two miles from job site.

The Expanded System lists cost per caisson. End-bearing loads vary from 200 Kips to 3,200 Kips. The dimensions of the caissons range from 2' x 50' to 7' x 200'.

Please see the reference section for further design and cost information.

System Components	QUANTITY	UNIT	COST EACH		
			MAT.	INST.	TOTAL
SYSTEM 1.4-200-2200					
CAISSON, STABLE GROUND, 3000 PSI CONC., 10KSF BRNG, 200K LOAD, 2'X50'					
Caissons, drilled, to 50', 24" shaft diameter, .116 C.Y./L.F.	50.000	V.L.F.	937.50	915	1,852.50
Reinforcing in place, columns, #3 to #7	.060	Ton	34.50	63	97.50
Caisson bell excavation and concrete, 4' diameter .444 CY	1.000	Ea.	49	173.50	222.50
Load & haul excess excavation, 2 miles	6.240	C.Y.		25.53	25.53
TOTAL			1,021	1,177.03	2,198.03

1.4-200	Caissons		COST EACH		
			MAT.	INST.	TOTAL
2200	Caisson, stable ground, 3000 PSI conc, 10KSF brng, 200K load, 2'x50'		1,025	1,175	2,200
2400	400K load, 2'-6"x50'-0"		1,600	1,900	3,500
2600	800K load, 3'-0"x100'-0"	R1.4	4,625	4,475	9,100
2800	1200K load, 4'-0"x100'-0"	-200	7,725	5,650	13,375
3000	1600K load, 5'-0"x150'-0"		17,200	8,725	25,925
3200	2400K load, 6'-0"x150'-0"		25,900	11,400	37,300
3400	3200K load, 7'-0"x200'-0"		45,800	16,700	62,500
5000	Wet ground, 3000 PSI conc., 10 KSF brng, 200K load, 2'-0"x50'-0"		860	1,650	2,510
5200	400K load, 2'-6"x50'-0"		1,425	2,825	4,250
5400	800K load, 3'-0"x100'-0"		3,950	7,600	11,550
5600	1200K load, 4'-0"x100'-0"		6,500	11,300	17,800
5800	1600K load, 5'-0"x150'-0"		14,800	24,400	39,200
6000	2400K load, 6'-0"x150'-0"		22,200	30,000	52,200
6200	3200K load, 7'-0"x200'-0"		38,300	48,200	86,500
7800	Soft rock, 3000 PSI conc., 10 KSF brng, 200K load, 2'-0"x50'-0"		855	8,925	9,780
8000	400K load, 2'-6"x50'-0"		1,450	14,600	16,050
8200	800K load, 3'-0"x100'-0"		3,900	38,100	42,000
8400	1200K load, 4'-0"x100'-0"		6,400	56,000	62,400
8600	1600K load, 5'-0"x150'-0"		14,700	114,500	129,200
8800	2400K load, 6'-0"x150'-0"		22,100	136,000	158,100
9000	3200K load, 7'-0"x200'-0"		38,300	216,500	254,800

Figure 5.15

Chapter 6

Roofing: Assemblies Estimate

The type of roofing system selected for a project is usually predicated on the knowledge and experience of the designer. There are times, however, when the owner has had contact with someone in the roofing industry who encourages the use of newer and perhaps unproven roofing assemblies. Also, local fire codes or the community's or owner's architectural or environmental guidelines may dictate one roofing assembly over another. The assembly selected should be one that has a demonstrated performance in the area where the project will be constructed. There should also exist in the construction location at least one qualified roofing contractor who can, if need be, make any necessary future repairs.

Selecting a Roofing Assembly

Presently, there are four basic groups of roof covering assemblies that are easily discernible from one another:

- Built-up
- Single-ply
- Metal
- Shingles and tiles

The roofing assembly should last the life of the building, provided it is specified, detailed, installed, and cared for properly and according to the manufacturer's instructions. It is important to remember that these are assemblies—a set or arrangement of items so closely related that they form a composite whole. Most often these assemblies consist of many different items made by different manufacturers—items that must be put together to form either a waterproofing membrane or a water shed. There are also related items that are required for the success of the roofing assembly. These items include flashing materials, roof deck insulation, roof hatches, skylights, gutters, and downspouts.

Built-Up Roofing Assemblies

A built-up roof is comprised of three different and distinct elements: felt, bitumen, and surfacing. The *felts*—which were traditionally organic (rag/paper) but now are almost exclusively inorganic (glass)—work much like reinforcing steel in concrete. Felts are necessary as tensile reinforcement to resist the extreme pulling forces on top of buildings. The felts, installed in layer fashion, also allow more bitumen to be applied in the whole assembly.

Bitumen, either coal tar pitch or asphalt, is the "glue" that holds the felts together and also acts as the waterproofing material in the roofing assembly. Felts in and of themselves do not waterproof; the layers of bitumen provide the waterproofing function.

The *surfacings* normally applied to built-up roofs are smooth, gravel or slag, mineral granules, and mineral coated cap sheet. Gravel, slag, and mineral granules may be embedded into the still-fluid flood coat. Gravel and slag serve as an excellent wearing surface to protect the membrane from mechanical damage. In some systems, a mineral-coated cap sheet is applied on top of the plies of felt and interply mappings of bitumen. This material is nothing more than a thicker or heavier ply of felt with a mineral granule surface hot mopped into place.

The most common built-up assemblies available contain two, three, or four plies of felt with either asphalt bitumen or coal tar pitch. Almost all assemblies are available for application to either nailable or non-nailable decks. Any of the assemblies may be applied to rigid deck insulation or directly to the structural deck.

See Figure 6.1 for a drawing of a typical built-up roof.

Single-Ply Roofing Assemblies

Since the early 1970s, the use of single-ply roofing membranes in the construction industry has been on the rise. Market surveys show that of all the single-ply assemblies being installed, about one in three are on new construction. Whether or not it is called "single-ply" or "elastomeric" is academic, because all the materials have the same idea in mind: reduce the time and difficulty of installing a waterproofing roof membrane by using new technology and materials. Many proponents claim that the old built-up

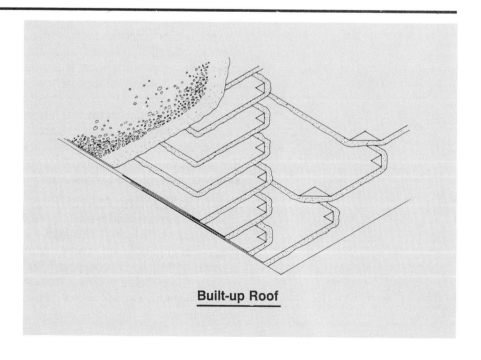

Built-up Roof

Figure 6.1

roof has too many problems associated with it to function properly on the buildings of today and those of the future.

Single-Ply Roofing Materials

Materials for the single-ply assemblies are more expensive than other, more conventional roofing systems; however, labor costs tend to be lower because installation is faster. Re-roofing represents the largest market today because of single-ply roofing's ease of application and because the hazards, odors, and time associated with built-up assemblies are avoided. One of the greatest advantages claimed by manufacturers of single-ply roofing materials is that the assemblies can be installed in a way that best suits the particular job. If the building structure cannot take the additional weight of stone ballast (sometimes as much as 12 pounds per square foot), the membrane can be fully adhered to the roof deck or insulation. This method is acceptable, provided that the membrane material is treated to withstand ozone and ultraviolet degradation.

Single-Ply Roofing Types

There are three categories of single-ply roofing assemblies:

- Thermo setting
- Thermo plastic
- Composites

Figure 6.2 is a guide to where each material falls. The chart also includes other useful information, such as compatible substrates, attachment methods, and sealing methods. The industry is expanding so rapidly that even the most conscientious contractors may have difficulty keeping up with the changes. Each of the materials and assemblies has unique requirements and performance characteristics.

Single-Ply Roofing Installation

The installation methods for single-ply assemblies generally follow the guidelines below. For accuracy, check with the system manufacturer for the proper application of their materials.

- The *loose-laid and ballasted* method is generally the easiest and quickest installation. Some special consideration must be given where flashings are attached. The membrane is stretched out flat, fused, or glued together at the side and end laps, and then fully ballasted with 3/4" to 1-1/2" stones (10 to 12 pounds per square foot) to prevent wind blow-off. This extra dead load must be considered during initial structural design stages. It is particularly important when re-roofing over an existing built-up roof that already weighs 10 to 15 pounds per square foot. A slip sheet or vapor retarder is sometimes required to separate the new roofing membrane from the old.

- The *partially adhered* method uses a series of bar or point attachments that mechanically attach the membrane to the substrate. The membrane manufacturer typically specifies the method of attachment based on the materials used and the substrate. Partially adhered systems do not require ballast material. A slip sheet may be required for some materials.

- The *fully adhered* method is by far the most time consuming because there is a requirement for contact cement, cold adhesive, or hot bitumen to uniformly adhere the membrane to the substrate below. Only manufacturer-approved roof deck insulation board should be used to receive the membrane. No ballast is required. A slip sheet may be required to separate the new and old materials.

Metal Roofing Assemblies

Metal roofing materials have been in use for a long time and may be divided into two groups: preformed metal and formed metal. The *preformed metal* types are available in sheets of varying lengths and widths, and are customarily in the corrugated shape. Aluminum, asphalt, fibrous glass, colored metal, and galvanized steel are the materials that are typically available. They are usually nailed or screwed into place. Preformed metal roofing is relatively economical, but can be used only where positive drainage is provided because the material, with its mechanical joints, is not watertight.

Formed metal roofing is practical on all sloped roofs from 1/4" per foot to vertical. The decision to use formed metal roofing is based more on

		Compatible Substrates						Attachment Method				Sealing Method				
Single-Ply Roofing Membrane Installation Guide — Generic Materials (Classification)		Slip-sheet req'd.	Concrete	Exist. asphalt memb.	Insulation board	Plywood	Spray urethane foam	Adhesive	Fully adhered	Loosely laid/ballast	Partially-adhered	Adhesive	Hot air gun	Self sealing	Solvent	Torch heating
Thermo Setting	EPDM (Ethylene, propylene)	•	•	•	•	•	•	•	•	•	•	•		•	•	
	Neoprene (Synthetic rubber)	•	•		•	•		•	•	•		•				
	PIB (Polyisobutylene)	•	•	•		•	•	•		•		•	•		•	
Thermo Plastic	CSPE (Chlorosulfenated polyethylene)	•	•		•	•	•	•	•	•	•	•				
	CPE (Chlorinated polyethylene)	•	•							•	•		•			
	PVC (Polyvinyl chloride)	•	•		•	•				•	•		•		•	
Composites	Glass reinforced EPDM/neoprene	•	•		•	•	•		•			•				
	Modified bitumen/polyester	•		•	•	•			•			•	•			•
	Modified bitumen/polyethylene & aluminum	•	•		•	•		•	•			•				•
	Modified bitumen/polyethylene sheet	•	•		•	•				•		•				•
	Modified CPE				•	•			•			•				
	Non-woven glass reinforced PVC							•	•	•		•				
	Nylon reinforced PVC		•		•	•			•				•		•	
	Nylon reinforced/butylorneoprene	•							•				•		•	
	Polyester reinforced CPE	•	•	•	•	•	•		•	•		•	•			
	Polyester reinforced PVC	•	•		•	•	•		•	•		•	•		•	
	Rubber asphalt/plastic sheet	•	•	•	•	•			•						•	

Figure 6.2

aesthetics than economics. The materials typically used are copper, lead, and zinc-copper alloy. Joining of the flat sheet is by either batten seam, flat seam, or standing seam.

Shingles and Tiles

For a very long time, shingles and tiles have been used as roofing materials on both residential and commercial buildings. They are not waterproofing materials, so they require a 3" to 4" slope per foot to perform properly.

Included in the category of *shingles* are composition asphalt strip shingles, wood shingles and shakes, aluminum shingles, and slate shingles. *Tiles* include concrete and clay materials, as well as aluminum.

The most significant design precautions are slope and weight. Each of the materials requires a minimum slope since it serves only as a watershed. Also, the heavier materials—concrete, clay, and slate—require structural systems that are capable of carrying the unusually large dead loads of these materials.

With a sloping roof, the dimensions from ridge to eave, as well as the eave and ridge lengths, must be determined before the actual roof area can be calculated. When the plan dimensions of the roof are known but the sloping dimensions are not, the actual roof area can be estimated if the approximate slope of the roof is known. See Figure 6.3. The multipliers shown may be used to convert slope to linear feet in the horizontal plan.

Factors for Converting Inclined to Horizontal

Roof Slope	Approx. Angle	Factor	Roof Slope	Approx. Angle	Factor
Flat	0°	1.000	12 in 12	45.0	1.414
1 in 12	4.8°	1.003	13 in 12	47.3	1.474
2 in 12	9.5°	1.014	14 in 12	49.4	1.537
3 in 12	14.0°	1.031	15 in 12	51.3	1.601
4 in 12	18.4°	1.054	16 in 12	53.1	1.667
5 in 12	22.6°	1.083	17 in 12	54.8	1.734
6 in 12	26.6°	1.118	18 in 12	56.3	1.803
7 in 12	30.3°	1.158	19 in 12	57.7	1.873
8 in 12	33.7°	1.202	20 in 12	59.0	1.943
9 in 12	36.9°	1.250	21 in 12	60.3	2.015
10 in 12	39.8°	1.302	22 in 12	61.4	2.088
11 in 12	42.5°	1.357	23 in 12	62.4	2.162

Figure 6.3

Figures 6.4 and 6.5 illustrate some examples for using the information in Figure 6.3.

Flashing

Flashings are necessary to prevent water from entering into the roofing system via roof discontinuities. The materials may be preformed or formed on the job to satisfy the particular requirements of the detail involved. The materials typically used for flashing include:

- Aluminum
- Copper
- Copper-coated lead
- Lead
- Polyvinyl chloride
- Butyl rubber
- Copper-clad stainless steel
- Stainless steel
- Galvanized metal

It's important to be aware of certain design precautions. Check to see that the flashing materials are compatible with the roofing assembly, and that the details are properly designed for the roofing assembly.

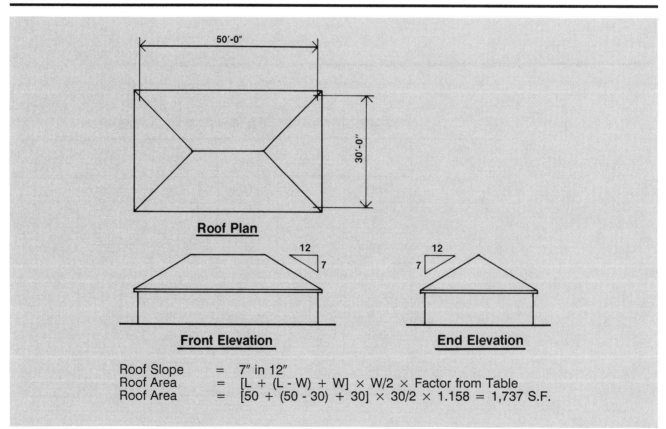

Roof Slope   = 7" in 12"
Roof Area   = [L + (L - W) + W] × W/2 × Factor from Table
Roof Area   = [50 + (50 - 30) + 30] × 30/2 × 1.158 = 1,737 S.F.

Figure 6.4

Roof Deck Insulation

Insulation is used to reduce heat transfer through the exterior enclosure of the building. The type and form of insulation varies according to where it is located in the structure. Major insulation types include mineral granules; glass fibers; plastic foams; organic vegetable fibers; solids; and composites of all of these.

Insulation is usually specified by "R" value. The designer and estimator must shop around to find the most economical material for the desired insulation capability. Labor costs vary from one insulation material to another because of fragility, weight, and bulkiness of the materials. Wood blocking and nailers to receive fasteners must match the insulation thickness on roof decks. As a result, a thick, inexpensive insulation board may far exceed the cost of a thinner, more expensive one when the thickness of the wood blocking and nailer is considered.

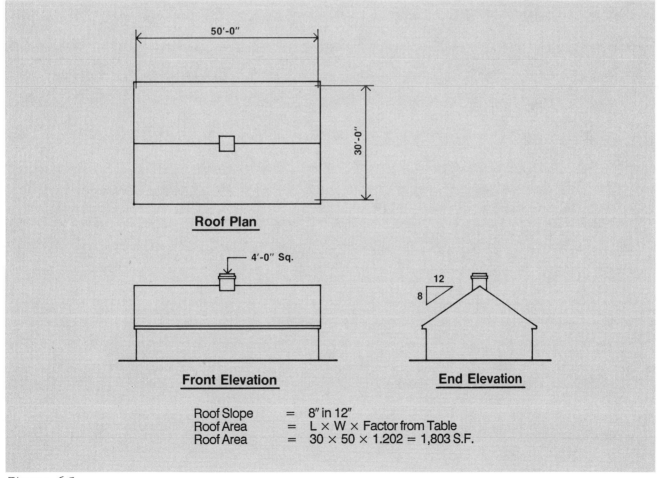

Roof Plan

— 4'-0" Sq.

Front Elevation   **End Elevation**

Roof Slope   =   8" in 12"
Roof Area   =   L × W × Factor from Table
Roof Area   =   30 × 50 × 1.202 = 1,803 S.F.

Figure 6.5

Conclusion

In making a cost-conscious selection of roof covering design, it is important to consider the total impact of each variable. Single-ply assemblies cost more for materials than built-up roofing, but are cheaper to apply. Ballast for single-ply assemblies adds weight, which requires more structural support. This cost must be compared to the cost and other features of a fully adhered system. Life cycle costs apply more to the roofing system than any other component of the building. Higher initial costs can be offset by savings in energy and re-roofing costs over the life of the building.

Roofing considerations frequently overlooked in preliminary estimates include:

- Flashing around mechanical equipment
- Insulation
- Hatches for access
- Skylights
- Gutters
- Downspouts
- Impact on storm drain systems
- Blocking
- Gravel stops

In a single-story building, the roof system, insulation, and roof covering costs have a significant impact on the total cost. The time spent considering alternatives for these during preliminary design is well worth it.

Chapter 7

Conveying: Assemblies Estimate

Conveying systems are used to transport people or things vertically, on an incline, or horizontally. These systems are typically provided and installed by the company that manufactures the system, or a certified installer. In construction they are very specialized and can have a dramatic effect on the overall cost of a building.

The list below indicates the many different types of conveying systems available. Each has individual design requirements and special applications.

- Ash hoist
- Book lifts
- Conveyors
- Dumbwaiters
- Elevators, freight
- Elevators, passenger
- Escalators
- Hoists & cranes
- Lifts
- Material handling systems
- Moving stairs & walks
- Pneumatic tube systems
- Vertical conveyors

Selecting a Conveying Assembly

There are specific factors that must be considered before selecting a particular assembly. These factors for some common conveying systems are listed below.

Dumbwaiters:

- Capacity
- Size
- Floors
- Stops
- Speed
- Finish

Elevators:

- Hydraulic or electric
- Geared or gearless

- Capacity
- Size
- Floors
- Number required
- Stops
- Speed
- Finish
- Machinery location
- Door type
- Signals
- Special requirements

Escalators:

- Capacity
- Size
- Floors
- Number required
- Story height
- Speed
- Finish
- Machinery location
- Incline angle
- Special requirements

Material Handling Systems:

- Automated
- Non-automated

Moving Stairs & Walks:

- Capacity
- Size
- Floors
- Number required
- Story height
- Speed
- Finish
- Machinery location
- Incline angle
- Special requirements

Pneumatic Tube Systems:

- Automatic
- Manual
- Size
- Stations
- Length
- Special requirements

Vertical Conveyers:

- Automatic
- Non-automatic

Elevators

This chapter focuses on elevators, because these are the most common conveying systems in construction today. Many of the points that are made apply to other systems found in this division of construction. Before any design details are finalized, manufacturers and consultants should be contacted for details.

Elevator Types

The two types of elevators in use today are hydraulic and electric. Each has distinct uses, advantages, and disadvantages. Hydraulic elevators are directly driven from below by a pump and cylinder mechanism. Electric elevators use motors and steel cables with counterweights, which work on the traction principle.

Hydraulic Elevators

Hydraulic elevators are used most often in low-rise buildings — 70′ maximum — with travel speeds between 25 and 150 feet per minute. Because they are rather slow and travel short distances, hydraulic elevators are most often used in low-rise commercial, industrial, and residential buildings, motels and hotels, and for moving freight.

The most significant factor to consider is the actual assembly operation. The cab rests on top of a hydraulic piston, which operates in a cylinder that extends as far into the ground as the cab rises. For example, a hydraulic elevator that travels 55′ up must have a piston and casing 55′ deep. The casing is installed by a drilling rig. Hydraulic elevators are not installed often in existing buildings because the drilling rig requires head room. Typically, hydraulic elevators cost less than their electric counterparts for a low-rise building. Refer to Figure 7.1 for a typical hydraulic elevator installation.

Electric Elevators

Electric elevators are typically used in commercial and institutional buildings that are mid- to high-rise (over 5 floors). Above 50 stories, they can attain speeds up to 1800 fpm. Though not thought of as freight haulers, electric elevators can handle capacities up to 10,000 pounds.

The lifting power in electric elevators is provided by traction to steel hoisting cables. One end of the cables is attached to the elevator cab and the other to counterweights, which can weigh up to 40% of the load capacity. The counterweights are needed to help reduce the power requirements of the motors.

There are two types of electric motor and drum assemblies: geared and gearless. The gearless assemblies are preferred for high-speed installations, while the geared assemblies are for low-speed applications.

Figure 7.2 is an illustration of a typical electric elevator.

Safety

Safety is a byword in elevator manufacturing and operation. Elevator systems are subject to many safety regulations and frequent safety inspections. As demanding as these safety requirements may seem, they are necessary, particularly when considering the heights and speeds that these machines must travel and the expected dependability.

Speed and Capacity

The speed and capacity of elevators vary depending on the intended use, travel height, overall quality of the building, and the building occupants. Normally, a small office building requires a minimum size elevator rated at 2,500 pounds capacity with a car measuring 6′-8″ x 4′-6″.

Elevator speeds should be sufficient to provide timely, efficient service. The maximum usable speed is limited by the building height, which includes an adequate allowance for acceleration and deceleration distances. A safe rule-of-thumb is that the rated elevator speed in fpm may be calculated as 1.6 times the rise in feet, plus 350.

For example:

A 20-story building (310 feet) requires several elevators. What is the maximum acceptable speed?

(1.6 × 310) + 350 = 846 fpm (maximum)

From this calculation, the elevators can be selected from manufacturer's data.

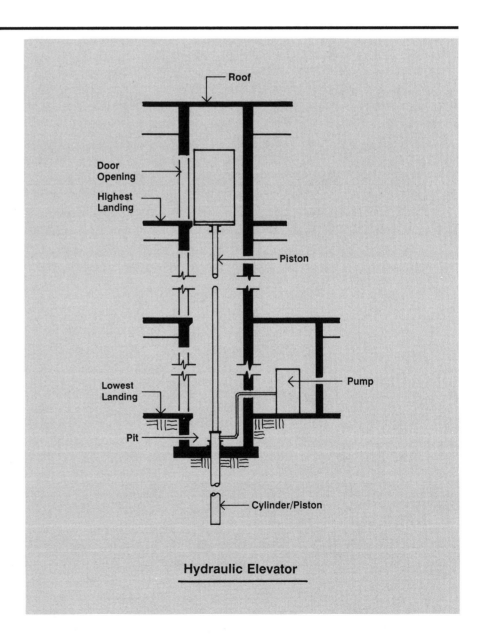

Hydraulic Elevator

Figure 7.1

The table in Figure 7.3 indicates the recommended elevator speed for office and general purpose buildings. There may be specific requirements by the local building code, so verify what is necessary and correct.

The table in Figure 7.4 shows relatively the same information; however, it indicates the height traveled based on speed and building type with recommended capacities. The same type of information is broken down even further in Figure 7.5.

Traffic

Other considerations in the use of elevators include size, number, and location for effective traffic patterns. All of these factors depend on the building function and population, building height and number of floors, car capacity, volume of traffic, and elevator speed.

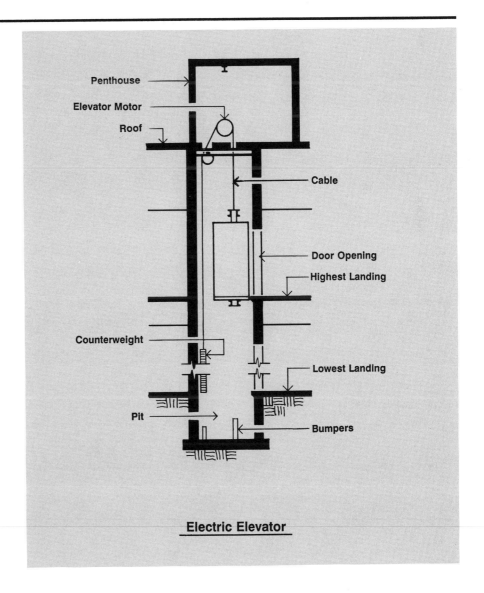

Electric Elevator

Figure 7.2

When computing elevator traffic, it is customary to use the peak or critical traffic periods. These periods vary greatly for different building types and occupancies. For example, office building traffic typically peaks three times: during the morning when most people are arriving for work, during lunch time, and in the late afternoon when they are leaving to go home. Hospital traffic usually reaches its peak during evening visiting hours.

Traffic is measured by the number of persons handled during a five-minute peak interval. When this figure is divided by the five-minute handling capacity of an elevator, the minimum number of elevators can be determined.

Handling capacity is computed based on a car's size (number of people carried) and its round-trip time. Car size is analyzed first.

Number of Floors	Recommended Speeds in Feet Per Minute			
	Small	Average	Prestige	Service
2-5	200-250	300-350	350-400	200
5-10	300-350	350-500	500	300
10-15	500	500-700	700	350-500
15-25	700	800	800	500
25-35	—	1000	1000	500
35-45	—	1000-2000	1200	700-800
45-60	—	1200-1400	1400-1600	800-1000
over 60	—	—	1800	1000

Figure 7.3

Elevator Speed vs. Height Requirements

Building Type and Elevator Capacities	Travel Speeds in Feet per Minute								
	100 fpm	200 fpm	250 fpm	350 fpm	400 fpm	500 fpm	700 fpm	800 fpm	1000 fpm
Apartments 1200 lb. to 2500 lb.	to 70	to 100	to 125	to 150	to 175	to 250	to 350		
Department Stores 1200 lb. to 2500 lb.		100		125	175	250	350		
Hospitals 3500 lb. to 4000 lb.	70	100	125	150	175	250	350		
Office Buildings 2000 lb. to 4000 lb.		100	125	150		175	250	to 350	over 350

Note: Vertical transportation capacity may be determined by code occupancy requirements of the number of square feet per person divided into the total square feet of building type. If we are contemplating an office building, we find that the Occupancy Code Requirement is 100 S.F. per person. For a 20,000 S.F. building, we would have a legal capacity of two hundred people. Elevator handling capacity is subject to the five minute evacuation recommendation, but it may vary from 11 to 18%. Speed required is a function of the travel height, number of stops and capacity of the elevator.

Figure 7.4

Sizing the Elevator

The first step in sizing the elevator is to determine the building population. For a commercial building with a diversified occupancy, this can be estimated by allowing 120 to 150 net square feet per person. For a similar building with a single purpose or common occupancy, the number may be 100 to 120 net square feet per person. The net square feet figure should include only rentable or occupied space and exclude toilets, lobbies, pipe chases, columns, etc. In apartments, hotels, and motels, the populations can be estimated by allowing two persons per sleeping room. If the floor layout is unknown, use 200 square feet per person. The five-minute capacity requirement for elevators is a percentage of the building population and is shown in Figure 7.6.

Under normal circumstances, the average person requires two square feet of space to feel comfortable in an elevator. This is the number for the average, healthy American. Change continents and there will be as many different requirements as the cultures visited. Social climate and regional differences determine the number to use, but for the sake of discussion we will use the two square feet figure. During rush hour, particularly in the densely populated areas of the Northeast, the amount of required space can be reduced to 1.3 square feet per person.

The chart in Figure 7.7 can be used to predict elevator passenger capacity during peak periods, provided the car load capacity (in pounds) is already

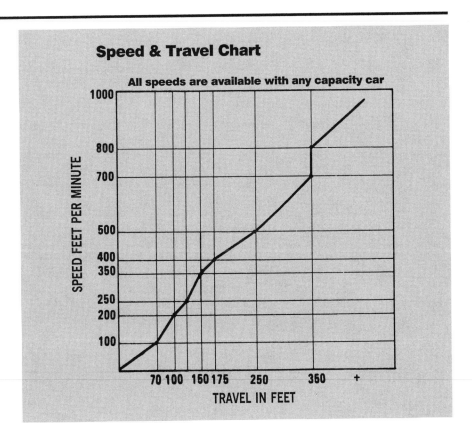

Figure 7.5

known. The round-trip time includes all the factors involved with a full speed, nonstop, round-trip journey as shown in Figure 7.8.

Generally speaking, waiting time between elevators should be 20 to 30 seconds. A waiting period that exceeds 35 seconds is unsatisfactory as far as most passengers are concerned. In residential buildings, passengers tend to be a bit more patient and will remain calm for up to 60 seconds.

Elevator, 5-Minute Capacity		
Building Type	No. Square Feet/Person	5-Minute Capacity
Commercial, Diversified Occupancy	120 to 150	10% to 12%*
Commercial, Single Purpose Occupancy	100 to 120	15% to 20%*
Apartment	200**	5% to 10%*
Hotel	200**	10% to 15%*
* Percentage of building occupancy.		
** Or 2 persons per sleeping room.		

Figure 7.6

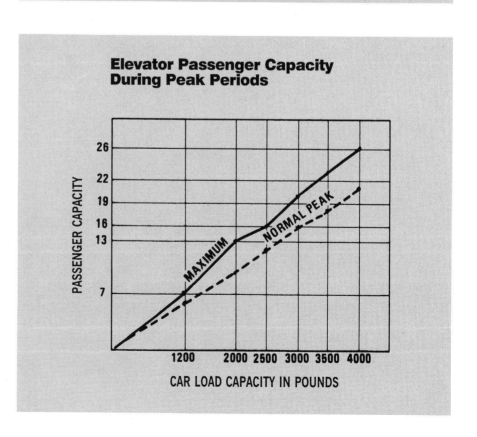

Figure 7.7

Elevator zoning has become popular recently, particularly with the advent of high-rise buildings. The elevators are initially sized by two or three groups, each group serving only a specified group of floors. The next group of cars serves the next group of floors, with no hall doors provided for the lower group of floors (except the lobby, of course). By omitting the hall doors at floors not assigned to the group, the cars run express or non-stop from the lobby to their lowest floor, and become local thereafter. Consider the following example:

> A proposed 30-story office tower is in the budget-setting stages. The architect has sized all the elevators based on the peak traffic periods and has determined that 10 passenger elevators are needed.
>
> Elevators: 1−3   Serve floors 1 through 10 and 30
>   4−6   Serve floors 1 and 11 through 21 and 30
>   7−9   Serve 1 and 22 through 30
>   10   Serves the 30th floor only
> The 30th floor contains the company snack bar and cafeteria.

This example may seem confusing at first glance, but it becomes clear after careful consideration that the time saved by having a series of express elevators can be realized in increased productivity.

Freight Elevators

Freight elevators are specifically designed to vertically transport equipment, materials, and goods rather than passengers. In low-rise buildings, hydraulic elevators are more appropriate, but electrical are perhaps better, economically, for lifts exceeding 50 feet.

Freight elevators have a more rugged, less finished appearance. They also have a lower speed, and greater capacity. Depending on the building height, standard speeds vary from 75 to 200 fpm, but the speeds may be greater for smaller capacities. Freight elevators are available in capacities from 2,500 to 25,000 pounds or more if needed.

Action During Round-Trip	Time (in seconds)
1. Passenger operates call button, doors open	3
2. Passenger enters car, operates floor button	2
3. Doors close	3
4. Car travels to selected floor	8
5. Doors reopen	2
6. Passenger leaves car	2
Total Time to Serve One Passenger	20
7. Doors close	3
8. Car returns to original floor	8
Total Round-Trip Time	31

Figure 7.8

The American National Standard Safety Code for Elevators classifies freight elevators in the following categories:

General freight loaded by hand truck. No single piece of freight to exceed one-quarter capacity load of the elevator, based on 50 pounds per square foot of platform area.

Motor vehicle or garage elevator. Automobiles and trucks only, based on capacity of 30 pounds per square foot of platform area.

Industrial truck loading. Must be capable of supporting impact load and weight of truck. Capacity not less than 50 pounds per square foot of platform.

Service Elevators

Service elevators are usually installed to handle oversized loads, trucks, and four-wheel carts. This type of system is usually found in commercial, industrial, apartment, hotel/motel and institutional buildings. Normally, service elevators are equipped with abuse-resistant materials and have horizontal doors that slide up and down.

Service elevators must comply with the code requirements for carrying passengers. The size is normally determined by the largest or bulkiest anticipated load.

Federal Requirements

The Americans with Disabilities Act (ADA) provides civil rights protection from discrimination to all individuals with disabilities. It contains specific requirements to ensure equal opportunity to disabled people in employment, public accommodations, and transportation. The ADA includes detailed requirements pertaining to elevators including construction, dimensions, controls, timing of signals, and access. These must be met by every elevator system and installation.

Conclusion

Conveying systems represent a significant cost in a new building. The more that is known about specific requirements, the closer the estimated cost of the system will be to the final installed cost. Codes must be complied with first. The next step is to decide between electric and hydraulic systems. When the installation is in a new, low-rise building, a hydraulic elevator is the most cost-effective choice.

There are a number of cost-effective trade-offs to consider when estimating the capacity, number of units, speed, and quality of elevators. The time spent evaluating the variables that affect the final system selection is well worth it. Work done in this area during the preliminary estimate can result in significant cost reduction.

Chapter 8

Mechanical: Assemblies Estimate

To properly develop mechanical assemblies in the assemblies square foot estimate format, it is necessary to first define certain parameters of the building, including:

- Building shape
- Number of floors
- Floor area
- Volume of building
- Expected occupancy (number of people and male/female ratio)
- Some idea of the systems desired

Once this is accomplished, the mechanical work can be examined on a system-by-system basis, making use of the above parameters to establish requirements.

Mechanical work is conveniently broken down into four distinct systems, each of which is discussed in this chapter:

- Plumbing
- Fire Protection
- Heating
- Air Conditioning

Plumbing

Once the building parameters are known, the first step for establishing the plumbing requirements is to define the necessary fixtures. Using the table in Figure 8.1, the plumbing fixture requirements can be determined by building type and occupancy. The table, which is based on model building codes, represents the minimum fixture requirements for various building types.

If the building population were unknown, the information in Figure 8.1 would be difficult to use. If the building size is known, Figure 8.2 can be used to determine the number of people that may occupy it.

Example One: Three-Story Office Building

Determine the amount of occupied space in a three-story office building with 10,000 square feet per floor. The Floor Area Ratios table in Figure 8.3 provides the data to make an approximation. From this table, Office Buildings have a .75 net to gross ratio. The remaining 25% includes hallways,

Minimum Plumbing Fixture Requirements

Type of Building/Use	Water Closets		Urinals		Lavatories		Bathtubs or Showers		Drinking Fountain	Other
	Persons	Fixtures	Persons	Fixtures	Persons	Fixtures	Persons	Fixtures	Fixtures	Fixtures
Assembly Halls Auditoriums Theater Public assembly	1-100 101-200 201-400	1 2 3	1-200 201-400 401-600	1 2 3	1-200 201-400 401-750	1 2 3			1 for each 1000 persons	1 service sink
	Over 400 add 1 fixt. for ea. 500 men; 1 fixt. for ea. 300 women		Over 600 add 1 fixture for each 300 men		Over 750 add 1 fixture for each 500 persons					
Assembly Public Worship	300 men 150 women	1 1	300 men	1	men women	1 1			1	
Dormitories	Men: 1 for each 10 persons Women: 1 for each 8 persons		1 for each 25 men; over 150 add 1 fixture for each 50 men		1 for ea. 12 persons 1 separate dental lav. for each 50 persons recom.		1 for ea. 8 persons For women add 1 additional for each 30. Over 150 persons add 1 for each 20.		1 for each 75 persons	Laundry trays 1 for each 50 serv. sink 1 for ea. 100
Dwellings Apartments and homes	1 fixture for each unit				1 fixture for each unit		1 fixture for each unit			
Hospitals Indiv. Room Ward Waiting room	8 persons	1 1 1			10 persons	1 1 1	20 persons	1 1	1 for 100 patients	1 service sink per floor
Industrial Mfg. plants Warehouses	1-10 11-25 26-50 51-75 76-100 1 fixture for each additional 30 persons	1 2 3 4 5	0-30 31-80 81-160 161-240	1 2 3 4	1-100 over 100	1 for ea. 10 1 for ea. 15	1 Shower for each 15 persons subject to excessive heat or occupational hazard		1 for each 75 persons	
Public Buildings Businesses Offices	1-15 16-35 36-55 56-80 81-110 111-150 1 fixture for ea. additional 40 persons	1 2 3 4 5 6	Urinals may be provided in place of water closets but may not replace more than 1/3 required number of men's water closets		1-15 16-35 36-60 61-90 91-125 1 fixture for ea. additional 45 persons	1 2 3 4 5			1 for each 75 persons	1 service sink per floor
Schools Elementary	1 for ea. 30 boys 1 for ea. 25 girls		1 for ea. 25 boys		1 for ea. 35 boys 1 for ea. 35 girls		For gym or pool shower room 1/5 of a class		1 for each 40 pupils	
Schools Secondary	1 for ea. 40 boys 1 for ea. 30 girls		1 for ea. 25 boys		1 for ea. 40 boys 1 for ea. 40 girls		For gym or pool shower room 1/5 of a class		1 for each 50 pupils	

Figure 8.1

Occupancy Determinations

Description		S.F. Required per Person		
		BOCA	SBC	UBC
Assembly Areas	Fixed Seats	**	6	7
	Movable Seats		15	15
	Concentrated	7		
	Unconcentrated	15		
	Standing Space	3		
Educational	Unclassified			
	Classrooms	20	40	20
	Shop Areas	50	100	50
Institutional	Unclassified		125	
	In-Patient Areas	240		
	Sleeping Areas	120		
Mercantile	Basement	30	30	20
	Ground Floor	30	30	30
	Upper Floors	60	60	50
Office		100	100	100

BOCA = Building Officials & Code Administrators
SBC = Southern Building Code
UBC = Uniform Building Code

** The occupancy load for assembly area with fixed seats shall be determined by the number of fixed seats installed.

Figure 8.2

Floor Area Ratios

Table below lists commonly used gross to net area and net to gross area ratios expressed in % for various building types.

Building Type	Gross to Net Ratio	Net to Gross Ratio	Building Type	Gross to Net Ratio	Net to Gross Ratio
Apartment	156	64	School Buildings (campus type)		
Bank	140	72	Administrative	150	67
Church	142	70	Auditorium	142	70
Courthouse	162	61	Biology	161	62
Department Store	123	81	Chemistry	170	59
Garage	118	85	Classroom	152	66
Hospital	183	55	Dining Hall	138	72
Hotel	158	63	Dormitory	154	65
Laboratory	171	58	Engineering	164	61
Library	132	76	Fraternity	160	63
Office	135	75	Gymnasium	142	70
Restaurant	141	70	Science	167	60
Warehouse	108	93	Service	120	83
			Student Union	172	59

The gross area of a building is the total floor area based on outside dimensions.

The net area of a building is the usable floor area for the function intended and excludes such items as stairways, corridors and mechanical rooms. In the case of a commercial building, it might be considered as the "leasable area."

Figure 8.3

lobbies, restrooms, janitor closets, and other non-rentable spaces. Effectively, the three-story office building has:

$$\frac{10,000 \text{ S.F./floor}}{\text{(gross)}} \times 75\% = \frac{7,500 \text{ S.F./floor}}{\text{(net)}}$$

Figure 8.2 can now be used to determine the building occupancy at 100 S.F./person. To determine the number of persons per floor, simply divide the net square feet by the square feet per person:

7,500 S.F./floor/100 S.F./person = 75 persons/floor

Now, using Figure 8.1, determine the minimum plumbing fixtures required on each floor for an office building with 75 persons per floor. See Figure 8.4.

When planning the restrooms, make certain that all meet ADA requirements. Figure 8.5 shows guidelines for toilet room layouts, including facilities for the handicapped. At least one of each type of fixture should be accessible in each toilet room.

Hot Water Requirements

Domestic water heating systems should be sized according to the consumption rates for maximum demand and average daily demand for the specific building type and occupancy. Water heating systems should be sized so that the maximum hourly demand anticipated can be met, in addition to an allowance for heat loss from the piping and storage tank.

Water heating systems can range from a small, under-the-counter unit heater, to the more common upright tank found in homes and offices, to a boiler in the building's mechanical room, to a central heating plant that serves several buildings in a complex. Each system has its own set of requirements and limitations. The discussion here focuses on unit or tank type water heating systems.

For a quick sizing method, use the table in Figure 8.6. Select the building type and size factor, then find the corresponding Maximum Hourly Demand and Average Day Demand.

For example, in the three-story office building with a population of 75 people/floor, from Figure 8.6, Office Buildings have a Maximum Hourly Demand of 0.4 gallons per person.

Fixture	Fixtures Per Floor			
	Minimum Required	Women	Men	Handicapped
Water Closets	4	2	1	2
Urinals	—	—	1	—
Lavatories	4	2	2	2
Drinking Fountain	1	—	—	—
Service Sink	1	—	—	—

Figure 8.4

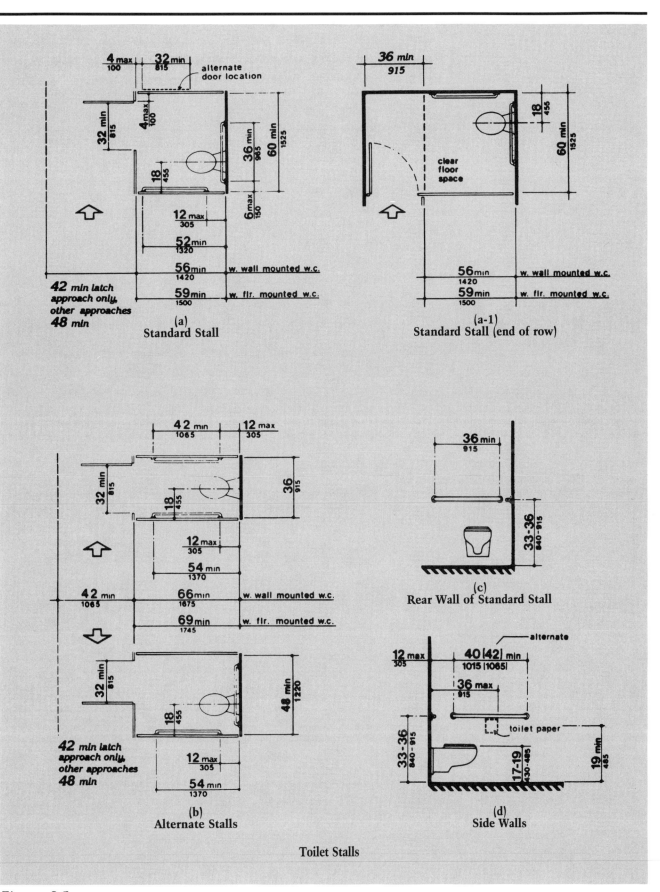

(a)
Standard Stall

4 max / 100   *32* min / 815   alternate door location

32 min / 815   *4* max / 100

60 min / 1525   *36* min / 965

18 / 455

6 max / 150

12 max / 305

52 min / 1320

56 min / 1420   w. wall mounted w.c.

59 min / 1500   w. flr. mounted w.c.

42 min latch approach only, other approaches *48* min

(a-1)
Standard Stall (end of row)

36 min / 915

18 / 455

60 min / 1525

clear floor space

56 min / 1420   w. wall mounted w.c.

59 min / 1500   w. flr. mounted w.c.

(b)
Alternate Stalls

42 min / 1065   *12* max / 305

32 min / 815

18 / 455

36 / 915

12 max / 305

54 min / 1370

42 min / 1065   *66* min / 1675   w. wall mounted w.c.

69 min / 1745   w. flr. mounted w.c.

32 min / 815

18 / 455

48 min / 1220

42 min latch approach only, other approaches *48* min

12 max / 305

54 min / 1370

(c)
Rear Wall of Standard Stall

36 min / 915

33-36 / 840-915

(d)
Side Walls

12 max / 305   *40 (42)* min / 1015 (1065)   alternate

36 max / 915

33-36 / 840-915   toilet paper

17-19 / 430-485

19 min / 485

Toilet Stalls

Figure 8.5

89

75 people/floor × 0.4 gallons/person = 30 gallons/floor

Water heaters may be sized using one per floor or as a total for all floors.

Example Two: Three-Story Office Building Roof Drainage
Size the roof drainage system for the three-story office building. For sizing purposes, assume the building is located in a city where the maximum rainfall is 4 inches/hour.

Step 1: Sketch the Roof
See Figures 8.7 and 8.8.

Step 2: Compute the Roof Area to be Drained

Roof: 100′ × 100′	=	10,000 S.F.	
Parapet: (100′ × 4) × 1′	=	400 S.F.	
Stair & elevator shaft walls	=	580 S.F.	
Total area to be drained	=	10,980 S.F.	or 11,000 S.F.

For simplicity, the roof is divided into four equal zones of 2,750 S.F./each.

Hot Water Consumption Rates

Type of Building	Size Factor	Maximum Hourly Demand	Average Day Demand
Apartment Dwellings	No. of Apartments:		
	Up to 20	12.0 Gal. per apt.	42.0 Gal. per apt.
	21 to 50	10.0 Gal. per apt.	40.0 Gal. per apt.
	51 to 75	8.5 Gal. per apt.	38.0 Gal. per apt.
	76 to 100	7.0 Gal. per apt.	37.0 Gal. per apt.
	101 to 200	6.0 Gal. per apt.	36.0 Gal. per apt.
	201 up	5.0 Gal. per apt.	35.0 Gal. per apt.
Dormitories	Men	3.8 Gal. per man	13.1 Gal. per man
	Women	5.0 Gal. per woman	12.3 Gal. per woman
Hospitals	Per bed	23.0 Gal. per patient	90.0 Gal. per patient
Hotels	Single room with bath	17.0 Gal. per unit	50.0 Gal. per unit
	Double room with bath	27.0 Gal. per unit	80.0 Gal. per unit
Motels	No. of units:		
	Up to 20	6.0 Gal. per unit	20.0 Gal. per unit
	21 to 100	5.0 Gal. per unit	14.0 Gal. per unit
	101 Up	4.0 Gal. per unit	10.0 Gal. per unit
Nursing Homes		4.5 Gal. per bed	18.4 Gal. per bed
Office buildings		0.4 Gal. per person	1.0 Gal. per person
Restaurants	Full meal type	1.5 Gal./max. meals/hr.	2.4 Gal. per meal
	Drive-in snack type	0.7 Gal./max. meals/hr.	0.7 Gal. per meal
Schools	Elementary	0.6 Gal. per student	0.6 Gal. per student
	Secondary & High	1.0 Gal. per student	1.8 Gal. per student

For evaluation purposes, recovery rate and storage capacity are inversely proportional. Water heaters should be sized so that the maximum hourly demand anticipated can be met in addition to allowance for the heat loss from the pipes and storage tank.

Figure 8.6

Step 3: Select the Leader and Horizontal Drain Size

Drains and leaders will be the same size for all areas.

> Leader (see Figure 8.9): 4″
> Horizontal Drain (see Figure 8.10): 4″ @ 1/2″ slope/foot

Two horizontal drains will join at a common vertical leader.

> The contributing area now becomes 5,500 S.F.
> Leader (Figure 8.9): 5″
> Horizontal Drain (Figure 8.10): 5″ @ 1/2″ slope/foot

The two horizontal drains will join at a common building horizontal drain, which will carry the roof drainage to the storm drainage system. The contributing area now becomes 11,000 S.F.

> Horizontal drain (Figure 8.10): Select 6″ @ 1/2″ slope/foot since it is so close to 11,000 S.F.

Determining Carrying Capacity of Leaders

When computing the carrying capacity of leaders, there are three factors to consider: dimensions, cross-sectional area, and inside perimeter (area of water contact).

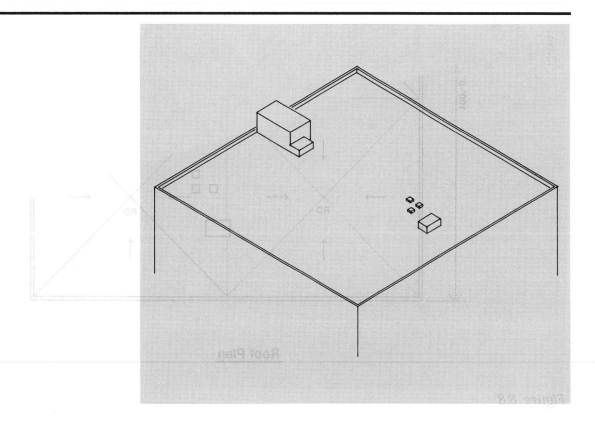

Figure 8.7

A rectangular leader, because of its four walls and corners, offers greater friction loss, thereby diminishing its carrying capacity. To compensate for this loss, a rectangular leader needs to be about 10% larger than a round leader to carry the same load.

The sizes for rectangular leaders (shown in Figure 8.11 as equivalents of the round leaders) include the 10% adjustment. In some cases, the rectangular sizes are more than 10% larger. Where the computation of the equivalent size has resulted in an unavailable size, the next larger stock size is given. Always use "Water Contact Area, Inches" as the basis of equivalency.

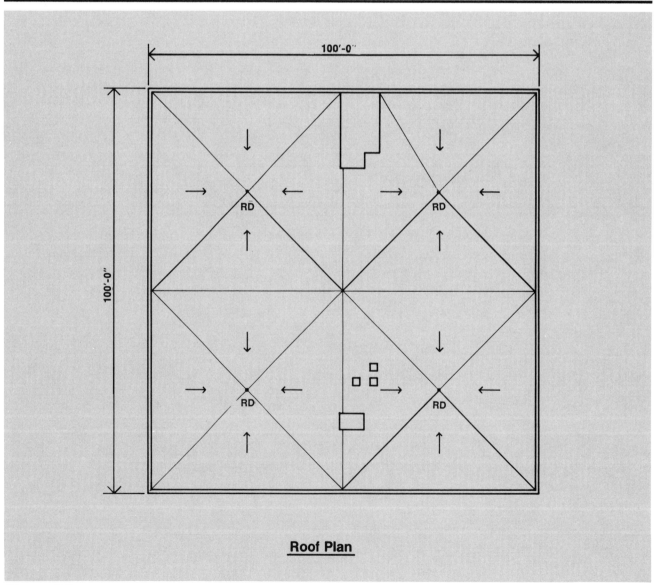

Roof Plan

Figure 8.8

Fire Protection

Fire protection has four specific priorities in the event of a fire:

- To provide for the safety of all building occupants, a route of evacuation, or a place of refuge. In relatively short and small buildings, evacuation is the normal method, but in high-rise construction, a place of refuge is also required.
- To provide for the safety of firefighters who may enter the burning building. This is accomplished by ensuring sufficient structural integrity in the building so that firefighters are not in danger of a building collapse.
- To protect adjacent property, which is accomplished by adequate building separation and by the proper use of exterior materials that do not have unreasonable fire risk.

Size of Vertical Leaders	
Size of Leader or Conductor, In.*	Maximum Projected Roof Area, S.F.
2	720
2-1/2	1,300
3	2,200
4	4,600
5	8,600
6	13,500
8	29,000

*The equivalent diameter of square or rectangular leader may be taken as the diameter of the circle that may be inscribed within the cross-sectional area of the leader.

Figure 8.9

Size of Horizontal Storm Drains			
Diameter of Drain, In.	Maximum Projected Roof Area for Drains at Various Slopes, S.F.		
	1/8″ Slope	1/4″ Slope	1/2″ Slope
3	822	1,160	1,644
4	1,880	2,650	3,760
5	3,340	4,720	6,680
6	5,350	7,550	10,700
8	11,500	16,300	23,000
10	20,700	29,200	41,400
12	33,300	47,000	66,600
15	59,500	84,000	119,000

Figure 8.10

- To preserve the building itself. It is this priority that will be dealt with in this book. The first three are normally taken care of during the planning, layout, and design of the building.

To correctly interpret the many model building codes that apply to any structure, the building must first be classified. As a minimum, the following information must be established:

- Fire zone
- Occupancy group
- Type of construction
- Number and height of floors
- Location on property
- Number of occupants
- Floor area

In most densely populated areas, cities are divided into three distinct fire zones. Certain characteristics are required of a building in each zone. For example, if Fire Zone 1 is the downtown, high-rise business district, local codes may require buildings located there to be of highly fire-resistive materials and construction.

To provide further guidance for fire resistance required in a building, the model codes have established specific occupancy requirements. Buildings have been broken down into broad groups as follows:

Equivalent Size of Round & Rectangular Leaders					
Round Leader			Rectangular Leader		
Diameter, Inches	Cross Sectional Area, Sq. In.	Water Contact Area, In.	Dimension, Inches	Cross Sectional Area, Sq. In.	Water Contact Area, In.
2	3.14	6.28	2 x 2 1-1/2 x 1-1/2	4 3.75	88
3	7.07	9.42	2 x 4 2-1/2 x 3	8 7.50	12 11
4	12.57	12.57	3 x 4-1/4 3-1/2 x 4	12.75 14	14.5 14
5	19.06	15.07	4 x 5 4-1/2 x 4-1/2	20 20.25	18 18
6	28.27	18.85	5 x 6 5-1/2 x 5-1/2	30 30.25	22 22

Circumference as Straight Line

Rectangle as Straight Line

Figure 8.11

Group	Occupancy Type
A	Assembly
E	Educational
I	Institutional
H	Hazardous
B	Business, including offices, factories, and mercantile
R	Residential
M	Miscellaneous structures

Buildings are further classified by the type of construction and degree of fire resistance. Fire resistance describes the hourly ratings for the major elements in each building type. These ratings are hourly increments from one to four hours. For example, the exterior bearing wall in a noncombustible, fire-resistive building requires four-hour construction. The structural frame in the same building requires three-hour construction.

Classes of Fires

There are four classes of fires:

Class A: Fires involving ordinary materials, including wood, paper, cloth, and rubber.

Class B: Flammable gases and liquids, such as natural gas, gasoline, and oil.

Class C: Electrical equipment.

Class D: Fires involving combustible metals such as sodium, potassium, and magnesium.

Water, foam, soda acid, and other water-based extinguishing agents should never be used on Class C or D fires. A system with relatively low toxicity using Halon has been developed to safely extinguish Class B and C fires.

Standpipes

In buildings where water may be used as the extinguishing agent, standpipes are used to carry the water in large diameter pipes throughout the building. There are three standpipe systems that are recognized by the model codes.

The *dry standpipe* is a large, normally empty pipe that rises from the lowest floor to the roof. At the street level is a Siamese connection that the fire department connects to a water supply. The dry standpipe is an extension of fire fighting equipment—once activated, the fire fighters can connect hoses to the standpipe and begin to extinguish the fire at any level. The dry standpipe is not designed to be used by any building occupants.

In buildings over four stories, the model codes require that a *wet standpipe* system be installed in the building. Occupants can use this system to help hold a fire until the fire department arrives. Wet standpipe systems must also be equipped with a Siamese fitting at the street level. The model codes state that wet pipe systems must be located so that every point of every floor is within 30 feet of the end of a 100-foot hose that is attached to a wet standpipe outlet.

Buildings higher than 150 feet must be provided with a combination standpipe in every stairway or smoke-proof enclosure. This system replaces the wet and dry standpipes normally required.

Properly designed and maintained, standpipe systems are an effective and valuable aid for extinguishing fires, especially in the upper stories of tall buildings, in the interior of large commercial or industrial malls, or in other areas where construction features or access make the laying of temporary hose lines time-consuming and hazardous.

There are three general classes of service for standpipes.

Class I: For use by fire department and personnel with special training for heavy streams (2-1/2" hose connections).

Class II: For general use by building occupants until the fire department arrives (1-1/2" hose connector with hose).

Class III: To be used by either fire department or trained building occupants (2-1/2" and 1-1/2" hose connections or a 2-1/2" hose valve with an easily removable 2-1/2" by 1-1/2" adapter).

Sprinkler Systems

Automatic sprinkler systems have been in use for many years in buildings as a very effective means of controlling and even extinguishing fires that are in the early stages. Because they are automatic, no action by building occupants is required for the systems to activate. The flow of water from a sprinkler system can be compared with a heavy rainfall.

The model codes require automatic sprinkler systems in some building types. In others, the systems are highly recommended. The codes generally require automatic sprinklers in the following occupancies. There are exceptions, so check the local code for specifics.

- Basements and cellars, except houses and garages.
- Theater backstage area, dressing rooms, workshops, and storage areas.
- Concealed areas above stairways in schools, hospitals, institutions such as prisons, and places of assembly like theaters, stadiums and schools.
- Most hazardous areas.
- All institutional occupancies, except jails.
- Retail sales areas over 12,000 S.F./floor or over 24,000 S.F. gross.
- Places of assembly over 12,000 S.F.

There are a number of different types of sprinker systems: wet pipe, dry pipe, preaction, deluge, and firecycle.

Wet Pipe Sprinkler Systems

The wet pipe sprinkler system employs automatic sprinklers attached to a piping system containing water and connected to a water supply so that when the sprinkler heads are opened by heat or fire, water discharges from the heads immediately. Benefits of the wet pipe system include quick response and a low initial cost.

Dry Pipe Sprinkler Systems

In a dry pipe sprinkler system, the water in the pipes has been replaced with compressed air. When the air is released by a sprinkler head opening, the dry pipe valve opens to allow water to flow to the open sprinkler head.

Preaction Sprinkler Systems

The preaction sprinkler system is a refinement of the dry pipe system. The system still employs automatic sprinklers connected to a piping system, which may or may not contain compressed air. To provide a second "look," a supplemental system of heat detectors with more sensitive characteristics is installed in the same areas as the sprinkler heads. When the heat rises above the threshold of the heat detectors, a valve opens, permitting the flow of water into the piping system. The water is discharged from any open heads. The advantage of this system occurs when a sprinkler head is accidentally damaged, because nothing happens — as long as the heat detectors are not activated, no water flows.

Deluge Sprinkler Systems

The deluge system is identical to the preaction system except that the sprinkler heads are always wide open. This system is normally used in high fire hazard areas.

Firecycle Sprinkler Systems

The firecycle system is a fixed fire protection sprinkler system that uses water as its extinguishing agent. It is a time-delayed, recycling, preaction system that automatically shuts off the flow of water when heat is reduced below the detector operating temperature. If the temperature is exceeded again, the water turns back on automatically. The system senses a fire condition through a closed circuit electrical detector system that automatically controls water flow to the fire. A battery system supplies up to 90 hours of emergency power to operate the system. The piping system is dry until water is required, and is monitored with pressurized air. Should any leak in the system occur, an alarm will sound, but water will not enter the system until heat is sensed by a firecycle detector.

Occupancy Hazards

To determine the permitted square feet of building floor area for each sprinkler head, three hazard levels for sprinklers have been established by the codes. Since the model building codes have varying rules regarding the installation of sprinkler systems, verification of the local code is an absolute must before proceeding too far into project design. Depending on the occupancy classification, buildings will fall into one of three categories based on expected hazards: light hazard occupancy, ordinary hazard occupancy, or extra hazard occupancy.

Light Hazard Occupancy

The protection area allotted per sprinkler should not exceed 200 square feet, with the maximum distance between sprinkler lines and sprinklers on lines being 15 feet. The sprinklers do not need to be staggered. Branch lines should not exceed eight sprinklers on either side of a cross main. Each large area that requires more than 100 sprinklers and does not have a subdividing partition should be supplied by feed mains or risers sized for ordinary hazard occupancy.

Included in this group are:
    Auditoriums
    Churches
    Clubs
    Educational buildings
    Hospitals
    Institutional buildings
    Libraries (except large stack rooms)
    Museums
    Nursing homes
    Offices
    Residential buildings
    Restaurants
    Schools
    Theaters

Ordinary Hazard Occupancy

The protection area allotted per sprinkler cannot exceed 130 square feet noncombustible ceiling and 120 square feet of combustible ceiling. The maximum allowable distance between sprinkler lines and sprinklers on line is 15 feet. Sprinklers cannot be staggered if the distance between heads exceeds 12 feet. Branch lines should not exceed eight sprinklers on either side of a cross main.

Included in this group are:

Automotive garages
Bakeries
Beverage manufacturing plants
Bleacheries
Boiler houses
Canneries
Cement plants
Clothing factories
Cold storage warehouses
Dairy products manufacturing plants
Distilleries
Dry cleaning operations
Electric generating stations
Feed mills
Grain elevators
Ice manufacturing plants
Laundries
Machine shops
Mercantiles
Paper mills
Printing and publishing houses
Shoe factories
Warehouses
Wood product assembly plants

Extra Hazard Occupancy

The protection area allotted per sprinkler cannot exceed 90 square feet of noncombustible ceiling and 80 square feet of combustible ceiling. The maximum allowable distance between sprinkler lines and sprinklers on line is 12 feet. Sprinklers on alternate lines have to be staggered if the distance between sprinklers on lines exceeds eight feet. Branch lines cannot exceed six sprinklers on either side of a cross main.

Included in this group are:

Aircraft hangars
Chemical works
Explosives manufacturing plants
Linoleum manufacturing plants
Linseed oil mills
Liquid manufacturing & use
Oil refineries
Paint shops
Shade cloth manufacturing
Solvent extracting
Varnish works
Volatile flammable

Heating

The basic function of a heating system is to heat an enclosed space to a desired temperature, and then to maintain that temperature within a reasonable range. To accomplish this, the selected system must be able to offset transmission losses resulting from the temperature difference between the interior and exterior of the enclosing walls and the roof/ceiling assembly, and the losses resulting from cold air infiltration through cracks and crevices around doors and windows. The amount of heat to be furnished depends on the building size, construction quality, temperature difference, air leakage, building use, shape, orientation, and exposure. Air circulation is also an important consideration. Circulation prevents stratification, a condition that can result in heat loss through uneven temperatures at various levels, not to mention discomfort for the occupants.

While the most accurate estimates of heating requirements are naturally based on detailed information, it is possible to approximate the cost using the following procedure:

1. Calculate the volume of the room or building.
2. Select the appropriate factor from Figure 8.12. Note that these factors apply only to the inside design temperatures listed in the first column and to a 0°F outside design temperature.
3. If the building has undesirable north and west exposures, multiply the heat loss factor by 1.1. Bad exposures would be severe shade on the north or west elevations and exposure to intense winter winds.
4. If the outside design temperature is other than 0°F, multiply the factor from Figure 8.12 by the appropriate factor from Figure 8.13.
5. Multiply the cubic volume by the factor selected from Figure 8.12. This will result in the estimated Btu/hr. heat loss that must be made up to maintain the inside design temperature.

Air Conditioning

The purpose of air conditioning is to control the environment of an enclosed space for human comfort or for equipment operation. System objectives must be defined and evaluated by several factors, including:

- Temperature control
- Humidity control
- Cleanliness
- Odor, smoke, and fumes
- Ventilation

The table in Figure 8.14 provides a rough rule of thumb for the air conditioning requirements of various building types. The numbers reveal the Btu's per hour per square foot of floor area and square feet per ton of air conditioning.

Conclusion

Mechanical assemblies costs are a function of a building's type of construction, shape, and use. Plumbing and fixture costs depend mostly on the use and occupancy of the building. Building codes should be used to determine minimum requirements, with attention given to the ADA. The scope of the plumbing work includes the work usually performed by plumbers, but does not include the utility work beyond 5' of the face of the building. Water heaters should be sized according to the maximum hourly anticipated demand, with an allowance for heat loss from the pipes and storage tank.

Factor for Determining Heat Loss for Various Types of Buildings

Building Type	Conditions	Qualifications	Loss Factor*
Factories & Industrial Plants General Office Areas at 70°F	One Story	Skylight in Roof	6.2
		No Skylight in Roof	5.7
	Multiple Story	Two Story	4.6
		Three Story	4.3
		Four Story	4.1
		Five Story	3.9
		Six Story	3.6
	All Walls Exposed	Flat Roof	6.9
		Heated Space Above	5.2
	One Long Warm Common Wall	Flat Roof	6.3
		Heated Space Above	4.7
	Warm Common Walls on Both Long Sides	Flat Roof	5.8
		Heated Space Above	4.1
Warehouses at 60°F	All Walls Exposed	Skylights in Roof	5.5
		No Skylight in Roof	5.1
		Heated Space Above	4.0
	One Long Warm Common Wall	Skylight in Roof	5.0
		No Skylight in Roof	4.9
		Heated Space Above	3.4
	Warm Common Walls on Both Long Sides	Skylight in Roof	4.7
		No Skylight in Roof	4.4
		Heated Space Above	3.0

*Note: This table tends to be conservative particularly for new buildings designed for minimum energy consumption.

Figure 8.12

Outside Design Temperature Correction Factor (for Degrees Fahrenheit)

Outside Design Temperature	50	40	30	20	10	0	−10	−20	−30
Correction Factor	0.29	0.43	0.57	0.72	0.86	1.00	1.14	1.28	1.43

Figure 8.13

Fire protection costs are a function of the type and use of a building. Both building materials and fire suppression systems are defined by building codes. Most buildings require standpipe risers and hose cabinets with related equipment. Many buildings require sprinklers, either by law or code. Savings in insurance premiums also make sprinkler installation worthwhile.

The selection of a heating and air conditioning system involves consideration of the fuel to be used, the building type, the occupancy, and quality (longevity and sophistication) desired. Costs can vary significantly between basic requirements and sophisticated temperature control systems. Even for a preliminary cost estimate, it is important to determine the owner's requirements in order to select the proper quality systems. Since mechanical assemblies typically comprise about 20% of the construction cost of a new building, each of the four divisions — Plumbing, Fire Protection, Heating, and Air Conditioning — requires careful attention by the estimator.

Air Conditioning Requirements

BTU's per hour per S.F. of floor area and S.F. per ton of air conditioning.

Type of Building	BTU per S.F.	S.F. per Ton	Type of Building	BTU per S.F.	S.F. per Ton	Type of Building	BTU per S.F.	S.F. per Ton
Apartments, Individual	26	450	Dormitory, Rooms	40	300	Libraries	50	240
Corridors	22	550	Corridors	30	400	Low Rise Office, Exterior	38	320
Auditoriums & Theaters	40	300/18*	Dress Shops	43	280	Interior	33	360
Banks	50	240	Drug Stores	80	150	Medical Centers	28	425
Barber Shops	48	250	Factories	40	300	Motels	28	425
Bars & Taverns	133	90	High Rise Office — Ext. Rms.	46	263	Office (small suite)	43	280
Beauty Parlors	66	180	Interior Rooms	37	325	Post Office, Individual Office	42	285
Bowling Alleys	68	175	Hospitals, Core	43	280	Central Area	46	260
Churches	36	330/20*	Perimeter	46	260	Residences	20	600
Cocktail Lounges	68	175	Hotel, Guest Rooms	44	275	Restaurants	60	200
Computer Rooms	141	85	Corridors	30	400	Schools & Colleges	46	260
Dental Offices	52	230	Public Spaces	55	220	Shoe Stores	55	220
Dept. Stores, Basement	34	350	Industrial Plants, Offices	38	320	Shop'g. Ctrs., Supermarkets	34	350
Main Floor	40	300	General Offices	34	350	Retail Stores	48	250
Upper Floor	30	400	Plant Areas	40	300	Specialty	60	200

*Persons per ton
12,000 BTU = 1 ton of air conditioning

Figure 8.14

Chapter 9

Electrical: Assemblies Estimate

The electrical assembly in a building is comprised of several subassemblies that can be initially estimated from limited information available in the early stages of a project. Before any subassembly sizing or estimating can be performed, the building electrical load or total number of watts required must first be determined. From this calculation, a second calculation must be done to find the total amperes for the service. The process requires simple math and a basic knowledge of the different components and where each fits into the overall system.

Estimating Procedure

Developing the building electrical loads is the first step, outlined below:

1. Determine the building size and anticipated use.
2. Develop the total load, in watts, for:
   - Lighting
   - Receptacles
   - Air conditioning
   - Elevators
   - Other power requirements
3. Determine the voltage available from the utility company.
4. Determine the size of the building service from formulas.
5. Determine costs for service, panels, and feeders.
6. Determine costs for above subsystems using loads.
   (Steps 5 and 6 will be discussed in Part 2 of this book.)

The conversion of watts to amperes is necessary to size the building service entrance. Once the number of watts required for the building is developed, ask the local utility company for the voltage available at or near the project site.

Example One: Three-Story Office Building

Perhaps the easiest way to understand the estimating procedure just outlined is to work through the same example used in the previous chapter, the Three-Story Office Building.

Power Requirements

Lighting (Figures 9.1 and 9.2):
3.0 watts/S.F.   30,000 S.F. × 3 W/S.F.   =   90,000 watts
Receptacles (Figure 9.1):
2.0 watts/S.F.   30,000 S.F. × 2 W/S.F.   =   60,000 watts
HVAC (Figures 9.1 and 9.3):
4.7 watts/S.F.   30,000 S.F. × 4.7 W/S.F.   =   141,000 watts
Misc. Motors & Power (Figure 9.1):
1.2 watts/S.F.   30,000 S.F. × 1.2 W/S.F.   =   36,000 watts

Elevators:
Figure 9.4: 2 @ 3000 lb. and 100 FPM = 30 hp
Figure 9.5: 30 hp, 460 volts = 40 amps

Formula:
Watts = 1.73 × volts × current × power factor × efficiency
Watts = 1.73 × 460 × 40 × 0.9 × 0.9   =   <u>25,784</u>
                                            352,784 watts

Voltage Available

Assume that the Utility Company has the following available at the site:

120/208 volt, 3 phase, 4 wire or 277/480 volt, 3 phase, 4 wire service.

Select the 277/480 volt service, as it allows smaller feeder to be used.

Size of Service

$$\text{Formula: Amperes} = \frac{\text{watts}}{\text{volts} \times \text{power factor} \times 1.73}$$

$$\text{Amperes} = \frac{352,784}{480 \times 0.8 \times 1.73} = 531 \text{ amps}$$

Use 600 amp service.

Conclusion

Energy codes set lighting limitations for many types of building occupancies. This wattage for lighting, combined with the power requirements for receptacles, air conditioning, elevators, and other systems, determines the total wattage. An approximation of all electrical systems of a building can be determined with a minimum of information. Because the electrical requirements for similar buildings are essentially the same, the costs per square foot are also the same. One branch bank or hospital will require nearly the same wattage as another branch bank or hospital. Except in special cases, developing a preliminary cost by using cost per fixture or per receptacle is cumbersome and no more accurate.

After the total wattage requirements are determined, simple formulas convert watts to amperes for the available voltages to size of service.

Costs for the various elements of electrical systems are developed in Part 2 of this book.

Nominal Watts Per S.F. for Electric Systems for Various Building Types

Type Construction	1. Lighting	2. Devices	3. HVAC	4. Misc.	5. Elevator	Total Watts
Apartment, luxury high rise	2	2.2	3	1		
Apartment, low rise	2	2	3	1		
Auditorium	2.5	1	3.3	.8		
Bank, branch office	3	2.1	5.7	1.4		
Bank, main office	2.5	1.5	5.7	1.4		
Church	1.8	.8	3.3	.8		
College, science building	3	3	5.3	1.3		
College, library	2.5	.8	5.7	1.4		
College, physical education center	2	1	4.5	1.1		
Department store	2.5	.9	4	1		
Dormitory, college	1.5	1.2	4	1		
Drive-in donut shop	3	4	6.8	1.7		
Garage, commercial	.5	.5	0	.5		
Hospital, general	2	4.5	5	1.3		
Hospital, pediatric	3	3.8	5	1.3		
Hotel, airport	2	1	5	1.3		
Housing for the elderly	2	1.2	4	1		
Manufacturing, food processing	3	1	4.5	1.1		
Manufacturing, apparel	2	1	4.5	1.1		
Manufacturing, tools	4	1	4.5	1.1		
Medical clinic	2.5	1.5	3.2	1		
Nursing home	2	1.6	4	1		
Office building, hi rise	3	2	4.7	1.2		
Office building, low rise	3	2	4.3	1.2		
Radio-TV studio	3.8	2.2	7.6	1.9		
Restaurant	2.5	2	6.8	1.7		
Retail store	2.5	.9	5.5	1.4		
School, elementary	3	1.9	5.3	1.3		
School, junior high	3	1.5	5.3	1.3		
School, senior high	2.3	1.7	5.3	1.3		
Supermarket	3	1	4	1		
Telephone exchange	1	.6	4.5	1.1		
Theater	2.5	1	3.3	.8		
Town Hall	2	1.9	5.3	1.3		
U.S. Post Office	3	2	5	1.3		
Warehouse, grocery	1	.6	0	.5		

Figure 9.1

Lighting Limit (Connected Load) for Listed Occupancies: New Building Proposed Energy Conservation Guideline

Type of Use	Maximum Watts per S.F.
Interior **Category A:** Classrooms, office areas, automotive mechanical areas, museums, conference rooms, drafting rooms, clerical areas, laboratories, merchandising areas, kitchens, examining rooms, book stacks, athletic facilities.	3.00
Category B: Auditoriums, waiting areas, spectator areas, restrooms, dining areas, transportation terminals, working corridors in prisons and hospitals, book storage areas, active inventory storage, hospital bedrooms, hotel and motel bedrooms, enclosed shopping mall concourse areas, stairways.	1.00
Category C: Corridors, lobbies, elevators, inactive storage areas.	0.50
Category D: Indoor parking.	0.25
Exterior **Category E:** Building perimeter: wall-wash, facade, canopy.	5.00 (per linear foot)
Category F: Outdoor parking.	0.10

Figure 9.2

Central Air Conditioning Watts per S.F., BTU's per Hour per S.F. of Floor Area and S.F. per Ton of Air Conditioning

Type Building	Watts per S.F.	BTUH per S.F.	S.F. per Ton	Type Building	Watts per S.F.	BTUH per S.F.	S.F. per Ton	Type Building	Watts per S.F.	BTUH per S.F.	S.F. per Ton
Apartments, Individual	3	26	450	Dormitory, Rooms	4.5	40	300	Libraries	5.7	50	240
Corridors	2.5	22	550	Corridors	3.4	30	400	Low Rise Office, Ext.	4.3	38	320
Auditoriums & Theaters	3.3	40	300/18*	Dress Shops	4.9	43	280	Interior	3.8	33	360
Banks	5.7	50	240	Drug Stores	9	80	150	Medical Centers	3.2	28	425
Barber Shops	5.5	48	250	Factories	4.5	40	300	Motels	3.2	28	425
Bars & Taverns	15	133	90	High Rise Off.-Ext. Rms.	5.2	46	263	Office (small suite)	4.9	43	280
Beauty Parlors	7.6	66	180	Interior Rooms	4.2	37	325	Post Office, Int. Office	4.9	42	285
Bowling Alleys	7.8	68	175	Hospitals, Core	4.9	43	280	Central Area	5.3	46	260
Churches	3.3	36	330/20*	Perimeter	5.3	46	260	Residences	2.3	20	600
Cocktail Lounges	7.8	68	175	Hotels, Guest Rooms	5	44	275	Restaurants	6.8	60	200
Computer Rooms	16	141	85	Public Spaces	6.2	55	220	Schools & Colleges	5.3	46	260
Dental Offices	6	52	230	Corridors	3.4	30	400	Shoe Stores	6.2	55	220
Dept. Stores, Basement	4	34	350	Industrial Plants, Offices	4.3	38	320	Shop'g. Ctrs., Sup. Mkts.	4	34	350
Main Floor	4.5	40	300	General Offices	4	34	350	Retail Stores	5.5	48	250
Upper Floor	3.4	30	400	Plant Areas	4.5	40	300	Specialty Shops	6.8	60	200

*Persons per ton   12,000 BTUH = 1 ton of air conditioning

Figure 9.3

106

Description: The table below shows approximate horsepower requirements for front opening elevators with 3-phase motors. Tabulations are for given car capacities at various travel speeds.

Watts = 1.73 × Volts × Current × Power Factor × Efficiency

$$\text{Horsepower} = \frac{\text{Volts} \times \text{Current} \times 1.73 \times \text{Power Factor}}{746 \text{ Watts}}$$

The power factor of electric motors varies from 80% to 90% in larger size motors. The efficiency likewise varies from 80% on a small motor to 90% on a large motor.

Horsepower Requirements For Elevators

Type	Maximum Travel Height	Travel Speeds In FPM	Capacity Of Cars In Lbs.								
			1200	1500	1800	2000	2500	3000	3500	4000	4500
Hydraulic	70 Ft.	70	10	15	15	15	20	20	20	25	30
		85	15	15	15	20	20	25	25	30	30
		100	15	15	20	20	25	30	30	40	40
		110	20	20	20	20	25	30	40	40	50
		125	20	20	20	25	30	40	40	50	50
		150	25	25	25	30	40	50	50	50	60
		175	25	30	30	40	50	50	60		
		200	30	30	40	40	50	60	60		
Geared Traction	300 Ft.	200				10	10	15	15		23
		350				15	15	23	23		36

Figure 9.4

Ampere Values Determined by Horsepower, Voltage and Phase Values

H.P.	Amperes					
	Single Phase		Three Phase			
	115V	230V	200V	230V	460V	575V
1/6	4.4A	2.2A				
1/4	5.8	2.9				
1/3	7.2	3.6				
1/2	9.8	4.9	2.3A	2.0A	1.0A	0.8A
3/4	13.8A	6.9	3.2	2.8	1.4	1.1
1	16	8	4.1	3.6	1.8	1.4
1-1/2	20	10	6.0	5.2	2.6	2.1
2	24	12	7.8	6.8	3.4	2.7
3	34	17	11.0	9.6	4.8	3.9
5			17.5	15.2	7.6	6.1
7-1/2			25.3	22	11	9
10			32.2	28	14	11
15			48.3	42	21	17
20			62.1	54	27	22
25			78.2	68	34	27
30			92.0	80	40	32
40			119.6	104	52	41
50			149.5	130	65	52
60			177	154	77	62
75			221	192	96	77
100			285	248	124	99
125			359	312	156	125
150			414	360	180	144
200			552	480	240	192

Figure 9.5

107

Chapter 10

Structural Load Example

Follow the procedures outlined in Chapters 3 and 4 to develop structural loads for the Three-Story Office Building used as an example for actual cost development in Part 2.

Specifications: Three-Story Office Building

Occupancy:	Office Building
Live Load:	Floor: 50 psf
	Roof: 30 psf (snow load)
Allowable Soil Bearing Capacity:	6 KSF
Footprint:	$100' \times 100' = 10,000$ S.F.
Floors:	3 floors
Floor to Floor Height:	12'-0"
Bay Size:	25'-0" x 25'-0"
Construction:	Floor: Open web joists and concrete slab on steel columns and beams.
	Roof: Open web joists and metal deck on steel columns and beams.

Superimposed Load

Floor:

Live Load	50 psf	BOCA Code	(Figure 4.2)
Partition	15 psf		(Figure 4.3)
Ceiling	5 psf		(Figure 4.3)
Mechanical	5 psf		(Figure 4.3)
Superimposed Floor Load:	75 psf		

Roof:

Snow Load	30 psf	BOCA Code	(Figure 4.2)
Ceiling	5 psf		(Figure 4.3)
Mechanical	5 psf		(Figure 4.3)
Superimposed Roof Load:	40 psf		

Calculate Total (Demand and Live) Loads

Enter the table in Figure 10.1 with 25'-0" x 25'-0" bays and a superimposed load of 75 psf. The result is:

Total Floor Load = 120 psf

Enter the table in Figure 10.2 with 25'-0" x 25'-0" bays and a superimposed load of 40 psf. The result is:

Total Roof Load = 60 psf

3.5-460 — Steel Joists, Beams & Slab on Columns

	BAY SIZE (FT.)	SUPERIMPOSED LOAD (P.S.F.)	DEPTH (IN.)	TOTAL LOAD (P.S.F.)	COLUMN ADD	COST PER S.F. MAT.	COST PER S.F. INST.	COST PER S.F. TOTAL
3900	20x25	65	26	110		4.65	3.28	7.93
4000					column	.48	.21	.69
4100	20x25	75	26	120		4.73	3.14	7.87
4200					column	.57	.25	.82
4300	20x25	100	26	145		5	3.28	8.28
4400					column	.57	.25	.82
4500	20x25	125	29	170		5.60	3.55	9.15
4600					column	.67	.29	.96
4700	25x25	40	23	84		4.58	3.22	7.80
4800					column	.46	.20	.66
4900	25x25	65	29	110		4.85	3.35	8.20
5000					column	.46	.20	.66
5100	25x25	75	26	120		5.25	3.36	8.61
5200					column	.54	.24	.78
5300	25x25	100	29	145		5.85	3.65	9.50
5400					column	.54	.24	.78
5500	25x25	125	32	170		6.15	3.80	9.95
5600					column	.59	.25	.84
5700	25x30	40	29	84		4.97	3.35	8.32
5800					column	.45	.20	.65
5900	25x30	65	29	110		5.20	3.49	8.69
6000					column	.45	.20	.65
6050	25x30	75	29	120		5.55	3.22	8.77
6100					column	.49	.21	.70
6150	25x30	100	29	145		6.05	3.41	9.46
6200					column	.49	.21	.70
6250	25x30	125	32	170		6.50	4.12	10.62
6300					column	.57	.25	.82
6350	30x30	40	29	84		5.15	3.06	8.21
6400					column	.41	.17	.58
6500	30x30	65	29	110		5.85	3.36	9.21
6600					column	.41	.17	.58
6700	30x30	75	32	120		6	3.40	9.40
6800					column	.47	.21	.68
6900	30x30	100	35	145		6.65	3.69	10.34
7000					column	.55	.24	.79
7100	30x30	125	35	172		7.30	4.51	11.81
7200					column	.61	.26	.87
7300	30x35	40	29	85		5.85	3.34	9.19
7400					column	.35	.16	.51
7500	30x35	65	29	111		6.55	4.17	10.72
7600					column	.46	.20	.66
7700	30x35	75	32	121		6.55	4.17	10.72
7800					column	.47	.20	.67
7900	30x35	100	35	148		7.10	3.83	10.93
8000					column	.57	.25	.82
8100	30x35	125	38	173		7.90	4.15	12.05
8200					column	.58	.25	.83
8300	35x35	40	32	85		6	3.41	9.41
8400					column	.41	.17	.58
8500	35x35	65	35	111		6.90	4.32	11.22
8600					column	.49	.21	.70
9300	35x35	75	38	121		7.05	4.39	11.44
9400					column	.49	.21	.70
9500	35x35	100	38	148		7.60	4.65	12.25
9600					column	.60	.26	.86

Figure 10.1

A3.7-420 **Steel Joists & Beams on Cols.**

Description: Table below lists the cost per S.F. for a roof system with steel columns, beams, and deck, using open web steel joists and 1-1/2" galvanized metal deck.

Design and Pricing Assumptions:
Columns are 18' high.
Building is 4 bays long by 4 bays wide.
Joists are 5'-0" O.C. and span the long direction of the bay.
Joists at columns have bottom chords extended and are connected to columns.
Column costs are not included but are listed separately per S.F. of floor.

Roof deck is 1-1/2", 22 gauge galvanized steel. Joist cost includes appropriate bridging. Deflection is limited to 1/240 of the span. Fireproofing is not included.

Design Loads	Min.		Max.	
Joists & Beams	3	PSF	5	PSF
Deck	2		2	
Insulation	3		3	
Roofing	6		6	
Misc.	6		6	
Total Dead Load	20	PSF	22	PSF

System Components	QUANTITY	UNIT	COST PER S.F.		
			MAT.	INST.	TOTAL
SYSTEM 3.7-420-1100					
METAL DECK AND JOISTS,15'X20' BAY,20 PSF S. LOAD					
Structural steel	.954	Lb.	.55	.19	.74
Open web joists	1.260	Lb.	.54	.27	.81
Metal decking, open, galvanized, 1-1/2" deep, 22 gauge	1.050	S.F.	.76	.37	1.13
TOTAL			1.85	.83	2.68

3.7-420					Steel Joists, Beams, & Deck on Columns			

	BAY SIZE (FT.)	SUPERIMPOSED LOAD (P.S.F.)	DEPTH (IN.)	TOTAL LOAD (P.S.F.)	COLUMN ADD	COST PER S.F.		
						MAT.	INST.	TOTAL
1100	15x20	20	16	40		1.85	.83	2.68
1200					columns	.90	.31	1.21
1300		30	16	50		2.03	.90	2.93
1400					columns	.90	.31	1.21
1500		40	18	60		2.07	.94	3.01
1600					columns	.90	.31	1.21
1700	20x20	20	16	40		1.99	.88	2.87
1800					columns	.67	.23	.90
1900		30	18	50		2.18	.95	3.13
2000					columns	.67	.23	.90
2100		40	18	60		2.43	1.05	3.48
2200					columns	.67	.23	.90
2300	20x25	20	18	40		2.10	.93	3.03
2400					columns	.54	.19	.73
2500		30	18	50		2.29	1.08	3.37
2600					columns	.54	.19	.73
2700		40	20	60		2.36	1.05	3.41
2800					columns	.72	.25	.97
2900	25x25	20	18	40		2.43	1.05	3.48
3000					columns	.43	.15	.58
3100		30	22	50		2.57	1.18	3.75
3200					columns	.57	.20	.77
3300		40	20	60		2.81	1.20	4.01
3400					columns	.57	.20	.77

Figure 10.2

Loading Diagram

Compute the loads to an interior column as shown in Figure 10.3. Note that steel columns are usually spliced in two-story jumps. The splice has been shown 3'-0" above the third floor.

Column Loads

Develop Column Loads as shown in Figure 10.3.

Column Loads:

Upper Columns	37.5 kips + 1.2 kips (incl. col and fireproofing)	= 38.7 kips
Lower Columns	38.7 + 75 + 1.2 + 75 + 1.2	= 191 kips

Foundation Loads

Interior Foundation Load	Figure 10.3	= 191 kips
Exterior Foundation Load	191 kips × .6	= 115 kips
Corner Foundation Load	191 kips × .45	= 86 kips

For spread footings with 6 KSF allowable soil bearing capacity, see Figure 10.4.

Interior Footing	191 kips (200 kips)	6'-0" sq x 20" deep
Exterior Footing	115 kips (125 kips)	5'-0" sq x 16" deep
Corner Footing	86 kips (100 kips)	4'-6" sq x 15" deep

As discussed in the Strip Footings section in Chapter 5, the load to the concrete wall and strip footing would be of no consequence.

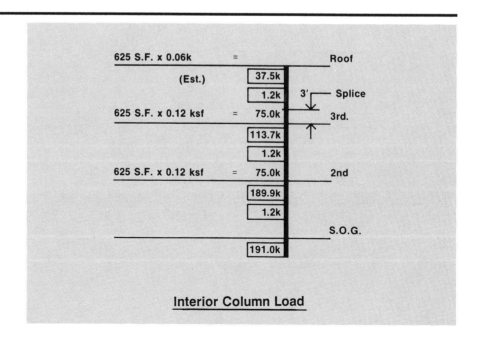

Interior Column Load

Figure 10.3

The Spread Footing System includes: excavation; backfill; forms (four uses); all reinforcement; 3,000 p.s.i. concrete (chute placed); and screed finish.

Footing systems are priced per individual unit. The Expanded System Listing at the bottom shows footings that range from 3' square x 12" deep, to 18' square x 52" deep. It is assumed that excavation is done by a truck mounted hydraulic excavator with an operator and oiler.

Backfill is with a dozer, and compaction by air tamp. The excavation and backfill equipment is assumed to operate at 30 C.Y. per hour.

Please see the reference section for further design and cost information.

System Components	QUANTITY	UNIT	COST EACH		
			MAT.	INST.	TOTAL
SYSTEM 1.1-120-7100					
SPREAD FOOTINGS, LOAD 25K, SOIL CAPACITY 3 KSF, 3' SQ X 12" DEEP					
Bulk excavation	.590	C.Y.		3.18	3.18
Hand trim	9.000	S.F.		4.77	4.77
Compacted backfill	.260	C.Y.		.55	.55
Formwork, 4 uses	12.000	S.F.	7.20	36.12	43.32
Reinforcing, fy = 60,000 psi	.006	Ton	3.45	4.50	7.95
Dowel or anchor bolt templates	6.000	L.F.	4.02	15	19.02
Concrete, f'c = 3,000 psi	.330	C.Y.	18.98		18.98
Place concrete, direct chute	.330	C.Y.		5	5
Screed finish	9.000	S.F.		2.97	2.97
TOTAL			33.65	72.09	105.74

1.1-120	Spread Footings		COST EACH		
			MAT.	INST.	TOTAL
7090	Spread footings, 3000 psi concrete, chute delivered				
7100	Load 25K, soil capacity 3 KSF, 3'-0" sq. x 12" deep		33.50	72	105.50
7150	Load 50K, soil capacity 3 KSF, 4'-6" sq. x 12" deep	R1.1 -120	69.50	125	194.50
7200	Load 50K, soil capacity 6 KSF, 3'-0" sq. x 12" deep		33.50	72	105.50
7250	Load 75K, soil capacity 3 KSF, 5'-6" sq. x 13" deep		109	177	286
7300	Load 75K, soil capacity 6 KSF, 4'-0" sq. x 12" deep		57	107	164
7350	Load 100K, soil capacity 3 KSF, 6'-0" sq. x 14" deep		138	212	350
7410	Load 100K, soil capacity 6 KSF, 4'-6" sq. x 15" deep		85.50	147	232.50
7450	Load 125K, soil capacity 3 KSF, 7'-0" sq. x 17" deep		217	305	522
7500	Load 125K, soil capacity 6 KSF, 5'-0" sq. x 16" deep		110	177	287
7550	Load 150K, soil capacity 3 KSF 7'-6" sq. x 18" deep		261	360	621
7610	Load 150K, soil capacity 6 KSF, 5'-6" sq. x 18" deep		145	223	368
7650	Load 200K, soil capacity 3 KSF, 8'-6" sq. x 20" deep		370	475	845
7700	Load 200K, soil capacity 6 KSF, 6'-0" sq. x 20" deep		189	275	464
7750	Load 300K, soil capacity 3 KSF, 10'-6" sq. x 25" deep		675	770	1,445
7810	Load 300K, soil capacity 6 KSF, 7'-6" sq. x 25" deep		355	465	820
7850	Load 400K, soil capacity 3 KSF, 12'-6" sq. x 28" deep		1,075	1,150	2,225
7900	Load 400K, soil capacity 6 KSF, 8'-6" sq. x 27" deep		490	605	1,095
7950	Load 500K, soil capacity 3 KSF, 14'-0" sq. x 31" deep		1,475	1,525	3,000
8010	Load 500K, soil capacity 6 KSF, 9'-6" sq. x 30" deep		675	785	1,460
8050	Load 600K, soil capacity 3 KSF, 16'-0" sq. x 35" deep		2,150	2,075	4,225
8100	Load 600K, soil capacity 6 KSF, 10'-6" sq. x 33" deep		905	1,025	1,930
8150	Load 700K, soil capacity 3 KSF, 17'-0" sq. x 37" deep		2,525	2,375	4,900
8200	Load 700K, soil capacity 6 KSF, 11'-6" sq. x 36" deep		1,150	1,250	2,400

Figure 10.4

Conclusion

Order of Magnitude costs, historical square foot costs, and model square foot costs are useful for developing rough, preliminary estimates and for the space planning of construction projects.

Order of Magnitude estimates can, in a short time, establish costs for many types of structures if the scope required or desired is known. With cost limitations in mind, divide the Order of Magnitude cost into the specified budget cost to obtain the project scope. The Order of Magnitude estimate should be within + or −20% of the actual construction cost of the facility.

The historical square foot estimate is also useful, provided the approximate size of the facility is known. Square foot costs can be used to develop the expected building area for the cost, the size building that can be constructed on a site with restrictions for parking and building height, and how much a building will cost if it contains a predetermined number of units. The square foot estimate, if modified for size and area, should be accurate within + or − 15% of the actual construction cost.

With a rough sketch of the building footprint, number of floors, and the floor to ceiling height, an assemblies square foot estimate can be accomplished in a reasonable amount of time. The time to complete the estimate will vary from project to project for several reasons. For example, changing structural characteristics, varying soil conditions, and the accuracy required of the estimate may cause the estimator to spend more time in the estimating process.

The assemblies square foot estimate has several advantages. With knowledge of the building's approximate bay sizes and required structural loads, the foundation and structural costs can be tailored to the most economical or preferred system. Exterior closure costs can be weighted to satisfy both the aesthetic and economic requirements. Interior construction costs can be quickly evaluated and compared with other alternatives. Realistic mechanical and electrical loads and costs can be determined, and should be within + or − 10% of the actual costs.

Assemblies square foot estimating shows the cost relationship of the major components of a structure. Although some components of the estimate may be priced using conventional unit or historical costs, other parts can be priced using the assemblies method to arrive at total building costs.

When computing structural loads for a building, make sketches. This approach not only helps to convey the ideas of the estimator, but also reduces omissions. Computations should be neat and orderly to reduce the chance of error. Show all floors in proper relationship when accumulating loads to avoid the possibility of omitting a floor.

Part 2 of the book shows how to transfer all the loads and assumptions developed in Part 1 into real estimated construction dollars. The loads and assumptions used are those that apply to a Three-Story Office Building. This example problem is fully explained in the introduction to Part 2.

Part 2

Cost Development — A Sample Estimate

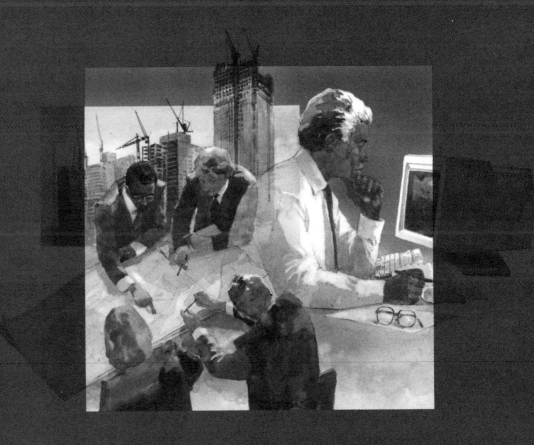

Introduction

For an estimate to be well organized, it must have a clear direction from the beginning. One of the biggest advantages of assemblies square foot estimating is the ease with which it guides a user through the complex task of combining the many different parts of a construction project.

Part 2 provides an inside look at how all the pieces of cost information for the estimate go together. It includes an example that closely follows the organizational pattern of the assemblies format. Essentially, the assemblies estimating method reorganizes the familiar 16 divisions of the Uniform Construction Index into 12 divisions that closely reflect the actual way a building is put together. There are 12 UniFormat divisions. Chapters 11 – 22 accurately reflect how the assemblies format breaks down the building construction project. For example, in Chapter 11: Foundations, the many different "unit price" elements of a foundation are placed in several basic groups, and the cost of an entire foundation is summarized in terms of assemblies.

For the estimator using this method, it's no longer necessary to know every detail to get a price for a foundation. The total cost for the entire system, the units involved, and the correct way to organize each section of the estimate are shown. The estimator needs to know only building dimensions and other relevant design information.

The principal example used here is the steel-frame Three-Story Office Building. Each chapter ends with a summary of the costs that have been assigned to the assemblies involved in that area of the job. Where necessary, additional examples have been included.

Beginning the Sample Estimate

This part of the book is arranged to follow, step by step, the process of preparing a cost estimate using assemblies square foot costs. The object of the sample estimate presented in the next 12 chapters is to explain where all the costs come from. Using this example, an estimate can be made based on square foot costs from any reliable source. A single source, *Means Assemblies Cost Data,* has been used exclusively for the sake of consistency.

The first step is to understand what factors affect the costs. For example, how do increased loads and varying soil conditions affect the total cost of a building? It's one thing to take the total project costs for a number of buildings, divide them by their square foot dimensions, and develop some ballpark figures for cost projections. It's an entirely different matter to know just why each of those buildings has its own unique price tag. Cost is not always directly related to size alone.

The estimator should be able to determine, based on practical experience, exactly how to develop the costs in an assemblies estimate. Although assemblies estimating eliminates the time-consuming procedures involved in a unit price estimate, it adds a new responsibility. Square foot estimates must be carefully organized and well thought out, or they simply will not work. Unless the correct costs are matched to each application, the estimate loses its accuracy.

Sample Project: Three-Story Office Building

The sample project estimate that begins in this chapter is based on a theoretical Three-Story Office Building that the client would like to have constructed at a budget of $1.7 million. Cost information for this example is based on 1996 prices from the 1996 edition of *Means Assemblies Cost Data*. Using the design criteria found in Part 1 of this book, the following 12 chapters assemble the building cost along typical project schedule guidelines, more or less from the bottom up, following the standard UniFormat division breakdown outlined below.

1. Foundations
2. Substructures
3. Superstructures
4. Exterior Closure
5. Roofing
6. Interior Construction
7. Conveying Systems
8. Mechanical
9. Electrical
10. General Conditions & Profit
11. Special Construction
12. Site Work

Since Excavation and Foundation work (UniFormat Division 1) are the initial activities in a construction project, they will be dealt with first.

In the real world, the column or wall loads must first be determined in order to design the foundation. With that in mind, the design and calculation work in Division 3, Superstructures, should be completed before starting an estimate of the excavation and foundation. What this means to the estimator is that when loads to the foundation are unknown, the loads for all elements in Division 3 must be determined before any work is begun on Division 1.

Using Means Assemblies Cost Data

Throughout the following chapters, which contain the estimate for the example Three-Story Office Building, pages from *Means Assemblies Cost Data* have been reproduced as a source of cost information. The next two pages after this Introduction explain how the editors of *Means Assemblies Cost Data* intend the information be used.

Each of the chapters in Part 2 ends with a Summary Form, which lists the costs that have been assigned to the assemblies in that particular chapter. The Summary Forms are cumulative, with new cost items added to the summary shown for the previous chapter. Following the "How To Use" pages is a page that explains how the Summary Forms are used.

How to Use the Assemblies Cost Tables

The following is a detailed explanation of a sample Assemblies Cost Table. Most Assembly Tables are separated into three parts: 1) an illustration of the system to be estimated; 2) the components and related costs of a typical system; and 3) the costs for similar systems with dimensional and/or size variations. Next to each bold number below is the item being described with the appropriate component of the sample entry following in parenthesis. In most cases, if the work is to be subcontracted, the general contractor will need to add an additional markup (R.S. Means suggests using 10%) to the "Total" figures.

System/Line Numbers (A3.5-140-1700)

Each Assemblies Cost Line has been assigned a unique identification number based on the UniFormat classification system.

UniFormat Division

3.5 140 1700

Means Subdivision
Means Major Classification
Means Individual Line Number

SUPERSTRUCTURES	A3.5-140	C.I.P. Flat Slab w/Drop Panels

General: Flat Slab: Solid uniform depth concrete two-way slabs with drop panels at columns and no column capitals.

Design and Pricing Assumptions:
Concrete f'c = 3 KSI, placed by concrete pump.
Reinforcement, fy = 60 KSI.
Forms, four use.
Finish, steel trowel.
Curing, spray on membrane.
Based on 4 bay x 4 bay structure.

System Components	QUANTITY	UNIT	COST PER S.F. MAT.	COST PER S.F. INST.	COST PER S.F. TOTAL
SYSTEM 3.5-140-1700					
15'X15' BAY 40 PSF S. LOAD, 12" MIN. COL. 6" SLAB, 1-1/2" DROP, 117 PSF					
Forms in place, flat slab with drop panels, to 15' high, 4 uses	.993	S.F.	1.02	3.52	4.54
Forms in place, exterior spandrel, 12" wide, 4 uses	.034	S.F.	.04	.21	.25
Reinforcing in place, elevated slabs #4 to #7	1.588	Lb.	.48	.43	.91
Concrete ready mix, regular weight, 3000 psi	.513	C.F.	1.09		1.09
Place and vibrate concrete, elevated slab, 6" to 10" pump	.513	C.F.		.41	.41
Finish floor, monolithic steel trowel finish for finish floor	1.000	S.F.		.61	.61
Cure with sprayed membrane curing compound	.010	C.S.F.	.02	.05	.07
TOTAL			2.65	5.23	7.88

3.5-140		Cast in Place Flat Slab with Drop Panels						
	BAY SIZE (FT.)	SUPERIMPOSED LOAD (P.S.F.)	MINIMUM COL. SIZE (IN.)	SLAB & DROP (IN.)	TOTAL LOAD (P.S.F.)	COST PER S.F. MAT.	COST PER S.F. INST.	COST PER S.F. TOTAL
1700		40	12	6 - 1-1/2	117	2.65	5.25	7.90
1720		75	12	6 - 2-1/2	153	2.70	5.25	7.95
1760		125	14	6 - 3-1/2	205	2.83	5.35	8.18
1780		200	16	6 - 4-1/2	281	2.98	5.50	8.48
1840	15 x 20	40	12	6-1/2 - 2	124	2.82	5.30	8.12
1860	R3.5 -100	75	14	6-1/2 - 4	162	2.94	5.40	8.34
1880		125	16	6-1/2 - 5	213	3.13	5.50	8.63
1900		200	18	6-1/2 - 6	293	3.21	5.60	8.81
1960	20 x 20	40	12	7 - 3	132	2.96	5.35	8.31
1980		75	16	7 - 4	168	3.14	5.55	8.69
		125						

2 Illustration

At the top of most assembly tables is an illustration, a brief description, and the design criteria used to develop the cost.

3 System Components

The components of a typical system are listed separately to show what has been included in the development of the total system price. The table below contains prices for other similar systems with dimensional and/or size variations.

4 Quantity

This is the number of line item units required for one system unit. For example, we assume that it will take 1.588 pounds of reinforcing on a square foot basis.

5 Unit of Measure for Each Item

The abbreviated designation indicates the unit of measure, as defined by industry standards, upon which the price of the component is based. For example, reinforcing is priced by lb. (pound) while concrete is priced by C.F. (cubic foot).

6 Unit of Measure for Each System (Cost per S.F.)

Costs shown in the three right hand columns have been adjusted by the component quantity and unit of measure for the entire system. In this example, "Cost per S.F." is the unit of measure for this system or "assembly."

7 Reference Number Information

R3.5 -100 You'll see reference numbers shown in bold squares at the beginning of some major classifications. These refer to related items in the Reference Section, visually identified by a vertical gray bar on the edge of pages.

The relation may be: (1) an estimating procedure that should be read before estimating, (2) an alternate pricing method, or (3) technical information.

The "R" designates the Reference Section. The numbers refer to the UniFormat classification system.

Example: The square number above is directing you to refer to the reference number R3.5-100. This particular reference number shows comparative costs of floor systems.

8 Materials (2.65)

This column contains the Materials Cost of each component. These cost figures are bare costs plus 10% for profit.

9 Installation (5.23)

Installation includes labor and equipment plus the installing contractor's overhead and profit. Equipment costs are the bare rental costs plus 10% for profit. The labor overhead and profit is defined on the inside back cover of this book.

10 Total (7.88)

The figure in this column is the sum of the material and installation costs.

Material Cost	+	Installation Cost	=	Total
$2.65	+	$5.23	=	$7.88

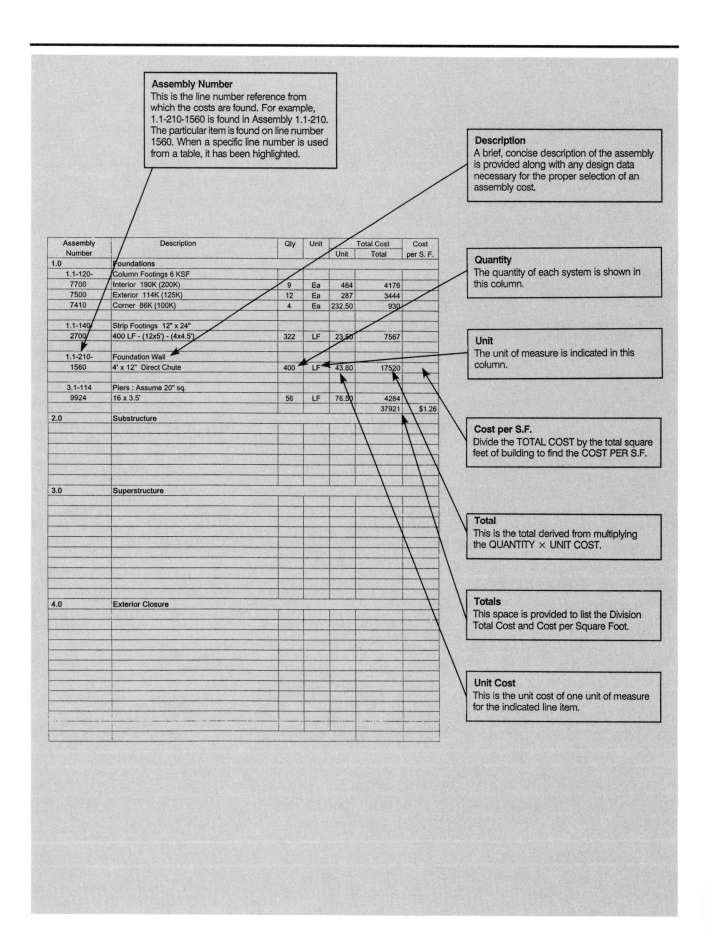

Assembly Number
This is the line number reference from which the costs are found. For example, 1.1-210-1560 is found in Assembly 1.1-210. The particular item is found on line number 1560. When a specific line number is used from a table, it has been highlighted.

Description
A brief, concise description of the assembly is provided along with any design data necessary for the proper selection of an assembly cost.

Quantity
The quantity of each system is shown in this column.

Unit
The unit of measure is indicated in this column.

Cost per S.F.
Divide the TOTAL COST by the total square feet of building to find the COST PER S.F.

Total
This is the total derived from multiplying the QUANTITY × UNIT COST.

Totals
This space is provided to list the Division Total Cost and Cost per Square Foot.

Unit Cost
This is the unit cost of one unit of measure for the indicated line item.

Assembly Number	Description	Qty	Unit	Total Cost Unit	Total Cost Total	Cost per S. F.
1.0	Foundations					
1.1-120-	Column Footings 6 KSF					
7700	Interior 190K (200K)	9	Ea	464	4176	
7500	Exterior 114K (125K)	12	Ea	287	3444	
7410	Corner 86K (100K)	4	Ea	232.50	930	
1.1-140	Strip Footings 12" x 24"					
2700	400 LF - (12x5') - (4x4.5')	322	LF	23.50	7567	
1.1-210-	Foundation Wall					
1560	4' x 12" Direct Chute	400	LF	43.80	17520	
3.1-114	Piers : Assume 20" sq.					
9924	16 x 3.5'	56	LF	76.50	4284	
					37921	$1.26
2.0	Substructure					
3.0	Superstructure					
4.0	Exterior Closure					

Chapter 11

Foundations

The foundation is the entire substructure below the first floor or frame of a building, including the footing on which the building rests. Once structural elements have been chosen and sized, loads to the foundation can be determined. With this information, foundation components can be selected and priced.

The sample project is the Three-Story Office Building described in Chapter 10.

Column Footings

Loads applied to the foundation and footing size (see "Foundation Loads" in Chapter 10) using a 6 KSF soil capacity:

Interior Columns	191 kips:	6'-0" sq. x 20" deep
Exterior Columns	115 kips:	5'-0" sq. x 16" deep
Corner Columns	86 kips:	4'-6" sq. x 15" deep

Costs:

Interior (Figure 10.4, line #7700)	(200 kips) 9 ea. @ $464	=	$4,176
Exterior (Figure 10.4, line #7500)	(125 kips) 12 ea. @ $287	=	3,444
Corners (Figure 10.4, line #7410)	(100 kips) 4 ea. @ $232.50	=	930
	Total for Column Footings		$8,550

Strip Footings

A strip footing is necessary around the perimeter of the building. To correctly find the total linear feet of strip footing, deduct the width of both exterior and corner column footings. See Figure 11.1.

Perimeter: 400 L.F.

Exterior Footings: 12 @ 5' each

Corner Footings: 4 @ 4.5' each

$400' - (12 \times 5') - (4 \times 4.5') = 322$ L.F.

Costs:

(Figure 11.2, line #2700) 322 L.F. @ $23.50 = $7,567

The load to the strip footing, as discussed in "Strip Footings" in Chapter 5, is negligible since the precast concrete fascia is hung from the columns and suspended floor slabs and accounted for in the column loads.

Foundation Wall

The foundation wall needed for the Three-Story Office Building is 4' deep x 12" wide.

$$100' \times 4 = 400 \text{ L.F.}$$

Cost:

(Figure 11.3, line #1560) 400 L.F. @ $43.80 = $17,520

Piers

To carry the column loads at the perimeter wall, provide a pier under each perimeter column to provide support directly to the footing instead of the slab on grade (see Figure 11.4). Assume a 20" square pier.

$$16 \times 3.5' = 56 \text{ L.F.}$$

Cost:

(Figure 11.5, line #9924) 56 L.F. @ $76.50 = $4,284

Excavation and Backfill

Excavation and fill (see Figure 11.6) is based on the following assumptions:

- 3' over-excavation in width is used for a work area at the bottom of excavation. The excavation has an appropriate side slope for the type of material being excavated.
- Backfill from excavation to the foundation wall is done with excavated material, except for clay (which is assumed to be all wasted).

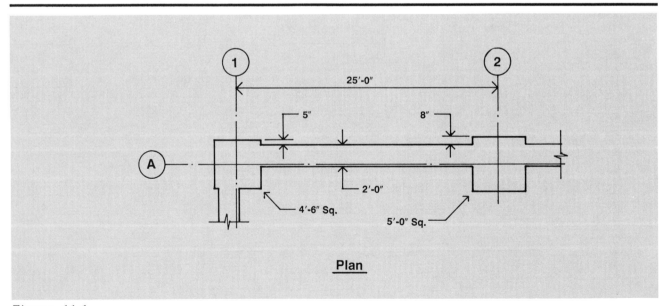

Plan

Figure 11.1

A1.1-140 | **Strip Footings**

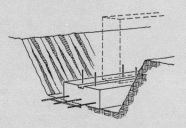

The Strip Footing System includes: excavation; hand trim; all forms needed for footing placement; forms for 2″ x 6″ keyway (four uses); dowels; and 3,000 p.s.i. concrete.

The footing size required varies for different soils. Soil bearing capacities are listed for 3 KSF and 6 KSF. Depths of the system range from 8″ to 24″. Widths range from 16″ to 96″. Smaller strip footings may not require reinforcement.

Please see the reference section for further design and cost information.

System Components	QUANTITY	UNIT	COST PER L.F.		
			MAT.	INST.	TOTAL
SYSTEM 1.1-140-2500					
STRIP FOOTING, LOAD 5.1KLF, SOIL CAP. 3 KSF, 24″WIDE X12″DEEP, REINF.					
Trench excavation	.148	C.Y.		.78	.78
Hand trim	2.000	S.F.		1.06	1.06
Compacted backfill	.074	C.Y.		.16	.16
Formwork, 4 uses	2.000	S.F.	.94	5.14	6.08
Keyway form, 4 uses	1.000	L.F.	.28	.64	.92
Reinforcing, fy = 60000 psi	3.000	Lb.	.90	1.11	2.01
Dowels	2.000	Ea.	2.32	3.94	6.26
Concrete, f'c = 3000 psi	.074	C.Y.	4.26		4.26
Place concrete, direct chute	.074	C.Y.		1.03	1.03
Screed finish	2.000	S.F.		.92	.92
TOTAL			8.70	14.78	23.48

1.1-140	Strip Footings	COST PER L.F.		
		MAT.	INST.	TOTAL
2100	Strip footing, load 2.6KLF,soil capacity 3KSF, 16″wide x 8″deep plain	4.27	8.80	13.07
2300	Load 3.9 KLF soil capacity, 3 KSF, 24″wide x 8″deep, plain	5.25	9.75	15
2500	Load 5.1KLF, soil capacity 3 KSF, 24″wide x 12″deep, reinf.	8.70	14.80	23.50
2700	Load 11.1KLF, soil capacity 6 KSF, 24″wide x 12″deep, reinf.	8.70	14.80	23.50
2900	Load 6.8 KLF, soil capacity 3 KSF, 32″wide x 12″deep, reinf.	10.45	16.15	26.60
3100	Load 14.8 KLF, soil capacity 6 KSF, 32″wide x 12″deep, reinf.	10.45	16.15	26.60
3300	Load 9.3 KLF, soil capacity 3 KSF, 40″wide x 12″deep, reinf.	12.20	17.50	29.70
3500	Load 18.4 KLF, soil capacity 6 KSF, 40″wide x 12″deep, reinf.	12.30	17.65	29.95
3700	Load 10.1KLF, soil capacity 3 KSF, 48″wide x 12″deep, reinf.	13.55	18.90	32.45
3900	Load 22.1KLF, soil capacity 6 KSF, 48″wide x 12″deep, reinf.	14.45	20	34.45
4100	Load 11.8KLF, soil capacity 3 KSF, 56″wide x 12″deep, reinf.	15.80	21	36.80
4300	Load 25.8KLF, soil capacity 6 KSF, 56″wide x 12″deep, reinf.	17.05	22.50	39.55
4500	Load 10KLF, soil capacity 3 KSF, 48″wide x 16″deep, reinf.	16.70	21.50	38.20
4700	Load 22KLF, soil capacity 6 KSF, 48″wide, 16″deep, reinf.	17.05	22	39.05
4900	Load 11.6KLF, soil capacity 3 KSF, 56″wide x 16″deep, reinf.	18.90	30.50	49.40
5100	Load 25.6KLF, soil capacity 6 KSF, 56″wide x 16″deep, reinf.	19.95	31.50	51.45
5300	Load 13.3KLF, soil capacity 3 KSF, 64″wide x 16″deep, reinf.	21.50	26	47.50
5500	Load 29.3KLF, soil capacity 6 KSF, 64″wide x 16″deep, reinf.	23	27.50	50.50
5700	Load 15KLF, soil capacity 3 KSF, 72″wide x 20″deep, reinf.	28.50	31	59.50
5900	Load 33KLF, soil capacity 6 KSF, 72″wide x 20″deep, reinf.	30	33	63
6100	Load 18.3KLF, soil capacity 3 KSF, 88″wide x 24″deep, reinf.	40	38.50	78.50
6300	Load 40.3KLF, soil capacity 6 KSF, 88″wide x 24″deep, reinf.	43.50	43	86.50
6500	Load 20KLF, soil capacity 3 KSF, 96″wide x 24″deep, reinf.	43.50	41.50	85
6700	Load 44 KLF, soil capacity 6 KSF, 96″ wide x 24″ deep, reinf.	46	44.50	90.50

R1.1 -140

Figure 11.2

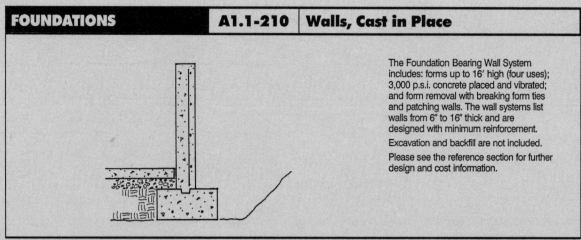

The Foundation Bearing Wall System includes: forms up to 16' high (four uses); 3,000 p.s.i. concrete placed and vibrated; and form removal with breaking form ties and patching walls. The wall systems list walls from 6" to 16" thick and are designed with minimum reinforcement.

Excavation and backfill are not included.

Please see the reference section for further design and cost information.

System Components	QUANTITY	UNIT	COST PER L.F.		
			MAT.	INST.	TOTAL
SYSTEM 1.1-210-1500 FOUNDATION WALL, CAST IN PLACE, DIRECT CHUTE, 4' HIGH, 6" THICK					
Formwork	8.000	SFCA	3.68	23.20	26.88
Reinforcing	3.300	Lb.	.95	.92	1.87
Concrete, 3,000 psi	.074	C.Y.	4.26		4.26
Place concrete, direct chute	.074	C.Y.		1.37	1.37
Finish walls, break ties and patch voids, one side	4.000	S.F.	.04	2.16	2.20
TOTAL			8.93	27.65	36.58

1.1-210		Walls, Cast in Place						
	WALL HEIGHT (FT.)	PLACING METHOD	CONCRETE (C.Y./L.F.)	REINFORCING (LBS./L.F.)	WALL THICKNESS (IN.)	COST PER L.F.		
						MAT.	INST.	TOTAL
1500	4'	direct chute	.074	3.3	6	8.95	27.50	36.45
1520			.099	4.8	8	10.80	28.50	39.30
1540			.123	6.0	10	12.50	29	41.50
1560	R1.1 -210		.148	7.2	12	14.30	29.50	43.80
1580			.173	8.1	14	16	30	46
1600			.197	9.44	16	17.75	31	48.75
1700	4'	pumped	.074	3.3	6	8.95	28.50	37.45
1720			.099	4.8	8	10.80	30	40.80
1740			.123	6.0	10	12.50	31	43.50
1760			.148	7.2	12	14.30	31.50	45.80
1780			.173	8.1	14	16	33	49
1800			.197	9.44	16	17.75	33.50	51.25
3000	6'	direct chute	.111	4.95	6	13.40	41.50	54.90
3020			.149	7.20	8	16.20	43	59.20
3040			.184	9.00	10	18.75	43.50	62.25
3060			.222	10.8	12	21.50	45	66.50
3080			.260	12.15	14	24	45.50	69.50
3100			.300	14.39	16	27	47	74
3200	6'	pumped	.111	4.95	6	13.40	43	56.40
3220			.149	7.20	8	16.20	45	61.20
3240			.184	9.00	10	18.75	46	64.75
3260			.222	10.8	12	21.50	48	69.50
3280			.260	12.15	14	24	49	73
3300			.300	14.39	16	27	50.50	77.50

Figure 11.3

- Costs are shown for material stored on- or off-site. Storage is within a two-mile haul distance.

Variables that affect the cost include:

- Availability and type of equipment.
- Haul distance to spoil area.
- Haul distance to borrow area as the source for fill.

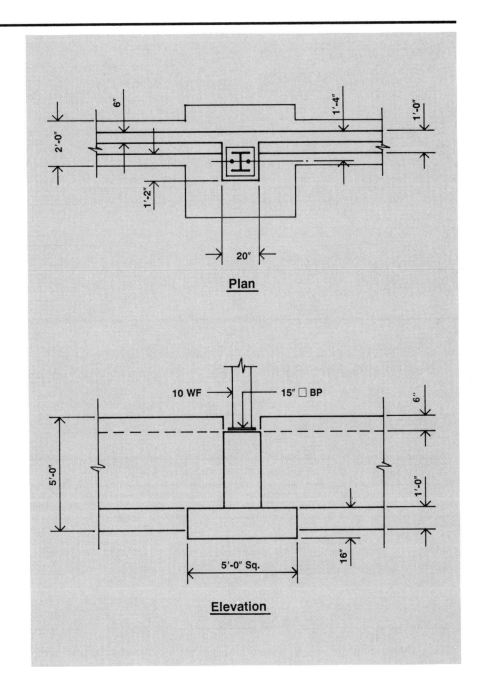

Figure 11.4

- Protection of the excavation, particularly in deep excavations where sheet piles may be necessary.

For pricing purposes, use Figure 11.6 with the "Footprint" or area of the excavation and the depth required. Determine the type of soil nearest the types listed in the table. Costs are listed per square foot of excavated area. Interpolate for depths and areas other than those shown.

Pricing excavation is not like pricing a ton of steel. When using cost tables, regardless of the source, review the assumptions made in compiling the data and temper the prices with judgment based on the anticipated variables at the project site.

Since there is no basement area in the Three-Story Office Building project, bulk excavation, as in Figure 11.6, is not needed. The only excavation necessary is for the perimeter foundation wall. This is similar to trench excavation and will be accounted for in Division 12, Site Work.

Waterproofing and Dampproofing

In the Three-Story Office Building, there is no basement; consequently, there is no need for waterproofing or dampproofing. However, if it were necessary, the table in Figure 11.7 could be used. Simply determine the total linear feet of perimeter foundation wall and then select the type of waterproofing or dampproofing needed. Multiply the cost per lineal foot times the total length of foundation wall. Note that each item in the table is defined for specific wall height. With some simple math, the costs can be translated into a cost per square foot for wall heights not listed.

C.I.P. Column, Square Tied

	LOAD (KIPS)	STORY HEIGHT (FT.)	COLUMN SIZE (IN.)	COLUMN WEIGHT (P.L.F.)	CONCRETE STRENGTH (PSI)	COST PER V.L.F.		
						MAT.	INST.	TOTAL
9800	1000	10	22	469	6000	26	64.50	90.50
9840		12	22	474	6000	26	64	90
9900		14	22	478	6000	26	65	91

C.I.P. Column, Square Tied-Minimum Reinforcing

	LOAD (KIPS)	STORY HEIGHT (FT.)	COLUMN SIZE (IN.)	COLUMN WEIGHT (P.L.F.)	CONCRETE STRENGTH (PSI)	COST PER V.L.F.		
						MAT.	INST.	TOTAL
9912	150	10-14	12	135	4000	11.35	33.50	44.85
9918	300	10-14	16	240	4000	15.15	43	58.15
9924	500	10-14	20	375	4000	20.50	56	76.50
9930	700	10-14	24	540	4000	28.50	69.50	98
9936	1000	10-14	28	740	4000	35	83	118
9942	1400	10-14	32	965	4000	47	96	143
9948	1800	10-14	36	1220	4000	56	110	166
9954	2300	10-14	40	1505	4000	63	122	185

Figure 11.5

Pricing Assumptions: Two-thirds of excavation is by 2-1/2 C.Y. wheel mounted front end loader and one-third by 1-1/2 C.Y. hydraulic excavator.

Two-mile round trip haul by 12 C.Y. tandem trucks is included for excavation wasted and storage of suitable fill from excavated soil. For excavation in clay, all is wasted and the cost of suitable backfill with two-mile haul is included.

Sand and gravel assumes 15% swell and compaction; common earth assumes 25% swell and 15% compaction; clay assumes 40% swell and 15% compaction (non-clay).

In general, the following items are accounted for in the costs in the table below.

1. Excavation for building or other structure to depth and extent indicated.
2. Backfill compacted in place.
3. Haul of excavated waste.
4. Replacement of unsuitable material with bank run gravel.

Note: Additional excavation and fill beyond this line of general excavation for the building (as required for isolated spread footings, strip footings, etc.) are included in the cost of the appropriate component systems.

System Components	QUANTITY	UNIT	COST PER S.F.		
			MAT.	INST.	TOTAL
SYSTEM 1.9-100-2220					
EXCAVATE & FILL, 1000 S.F., 4' DEEP, SAND, ON SITE STORAGE					
Excavating bulk shovel, 1.5 C.Y. bucket, 160 cy/hr	.108	C.Y.		.13	.13
Excavation, front end loader, 2-1/2 C.Y.	.216	C.Y.		.28	.28
Haul earth, 12 C.Y. dump truck	.170	C.Y.		.59	.59
Backfill, dozer bulk push 300', including compaction	.203	C.Y.		.43	.43
TOTAL				1.43	1.43

1.9-100	Building Excavation and Backfill		COST PER S.F.		
			MAT.	INST.	TOTAL
2220	Excav & fill, 1000 S.F. 4' sand, gravel, or common earth, on site storage			1.43	1.43
2240	Off site storage			2.85	2.85
2260	Clay excavation, bank run gravel borrow for backfill		.76	2.71	3.47
2280	8' deep, sand, gravel, or common earth, on site storage	R12.0 -010		3.37	3.37
2300	Off site storage			7.25	7.25
2320	Clay excavation, bank run gravel borrow for backfill		1.83	6.20	8.03
2340	16' deep, sand, gravel, or common earth, on site storage	R12.1 -400		9.55	9.55
2350	Off site storage			18.05	18.05
2360	Clay excavation, bank run gravel borrow for backfill		5.05	15.95	21
3380	4000 S.F.,4' deep, sand, gravel, or common earth, on site storage	R12.3 -100		1.09	1.09
3400	Off site storage			1.73	1.73
3420	Clay excavation, bank run gravel borrow for backfill		.35	1.66	2.01
3440	8' deep, sand, gravel, or common earth, on site storage			2.37	2.37
3460	Off site storage			4.09	4.09
3480	Clay excavation, bank run gravel borrow for backfill		.83	3.68	4.51
3500	16' deep, sand, gravel, or common earth, on site storage			5.85	5.85
3520	Off site storage			11.25	11.25
3540	Clay, excavation, bank run gravel borrow for backfill		2.23	8.80	11.03
4560	10,000 S.F., 4' deep, sand gravel, or common earth, on site storage			.96	.96
4580	Off site storage			1.36	1.36
4600	Clay excavation, bank run gravel borrow for backfill		.22	1.33	1.55
4620	8' deep, sand, gravel, or common earth, on site storage			2.04	2.04
4640	Off site storage			3.06	3.06
4660	Clay excavation, bank run gravel borrow for backfill		.51	2.84	3.35
4680	16' deep, sand, gravel, or common earth, on site storage			4.75	4.75
4700	Off site storage			7.90	7.90
4720	Clay excavation, bank run gravel borrow for backfill		1.34	6.50	7.84
5740	30,000 S.F., 4' deep, sand, gravel, or common earth, on site storage			.89	.89
5760	Off site storage			1.11	1.11
5780	Clay excavation, bank run gravel borrow for backfill		.12	1.08	1.20
5800	8' deep, sand & gravel, or common earth, on site storage			1.82	1.82

Figure 11.6

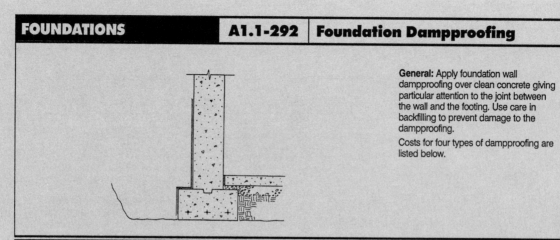

General: Apply foundation wall dampproofing over clean concrete giving particular attention to the joint between the wall and the footing. Use care in backfilling to prevent damage to the dampproofing.

Costs for four types of dampproofing are listed below.

System Components		QUANTITY	UNIT	COST PER L.F.		
				MAT.	INST.	TOTAL
SYSTEM 1.1-292-1000 FOUNDATION DAMPPROOFING, BITUMINOUS, 1 COAT, 4' HIGH						
	Bituminous asphalt dampproofing brushed on below grade, 1 coat	4.000	S.F.	.24	1.92	2.16
	Labor for protection of dampproofing during backfilling	4.000	S.F.		.76	.76
	TOTAL			.24	2.68	2.92

1.1-292	Foundation Dampproofing	COST PER L.F.		
		MAT.	INST.	TOTAL
1000	Foundation dampproofing, bituminous, 1 coat, 4' high	.24	2.68	2.92
1400	8' high	.48	5.35	5.83
1800	12' high	.72	8.30	9.02
2000	2 coats, 4' high	.44	3.28	3.72
2400	8' high	.88	6.55	7.43
2800	12' high	1.32	10.10	11.42
3000	Asphalt with fibers, 1/16" thick, 4' high	.56	3.28	3.84
3400	8' high	1.12	6.55	7.67
3800	12' high	1.68	10.10	11.78
4000	1/8" thick, 4' high	1.20	3.92	5.12
4400	8' high	2.40	7.85	10.25
4800	12' high	3.60	12	15.60
5000	Asphalt coated board and mastic, 1/4" thick, 4' high	1.68	3.56	5.24
5400	8' high	3.36	7.10	10.46
5800	12' high	5.05	10.95	16
6000	1/2" thick, 4' high	2.52	4.96	7.48
6400	8' high	5.05	9.90	14.95
6800	12' high	7.55	15.15	22.70
7000	Metallic coating, on walls, w/ 1/8" thick cement. coating, 4' high	7.20	26.50	33.70
7400	8' high	14.40	53	67.40
7800	12' high	21.50	79	100.50
8000	3 coat system on slabs, no protection course, 2' wide	.44	1.68	2.12
8400	4' wide	.88	3.36	4.24
8800	6' wide	1.32	5.05	6.37

Figure 11.7

Foundation Underdrain

Whenever waterproofing is used on a basement wall, it is necessary to provide positive foundation underdrains or footing drains, as they are sometimes called, to carry away any free water that may accumulate around the wall or strip footing.

Grade Beams

Grade beams are used primarily to support walls or slabs when footings are at a greater depth (because of poor soil conditions) than the required wall or floor.

To determine the cost of a grade beam, first determine the load per linear foot transmitted to the grade beam, whether it is from a load-bearing or non-load-bearing wall. If the building is to be constructed in an area that has frozen earth through the winter months, frost protection may be necessary. Choose a grade beam with a width no less than that of the supported wall. Also take into consideration the load supported and the minimum required depth. A typical grade beam system cost per linear foot is derived from designing a simply supported beam such as that shown in Figure 11.8, using design assumptions and the following items:

- Excavation
- Reinforcing steel

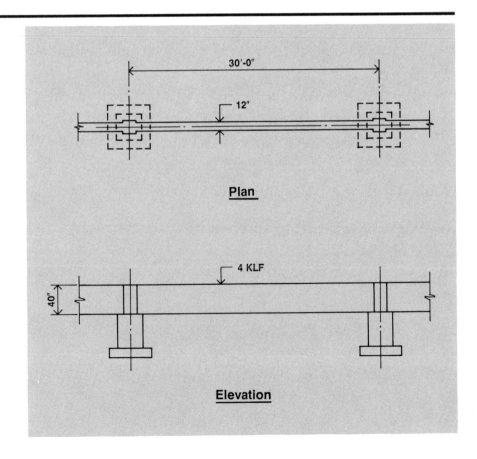

Figure 11.8

- Backfill
- Concrete
- Formwork

The entire assemblies cost of grade beams can be determined quickly once the loading is determined. The following example, although not part of the Three-Story Office Building, is representative of the method used to calculate the cost of grade beams.

Example: Grade Beam Cost

Using Figure 11.9:

30' Span, 4 KLF Loading

Grade Beam 40" Deep, 12" Wide $63.50/L.F.

$63.50/L.F. $\times$ L.F. of Grade Beam = Cost

Figure 11.10 shows the summary of Division 1, Foundation costs for this building. Note that the total cost is reached by adding the costs of each individual item as well as the cost per square foot, which is the total division cost divided by 30,000 square feet.

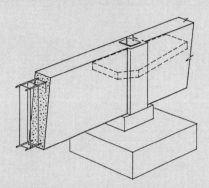

The Grade Beam System includes: excavation with a truck mounted backhoe; hand trim; backfill; forms (four uses); reinforcing steel; and 3,000 p.s.i. concrete placed from chute.

Superimposed loads vary in the listing from 8 Kips per linear foot (KLF) to 50 KLF. In the Expanded System Listing, the span of the beams varies from 15' to 40'. Depth varies from 28" to 52". Width varies from 12" to 28".

Please see the reference section for further design and cost information.

System Components	QUANTITY	UNIT	COST PER L.F.		
			MAT.	INST.	TOTAL
SYSTEM 1.1-230-2220					
GRADE BEAM, 15' SPAN, 28" DEEP, 12" WIDE, 8 KLF LOAD					
Excavation, trench, hydraulic backhoe, 3/8 CY bucket	.260	C.Y.		1.39	1.39
Trim sides and bottom of trench, regular soil	2.000	S.F.		1.06	1.06
Backfill, by hand, compaction in 6" layers, using vibrating plate	.170	C.Y.		.91	.91
Forms in place, grade beam, 4 uses	4.700	SFCA	3.10	14.99	18.09
Reinforcing in place, beams & girders, #8 to #14	.019	Ton	10.64	11.12	21.76
Concrete ready mix, regular weight, 3000 psi	.090	C.Y.	5.18		5.18
Place and vibrate conc. for grade beam, direct chute	.090	C.Y.		1	1
TOTAL			18.92	30.47	49.39

1.1-230	Grade Beams		COST PER L.F.		
			MAT.	INST.	TOTAL
2220	Grade beam, 15' span, 28" deep, 12" wide, 8 KLF load		18.90	30.50	49.40
2240	14" wide, 12 KLF load		19.50	30.50	50
2260	40" deep, 12" wide, 16 KLF load	R1.1 -230	19.70	37.50	57.20
2280	20 KLF load		22.50	40.50	63
2300	52" deep, 12" wide, 30 KLF load		26	51	77
2320	40 KLF load		31.50	57	88.50
2340	50 KLF load		37.50	62.50	100
3360	20' span, 28" deep, 12" wide, 2 KLF load		12.20	23.50	35.70
3380	16" wide, 4 KLF load		15.85	26	41.85
3400	40" deep, 12" wide, 8 KLF load		19.70	37.50	57.20
3420	12 KLF load		25	42.50	67.50
3440	14" wide, 16 KLF load		30	47.50	77.50
3460	52" deep, 12" wide, 20 KLF load		32.50	57.50	90
3480	14" wide, 30 KLF load		44	68.50	112.50
3500	20" wide, 40 KLF load		51.50	72	123.50
3520	24" wide, 50 KLF load		63.50	82.50	146
4540	30' span, 28" deep, 12" wide, 1 KLF load		12.75	24	36.75
4560	14" wide, 2 KLF load		20	33.50	53.50
4580	40" deep, 12" wide, 4 KLF load		23.50	40	63.50
4600	18" wide, 8 KLF load		33.50	48.50	82
4620	52" deep, 14" wide, 12 KLF load		42.50	66.50	109
4640	20" wide, 16 KLF load		51.50	73	124.50
4660	24" wide, 20 KLF load		63.50	83	146.50
4680	36" wide, 30 KLF load		90.50	104	194.50
4700	48" wide, 40 KLF load		119	127	246
5720	40' span, 40" deep, 12" wide, 1 KLF load		17.50	35.50	53
5740	2 KLF load		21.50	39.50	61
5760	52" deep, 12" wide, 4 KLF load		31.50	57	88.50

Figure 11.9

133

Assembly Number	Description	Qty	Unit	Total Cost		Cost per S. F.
				Unit	Total	
1.0	**Foundations**					
1.1-120-	Column Footings 6 KSF					
7700	Interior 190K (200K)	9	Ea	464	4176	
7500	Exterior 114K (125K)	12	Ea	287	3444	
7410	Corner 86K (100K)	4	Ea	232.50	930	
1.1-140-	Strip Footings 12" x 24"					
2700	400 LF - (12x5') - (4x4.5')	322	LF	23.50	7567	
1.1-210-	Foundation Wall					
1560	4' x 12" Direct Chute	400	LF	43.80	17520	
3.1-114	Piers : Assume 20" sq.					
9924	16 x 3.5'	56	LF	76.50	4284	
					37921	$1.26
2.0	**Substructure**					
3.0	**Superstructure**					
4.0	**Exterior Closure**					

Figure 11.10

Chapter 12

Substructures

Slab on Grade

When estimating a slab on grade, its intended use must be considered. The local building code may prescribe the required floor loading per square foot, which in turn defines the thickness and amount of reinforcement necessary to satisfy the loading conditions. Slabs may be categorized in three distinct groups according to their loading conditions: non-industrial; light industrial; and heavy industrial. Each can be divided further into reinforced and nonreinforced.

If the specifications or dimensions are predetermined by the project requirements or the estimator's past experience, it may be possible to develop a unit price estimate of the cost of the slab on grade to include such items as:

- Base course
- Topping
- Vapor barrier
- Finish
- Concrete
- Cure
- Reinforcement

However, it is even easier to do an assemblies estimate using either assemblies costs developed by the estimator or some other source. Figure 12.1 shows the assemblies costs with all components listed for reinforced and non-reinforced slabs for different slab thicknesses. Using a table such as this is certainly acceptable; however, the national average cost of concrete is $54.00/cubic yard. Using the system shown in Figure 12.1 (line number 2220) as an example, what is the effect of a delivered concrete price of $72 per cubic yard?

There is 0.012 C.Y./S.F. in a 4″ thick slab.

$$0.33/27 = 0.012$$
$$.012 \times (\$72 - \$54) = \$.22/S.F. + \$2.55/S.F. = \$2.77/S.F.$$

The system price now is $2.77/S.F. based on the new material prices instead of $2.55/S.F., which used $54/C.Y. as its basis.

Slab on Grade Selection and Cost Analysis

Using Figure 12.1, select the appropriate slab thickness and the proper usage. Multiply the system cost per square foot by the total area of slab to obtain the total cost for slabs on grade. For the area of slab, use the out-to-out dimension. Doing so will provide a built-in allowance for thickened slab under partitions and at the perimeter, as well as other areas where some extra quantity is needed.

Assembly Cost S.F. × Area of Slab = Total Cost for Slab on Grade

Example: Three-Story Office Building

For this particular application, a 4″ thick, non-industrial, reinforced slab grade is sufficient.

Cost:

(Figure 12.1, line #2240)   10,000 S.F. @ $2.86 = $28,600
Total Cost per Square Foot $28,600/30,000 = $0.95/S.F.

Refer to Figure 12.2 for the summary of the costs for this Division.

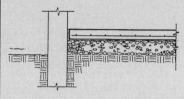

SUBSTRUCTURES | A2.1-200 | Plain & Reinforced Slab-on-Grade

There are four types of Slab on Grade Systems listed: Non-industrial, Light industrial, Industrial and Heavy industrial. Each type is listed two ways: reinforced and non-reinforced. A Slab on Grade system includes three passes with a grader; 6" of compacted gravel fill; polyethylene vapor barrier; 3500 p.s.i. concrete placed by chute; bituminous fibre expansion joint; all necessary edge forms (4 uses); steel trowel finish; and sprayed-on membrane curing compound.

The Expanded System Listing shows costs on a per square foot basis. Thicknesses of the slabs range from 4" to 8". Non-industrial applications are for foot traffic only with negligible abrasion. Light industrial applications are for pneumatic wheels and light abrasion. Industrial applications are for solid rubber wheels and moderate abrasion. Heavy industrial applications are for steel wheels and severe abrasion. All slabs are either shown unreinforced or reinforced with welded wire fabric.

System Components	QUANTITY	UNIT	MAT.	INST.	TOTAL
SYSTEM 2.1-200-2220					
SLAB ON GRADE, 4" THICK, NON INDUSTRIAL, NON REINFORCED					
Fine grade, 3 passes with grader and roller	.110	S.Y.		.16	.16
Gravel under floor slab, 6" deep, compacted	1.000	S.F.	.12	.18	.30
Polyethylene vapor barrier, standard, .006" thick	1.000	S.F.	.04	.09	.13
Concrete ready mix, regular weight, 3500 psi	.012	C.Y.	.71		.71
Place and vibrate concrete for slab on grade, 4" thick, direct chute	.012	C.Y.		.18	.18
Expansion joint, premolded bituminous fiber, 1/2" x 6"	.100	L.F.	.13	.19	.32
Edge forms in place for slab on grade to 6" high, 4 uses	.030	L.F.	.01	.06	.07
Cure with sprayed membrane curing compound	1.000	S.F.	.02	.05	.07
Finishing floor, monolithic steel trowel	1.000	S.F.		.61	.61
TOTAL			1.03	1.52	2.55

2.1-200	Plain & Reinforced	MAT.	INST.	TOTAL
2220	Slab on grade, 4" thick, non industrial, non reinforced	1.03	1.52	2.55
2240	Reinforced	1.11	1.75	2.86
2260	Light industrial, non reinforced	1.28	1.91	3.19
2280	Reinforced	1.36	2.14	3.50
2300	Industrial, non reinforced	1.64	3.15	4.79
2320	Reinforced	1.72	3.38	5.10
3340	5" thick, non industrial, non reinforced	1.21	1.57	2.78
3360	Reinforced	1.29	1.80	3.09
3380	Light industrial, non reinforced	1.46	1.96	3.42
3400	Reinforced	1.54	2.19	3.73
3420	Heavy industrial, non reinforced	2.04	3.70	5.74
3440	Reinforced	2.09	3.96	6.05
4460	6" thick, non industrial, non reinforced	1.44	1.53	2.97
4480	Reinforced	1.56	1.84	3.40
4500	Light industrial, non reinforced	1.70	1.92	3.62
4520	Reinforced	1.89	2.34	4.23
4540	Heavy industrial, non reinforced	2.29	3.75	6.04
4560	Reinforced	2.41	4.06	6.47
5580	7" thick, non industrial, non reinforced	1.64	1.58	3.22
5600	Reinforced	1.81	1.90	3.71
5620	Light industrial, non reinforced	1.90	1.97	3.87
5640	Reinforced	2.07	2.29	4.36
5660	Heavy industrial, non reinforced	2.48	3.69	6.17
5680	Reinforced	2.60	4	6.60
6700	8" thick, non industrial, non reinforced	1.82	1.61	3.43
6720	Reinforced	1.97	1.88	3.85

R2.1-200

Figure 12.1

137

Assembly Number	Description	Qty	Unit	Total Cost		Cost per S. F.
				Unit	Total	
1.0	**Foundations**					
1.1-120-	Column Footings 6 KSF					
7700	Interior 190K (200K)	9	Ea	464	4176	
7500	Exterior 114K (125K)	12	Ea	287	3444	
7410	Corner 86K (100K)	4	Ea	232.50	930	
1.1-140-	Strip Footings 12" x 24"					
2700	400 LF - (12x5') - (4x4.5')	322	LF	23.50	7567	
1.1-210-	Foundation Wall					
1560	4' x 12" Direct Chute	400	LF	43.80	17520	
3.1-114	Piers : Assume 20" sq.					
9924	16 x 3.5'	56	LF	76.50	4284	
					$37,921	$1.26
2.0	**Substructure**					
2.1-200-	4" non-industrial slab-on-grade					
2240	Reinforced	10000	SF	2.86	28600	
					$28,600	$0.95
3.0	**Superstructure**					
4.0	**Exterior Closure**					

Figure 12.2

Chapter 13

Superstructures

The superstructure is the part of the building or other structure above the foundation. The assemblies approach to estimating the cost of a superstructure must be extremely versatile to be accurate. To truly analyze the merits of each combination, data must be available for comparison and substitution of systems and combinations of systems. Basically, designs have to be determined and costs provided for:

- Suspended floor systems
- Roof framing systems
- Columns
- Exterior structural components
- Stairs

Exterior load-bearing walls are not included in this division; they are found in Division 4, Exterior Closure.

Suspended Floor Assemblies

The design live loads for suspended floor assemblies are determined by the proposed use of the building and the applicable building code. Once the loads have been determined, the type of assembly must be selected from the proper assemblies table. Depending on the assemblies selected, columns may or may not be included in the cost shown. Be certain that any design assumptions or conditions shown on the assemblies table are read and understood before selecting a price. With these costs available, it is possible to quickly compare the following possibilities:

- Alternate materials using the same dimensions
- Alternate dimensions using the same materials

Quick comparisons like these are what make assemblies estimating so effective. Not only is the designer able to make proper choices based on cost, but the owner is able to see more clearly how different systems and materials can affect the final cost of the project.

Assemblies Comparison

At this point in the estimate, the superimposed loads and the anticipated bay spacing can be established. It would be tremendously helpful if one could know which assemblies are most economical to use without having to spend a lot of time looking at page after page of tables. Figure 13.1 provides a quick comparison of assemblies before any more time is spent in the estimating process.

Comparative Costs ($/SF) of Floor Systems/Type, (Table Number), Bay Size, & Load

Bay Size	Cast-In-Place Concrete						Precast Concrete			Structural Steel					Wood	
	1 Way BM & Slab 3.5-120	2 Way BM & Slab 3.5-130	Flat Slab 3.5-140	Flat Plate 3.5-150	Joist Slab 3.5-160	Waffle Slab 3.5-170	Beams & Hollow Core Slabs 3.5-242	Beams & Hollow Core Slabes Topped 3.5-244	Beams & Double Tees Topped 3.5-254	Bar Joists on Cols. & Brg. Walls 3.5-440	Bar Joists Beams on cols. 3.5-460	Composite Beams & C.I.P. Slab 3.5-520	Wide Flange Beams Composite 3.5-530	Composite Beam & DK., Lt. Wt. Slab 3.5-540	Wood Beams & Joists 3.5-720	Laminated Wood Beams & Joists 3.5-730
Superimposed Load = 40 PSF																
15 x 15	9.46	8.92	7.90	7.52	9.25	–	–	–	–	–	–	–	–	–	7.09	7.32
15 x 20	9.47	9.17	8.12	7.91	9.29	–	–	–	–	5.28	6.14	–	9.48	–	9.49	7.44
20 x 20	9.52	9.57	8.31	7.91	9.21	10.71	11.57	13.29	–	5.39	6.55	–	10.05	–	9.26	7.62
20 x 25	9.61	10.08	8.91	8.61	9.30	10.80	11.05	12.68	–	6.05	7.35	611.00	10.45	9.11	–	–
25 x 25	9.60	10.32	9.11	8.77	9.14	11.01	11.70	13.71	–	6.37	7.8	11.00	11.43	9.20	–	–
25 x 30	9.76	10.69	9.61	–	9.66	11.10	11.25	12.92	13.13	6.78	8.32	11.85	11.26	9.26	–	–
30 x 30	10.79	11.51	10.11	–	9.89	11.55	11.60	13.76	13.96	6.59	8.21	11.85	12.25	9.09	–	–
30 x 35	11.26	12.33	10.66	–	10.29	11.70	11.67	13.77	–	7.37	9.19	12.25	13.15	9.44	–	–
35 x 35	11.87	12.85	10.83	–	10.30	12.20	12.37	14.16	–	7.56	9.41	12.70	13.60	10.12	–	–
35 x 40	12.14	13.50	–	–	10.64	12.50	12.58	14.72	12.47	–	–	13.65	13.90	11.05	–	–
40 x 40	–	–	–	–	10.97	12.80	13.23	15.36	14.13	–	–	–	–	–	–	–
40 x 45	–	–	–	–	11.36	13.10	–	–	–	–	–	–	–	–	–	–
40 x 50	–	–	–	–	–	–	–	–	13.37	–	–	–	–	–	–	–
Superimposed Load = 75 PSF																
15 x 15	9.56	9.10	7.95	7.54	9.32	–	–	–	–	–	–	–	–	–	8.98	9.07
15 x 20	9.78	9.88	8.34	8.18	9.46	–	–	–	–	5.71	6.93	–	10.22	–	11.41	9.64
20 x 20	10.21	10.39	8.69	8.27	9.72	10.84	12.12	13.84	–	6.07	7.46	–	11.16	–	11.36	9.57
20 x 25	10.51	11.18	9.36	8.77	9.80	11.00	11.05	13.21	–	6.85	7.87	11.80	12.40	9.32	–	–
25 x 25	10.33	11.14	9.51	9.06	9.74	11.26	12.15	13.71	–	6.91	8.61	12.50	12.85	9.79	–	–
25 x 30	10.50	11.58	10.06	–	10.00	11.38	11.76	13.88	13.13	7.05	8.77	13.20	13.35	9.71	–	–
30 x 30	11.57	12.42	10.63	–	10.21	11.75	12.10	14.56	13.96	7.56	9.4	13.05	13.95	9.74	–	–
30 x 35	11.75	12.91	11.21	–	10.39	11.70	12.37	13.78	–	8.73	10.72	13.90	14.95	10.05	–	–
35 x 35	12.95	13.35	11.48	–	10.77	12.40	12.97	15.23	–	9.31	11.44	14.35	15.40	11.24	–	–
35 x 40	13.13	13.95	–	–	11.09	12.80	–	–	13.29	–	–	15.20	15.85	11.73	–	–
40 x 40	–	–	–	–	11.27	13.25	–	–	14.59	–	–	–	–	–	–	–
40 x 45	–	–	–	–	11.50	13.50	–	–	–	–	–	–	–	–	–	–
40 x 50	–	–	–	–	–	–	–	–	13.36	–	–	–	–	–	–	–
Superimposed Load = 125 PSF																
15 x 15	9.68	9.43	8.18	7.69	9.44	–	–	–	–	–	–	–	–	–	13.19	13.04
15 x 20	10.14	10.75	8.63	8.63	9.97	–	–	–	–	6.55	8.03	–	11.51	–	15.44	14.27
20 x 20	10.80	10.75	9.23	8.64	9.85	11.04	12.12	–	–	7.33	8.48	–	12.35	–	18.82	14.07
20 x 25	11.17	11.50	9.92	9.18	10.34	11.19	11.80	–	–	7.50	9.15	13.20	14.00	11.30	–	–
25 x 25	12.03	12.04	9.97	9.43	10.78	11.54	12.15	–	–	8.06	9.95	14.10	14.80	10.41	–	–
25 x 30	12.05	12.52	10.44	–	10.73	11.61	12.47	–	–	8.64	10.62	15.35	14.75	11.07	–	–
30 x 30	12.30	13.06	10.97	–	10.80	11.95	13.42	–	–	9.53	11.81	15.25	16.10	11.42	–	–
30 x 35	13.06	14.05	11.50	–	10.93	12.20	–	–	–	9.78	12.05	16.90	17.10	11.71	–	–
35 x 35	13.75	14.55	11.80	–	10.81	12.65	–	–	–	10.88	12.49	16.45	18.05	13.01	–	–
35 x 40	13.85	14.60	–	–	11.06	13.20	–	–	–	–	–	17.70	18.35	13.28	–	–
40 x 40	–	–	–	–	11.72	13.40	–	–	–	–	–	–	–	–	–	–
40 x 45	–	–	–	–	11.92	13.70	–	–	–	–	–	–	–	–	–	–
40 x 50	–	–	–	–	–	–	–	–	–	–	–	–	–	–	–	–
Superimposed Load = 200 PSF																
15 x 15	10.09	10.06	8.48	–	9.75	–	–	–	–	–	–	–	–	–	23.10	20.25
15 x 20	10.96	11.51	8.81	–	10.22	–	–	–	–	–	–	–	13.95	–	21.05	18.30
20 x 20	11.70	11.68	9.34	–	10.36	11.49	–	–	–	–	–	–	14.65	–	–	19.30
20 x 25	11.98	12.52	10.23	–	10.89	11.62	–	–	–	–	–	16.00	15.55	12.49	–	–
25 x 25	13.00	13.58	10.31	–	11.23	11.80	–	–	–	–	–	16.60	16.80	12.95	–	–
25 x 30	13.08	13.65	10.93	–	11.33	12.40	–	–	–	–	–	17.55	17.95	13.01	–	–
30 x 30	13.64	14.00	11.50	–	11.35	12.72	–	–	–	–	–	18.40	21.10	13.50	–	–
30 x 35	13.80	14.80	–	–	11.75	13.25	–	–	–	–	–	19.85	20.60	13.50	–	–
35 x 35	14.80	15.35	–	–	11.66	13.45	–	–	–	–	–	20.20	22.10	14.45	–	–
35 x 40	14.90	15.65	–	–	11.90	14.25	–	–	–	–	–	21.70	22.80	15.40	–	–
40 x 40	–	–	–	–	–	–	–	–	–	–	–	–	–	–	–	–
40 x 45	–	–	–	–	–	–	–	–	–	–	–	–	–	–	–	–
40 x 45	–	–	–	–	–	–	–	–	–	–	–	–	–	–	–	–

Figure 13.1

Whatever assembly is chosen, ensure that compatible materials are used throughout. For instance, if steel is used for the floor and roof framing systems, then steel columns should be used. Concrete columns should be used with concrete framed buildings. The same holds true for wood framing and wood columns, although in some instances steel pipe columns are used in wood frame buildings.

Example One: Roof and Floor Assembly Selection

With the size of the building, bay size, and number of floors determined, select a floor and roof assembly. Assume the example and loads illustrated in Figure 13.2.

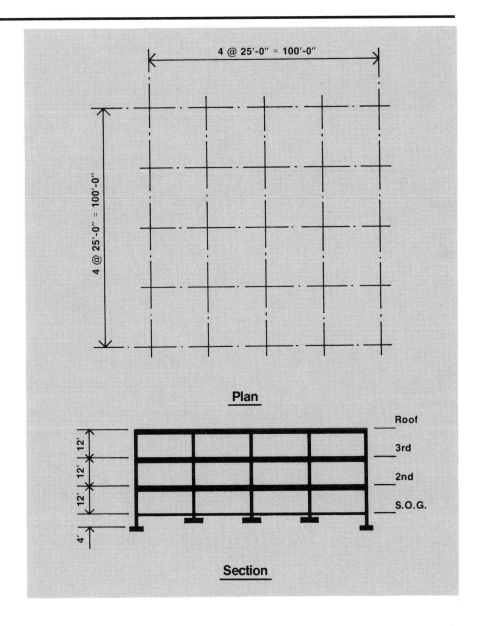

Figure 13.2

The examples that follow are not part of the Three-Story Office Building. They are presented to demonstrate some of the many uses for the tables in *Means Assemblies Cost Data.*

Floor:

Live Load Office Building	50 psf	BOCA Code
Partitions	15	
Ceiling	5	(Assume)
Mechanical	5	
	75 psf	Total Superimposed Load

Roof:

Snow Load	30 psf	BOCA Code
Ceiling	5	
Mechanical	5	
	40 psf	Total Superimposed Load

25′ x 25′ Bay Spacing

3 stories with 2 suspended floors and a roof.

Step One: Select the Floor Assembly
See Figure 13.1:

> Open Web Joists and Beams on Columns: $8.61/S.F.
> Concrete Flat Plate: $9.06/S.F.

For this example, assume a concrete frame building with flat plate construction. Ceilings can be eliminated in some areas for economy or architectural effect.

Step Two: Price the Floor Assembly
With the reference from Figure 13.1, go to the specific table for the floor assembly chosen.

See Figure 13.3:

> 25′ x 25′ bay; 75 psf Superimposed Load
> Total Load = 194 psf
> System Cost = $9.06/S.F.

Step Three: Note Minimum Required Column Size
See Figure 13.3:

> 25′ x 25′ bay; 75 psf Superimposed Load
> Column Size: 24″ square (for concrete floor systems only)

Step Four: Select a Roof Assembly
See Figure 13.3:

> 25′ x 25′ bay; 40 psf Superimposed Load
> Total Load = 152 psf
> System Cost = $8.77/S.F.

Step Five: Summarize Column Loads
The loads shown in Figure 13.4 were developed in Example Two in Chapter 4 for this structure.

The loads are as follows:

Minimum Column Load	=	102 kips
Maximum Column Load	=	358 kips
Load to the Foundation	=	358 kips

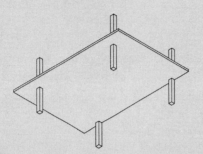

General: Flat Plates: Solid uniform depth concrete two-way slab without drops or interior beams. Primary design limit is shear at columns.

Design and Pricing Assumptions:
Concrete f'c to 4 KSI, placed by concrete pump.
Reinforcement, fy = 60 KSI.
Forms, four use.
Finish, steel trowel.
Curing, spray on membrane.
Based on 4 bay x 4 bay structure.

System Components	QUANTITY	UNIT	COST PER S.F.		
			MAT.	INST.	TOTAL
SYSTEM 3.5-150-2000					
15'X15' BAY 40 PSF S. LOAD, 12" MIN. COL.					
Forms in place, flat plate to 15' high, 4 uses	.992	S.F.	.83	3.41	4.24
Edge forms to 6" high on elevated slab, 4 uses	.065	L.F.	.03	.16	.19
Reinforcing in place, elevated slabs #4 to #7	1.706	Lb.	.51	.46	.97
Concrete ready mix, regular weight, 3000 psi	.459	C.F.	.98		.98
Place and vibrate concrete, elevated slab less than 6", pump	.459	C.F.		.44	.44
Finish floor, monolithic steel trowel finish for finish floor	1.000	S.F.		.61	.61
Cure with sprayed membrane curing compound	.010	C.S.F.	.02	.05	.07
TOTAL			2.37	5.13	7.50

3.5-150	Cast in Place Flat Plate							
	BAY SIZE (FT.)	SUPERIMPOSED LOAD (P.S.F.)	MINIMUM COL. SIZE (IN.)	SLAB THICKNESS (IN.)	TOTAL LOAD (P.S.F.)	COST PER S.F.		
						MAT.	INST.	TOTAL
2000	15 x 15	40	12	5-1/2	109	2.37	5.15	7.52
2200	R3.5 -010	75	14	5-1/2	144	2.39	5.15	7.54
2400		125	20	5-1/2	194	2.49	5.20	7.69
2600		175	22	5-1/2	244	2.54	5.20	7.74
3000	15 x 20	40	14	7	127	2.71	5.20	7.91
3400	R3.5 -100	75	16	7-1/2	169	2.88	5.30	8.18
3600		125	22	8-1/2	231	3.18	5.45	8.63
3800		175	24	8-1/2	281	3.20	5.45	8.65
4200	20 x 20	40	16	7	127	2.71	5.20	7.91
4400		75	20	7-1/2	175	2.92	5.35	8.27
4600		125	24	8-1/2	231	3.19	5.45	8.64
5000		175	24	8-1/2	281	3.21	5.50	8.71
5600	20 x 25	40	18	8-1/2	146	3.16	5.45	8.61
6000		75	20	9	188	3.27	5.50	8.77
6400		125	26	9-1/2	244	3.53	5.65	9.18
6600		175	30	10	300	3.68	5.70	9.38
7000	25 x 25	40	20	9	152	3.27	5.50	8.77
7400		75	24	9-1/2	194	3.46	5.60	9.06
7600		125	30	10	250	3.68	5.75	9.43
8000								

Figure 13.3

Step Six: Determine the Concrete Column Cost
See Figure 13.5:

> 24" square; 358 kips Working Load
> First check the "Minimum Reinforcing" section of Figure 13.5.
> Line # 9930: 24" sq. column with minimum reinforcing will support a
> total load of 700 kip (358 kips required).
> Column Cost = $98.00/V.L.F.

If the load accumulated had been in excess of of 700 kips, then the first portion of the table in Figure 13.5 would have been used.

Example: Proceed down the "Column Size" section to the first 24" square column @ 12' story height.

> Total Load = 900 kips
> Column Cost = $109.50/V.L.F.
> To determine the average column cost per vertical linear foot, proceed as follows:

$$\frac{\$98.00/V.L.F. + \$109.50/V.L.F.}{2} = \$103.75/V.L.F.$$

The result is more accurate for estimating since concrete columns at the upper floors, in this example, are not as costly as those at the ground level.

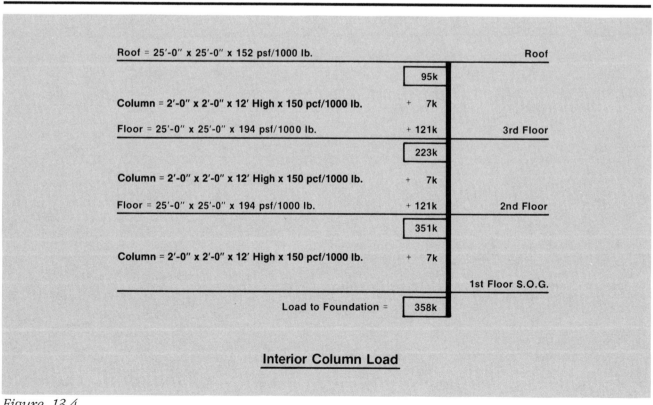

Interior Column Load

Figure 13.4

CONCRETE COLUMNS

General: It is desirable for purposes of consistency and simplicity to maintain constant column sizes throughout the building height. To do this, concrete strength may be varied (higher strength concrete at lower stories and lower strength concrete at upper stories), as well as varying the amount of reinforcing.

The first portion of the table provides probable minimum column sizes with related costs and weights per lineal foot of story height for bottom level columns.

The second portion of the table provides costs by column size for top level columns with minimum code reinforcement. Probable maximum loads for these columns are also given.

How to Use Table:

1. Enter the second portion (minimum reinforcing) of the table with the minimum allowable column size from the selected cast in place floor system.

   If the total load on the column does not exceed the allowable working load shown, use the cost per L.F. multiplied by the length of columns required to obtain the column cost.

2. If the total load on the column exceeds the allowable working load shown in the second portion of the table, enter the first portion of the table with the total load on the column and the minimum allowable column size from the selected cast in place floor system.

Select a cost per L.F. for bottom level columns by total load or minimum allowable column size.

Select a cost per L.F. for top level columns using the column size required for bottom level columns from the second portion of the table.

$$\frac{\text{Btm. + Top Col. Costs/L.F.}}{2} = \text{Avg. Col. Cost/L.F}$$

Column Cost = Average Col. Cost/L.F. x Length of Cols. Required.

See reference section in back of book to determine total loads.

Design and Pricing Assumptions:
Normal wt. concrete, f'c = 4 or 6 KSI, placed by pump.
Steel, fy = 60 KSI, spliced every other level.
Minimum design eccentricity of 0.1t.
Assumed load level depth is 8″ (weights prorated to full story basis).
Gravity loads only (no frame or lateral loads included).

Please see the reference section for further design and cost information.

System Components	QUANTITY	UNIT	COST PER V.L.F.		
			MAT.	INST.	TOTAL
SYSTEM 3.1-114-0640					
SQUARE COLUMNS, 100K LOAD,10′ STORY, 10″ SQUARE					
Forms in place, columns, plywood, 10″ x 10″, 4 uses	3.323	SFCA	3.13	18.47	21.60
Chamfer strip,wood, 3/4″ wide	4.000	L.F.	2.72	2.44	5.16
Reinforcing in place, columns, #3 to #7	3.653	Lb.	1.50	2.60	4.10
Reinforcing in place, column ties	1.405	Lb.	.30	.52	.82
Concrete ready mix, regular weight, 4000 psi	.026	C.Y.	1.57		1.57
Placing concrete, incl. vibrating, 12″ sq./round columns, pumped	.026	C.Y.		1.66	1.66
Finish, break ties, patch voids, burlap rub w/grout	3.323	S.F.	.27	2.17	2.44
TOTAL			9.49	27.86	37.35

3.1-114			C.I.P. Column, Square Tied					
	LOAD (KIPS)	STORY HEIGHT (FT.)	COLUMN SIZE (IN.)	COLUMN WEIGHT (P.L.F.)	CONCRETE STRENGTH (PSI)	COST PER V.L.F.		
						MAT.	INST.	TOTAL
0640	100	10	10	96	4000	9.50	28	37.50
0680	R3.1 -112	12	10	97	4000	9.35	27.50	36.85
0700		14	12	142	4000	11.50	33.50	45
0710								
0740	150	10	10	96	4000	9.50	28	37.50
0780		12	12	142	4000	11.55	33.50	45.05
0800		14	12	143	4000	11.50	33.50	45

Figure 13.5

|

| 3.1-114 | | C.I.P. Column, Square Tied | | | | | | |

	LOAD (KIPS)	STORY HEIGHT (FT.)	COLUMN SIZE (IN.)	COLUMN WEIGHT (P.L.F.)	CONCRETE STRENGTH (PSI)	COST PER V.L.F.		
						MAT.	INST.	TOTAL
0840	200	10	12	140	4000	11.60	34	45.60
0860		12	12	142	4000	11.55	33.50	45.05
0900		14	14	196	4000	13.55	39.50	53.05
0920	300	10	14	192	4000	13.85	40	53.85
0960		12	14	194	4000	13.75	39.50	53.25
0980		14	16	253	4000	15.25	43.50	58.75
1020	400	10	16	248	4000	16.35	45.50	61.85
1060		12	16	251	4000	16.25	45	61.25
1080		14	16	253	4000	16.15	45	61.15
1200	500	10	18	315	4000	20.50	54	74.50
1250		12	20	394	4000	21	56	77
1300		14	20	397	4000	21	56	77
1350	600	10	20	388	4000	24	61.50	85.50
1400		12	20	394	4000	24	61	85
1600		14	20	397	4000	23.50	61	84.50
1900	700	10	20	388	4000	24.50	62	86.50
2100		12	22	474	4000	24.50	61.50	86
2300		14	22	478	4000	24	61.50	85.50
2600	800	10	22	388	4000	33.50	77.50	111
2900		12	22	474	4000	33.50	77.50	111
3200		14	22	478	4000	33	77	110
3400	900	10	24	560	4000	33	78	111
3800		12	24	567	4000	32.50	77	109.50
4000		14	24	571	4000	32.50	76.50	109
4250	1000	10	24	560	4000	39.50	89	128.50
4500		12	26	667	4000	36.50	80	116.50
4750		14	26	673	4000	36	79.50	115.50
5600	100	10	10	96	6000	9.60	28	37.60
5800		12	10	97	6000	9.45	28	37.45
6000		14	12	142	6000	11.55	33.50	45.05
6200	150	10	10	96	6000	9.50	28	37.50
6400		12	12	98	6000	11.55	33.50	45.05
6600		14	12	143	6000	11.55	33.50	45.05
6800	200	10	12	140	6000	11.95	34.50	46.45
7000		12	12	142	6000	11.55	33.50	45.05
7100		14	14	196	6000	13.55	39.50	53.05
7300	300	10	14	192	6000	13.70	39.50	53.20
7500		12	14	194	6000	13.65	39.50	53.15
7600		14	14	196	6000	13.55	39.50	53.05
7700	400	10	14	192	6000	13.70	39.50	53.20
7800		12	14	194	6000	13.65	39.50	53.15
7900		14	16	253	6000	16.10	44.50	60.60
8000	500	10	16	248	6000	16.35	45.50	61.85
8050		12	16	251	6000	16.25	45	61.25
8100		14	16	253	6000	16.15	45	61.15
8200	600	10	18	315	6000	19.20	52.50	71.70
8300		12	18	319	6000	19.05	52	71.05
8400		14	18	321	6000	18.95	52	70.95
8500	700	10	18	315	6000	20.50	55	75.50
8600		12	18	319	6000	20.50	54.50	75
8700		14	18	321	6000	20	54	74
8800	800	10	20	388	6000	21	56.50	77.50
8900		12	20	394	6000	21	56.50	77.50
9000		14	20	397	6000	21	56	77
9100	900	10	20	388	6000	29	70	99
9300		12	20	394	6000	28.50	69.50	98
9600		14	20	397	6000	28	69	97

Figure 13.5 (cont.)

Step Seven: Determine the Superstructure Cost

Roof: 100′ × 100′ × $8.77/S.F.	=	$ 87,700
Suspended Floors: 2 × 100′ × 100′ × $9.06/S.F.	=	181,200
Columns: 25 each × 36′ high × $98.00/V.L.F.	=	88,200
Total Cost		$357,100
Total Cost per Square Foot	=	$11.90/S.F.

Example Two: Three-Story Office Building

The total superimposed loads, from the "Superimposed Load" section in Chapter 10, are:

Floor: 75 psf
Roof: 40 psf
25′ x 25′ Bay Spacing
3 stories with 2 suspended floors and a roof

Step One: Select the Floor System

See Figure 13.1:

Open Web Joists and Beams on Columns: $8.61/S.F.
Cast in Place Concrete Flat Plate: $9.06/S.F.

The designer and owner agree that the steel framing system is more desirable because of cost, speed of erection, availability, and the fact that structural elements will not be exposed to view.

Note: A concrete structure is fireproof. Steel and open web joist construction requires fireproofing. Therefore, the cost comparison at this point is not complete.

SUPERSTRUCTURES		A3.1-114	C.I.P. Column, Square Tied					
3.1-114			**C.I.P. Column, Square Tied**					
	LOAD (KIPS)	STORY HEIGHT (FT.)	COLUMN SIZE (IN.)	COLUMN WEIGHT (P.L.F.)	CONCRETE STRENGTH (PSI)	COST PER V.L.F.		
						MAT.	INST.	TOTAL
9800	1000	10	22	469	6000	26	64.50	90.50
9840		12	22	474	6000	26	64	90
9900		14	22	478	6000	26	65	91
3.1-114			**C.I.P. Column, Square Tied-Minimum Reinforcing**					
	LOAD (KIPS)	STORY HEIGHT (FT.)	COLUMN SIZE (IN.)	COLUMN WEIGHT (P.L.F.)	CONCRETE STRENGTH (PSI)	COST PER V.L.F.		
						MAT.	INST.	TOTAL
9912	150	10-14	12	135	4000	11.35	33.50	44.85
9918	300	10-14	16	240	4000	15.15	43	58.15
9924	500	10-14	20	375	4000	20.50	56	76.50
9930	700	10-14	24	540	4000	28.50	69.50	98
9936	1000	10-14	28	740	4000	35	83	118
9942	1400	10-14	32	965	4000	47	96	143
9948	1800	10-14	36	1220	4000	56	110	166
9954	2300	10-14	40	1505	4000	63	122	185

Figure 13.5 (cont.)

Step Two: Price the Floor System

Choose the appropriate table for an Open Web Joist and Beams on Columns floor assembly.

> (Figure 13.6 #5100)   25' x 25' Bay; 75 psf Superimposed Load
> Total Load   =   120 psf
> System Cost   = $8.61/S.F.

Step Three: Select a Roof System

Now select a roof system and price it.

See Figure 13.7:

> 25' x 25' Bay; 40 psf Superimposed Load
> Total Load   =   60 psf
> System Cost   = $4.01/S.F.

Step Four: Summarize Loads

The loads depicted in Figure 13.8 were developed in Chapter 10 for this structure. The loads are as follows:

> Minimum Column Load   =   39
> Maximum Column Load   =   191
> Load to the Foundation   =   191

Step Five: Price the Columns

Select the column type and determine the cost.

See Figure 13.9:

> 200 kips; 10' unsupported length.
> Use a W-shape.
> Column Cost: $37.85/V.L.F. (1st & 2nd floor)

See Figure 13.9:

> 50 kips; 10' unsupported length.
> Use a W-shape.
> Column Cost: $17.60/V.L.F. (3rd floor)

Step Six: Determine Total Superstructure Cost

Roof: 100' × 100' × $4.01/S.F.	=	$ 40,100
Suspended Floors: 2 × 100' × 100' × $8.61/S.F.	=	172,200
Columns: 25 ea. × 27.33' high × $37.85/V.L.F.	=	25,861
Columns: 25 ea. × 9.00' high × $17.60/V.L.F.	=	3,960
Total Cost:		$242,121
Cost per Square Foot: $242,121/30,000 S.F.	=	$8.07/S.F.

Floor Assembly Commentary

If the pages of assemblies in *Means Assemblies Cost Data* were used without regard to some design assumptions, the user could have some difficulty. Although the pages are not reproduced here, what follows is a brief recommendation on how to use each assembly.

Cast-in-Place Slabs, One Way: May be used in conjunction with precast concrete beams, steel framing, or bearing walls.

Cast-in-Place Beams and Slabs, One Way: Effective system for heavy or concentrated loads, with spans under 20'. These systems have relatively deep floors and complex framing.

Cast-in-Place Beams and Slabs, Two Way: Efficient for heavy or concentrated loads under 30' spans and fairly square bays.

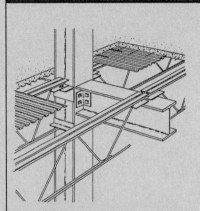

Table below lists costs for a floor system on steel columns and beams using open web steel joists, galvanized steel slab form, and 2-1/2″ concrete slab reinforced with welded wire fabric.

Design and Pricing Assumptions:
Structural Steel is A36.
Concrete f'c = 3 KSI placed by pump.
WWF 6 x 6 — W1.4 x W1.4 (10 x 10)
Columns are 12′ high.
Building is 4 bays long by 4 bays wide.
Joists are 2′ O.C. ± and span the long direction of the bay.

Joists at columns have bottom chords extended and are connected to columns.

Slab form is 28 gauge galvanized. Column costs in table are for columns to support 1 floor plus roof loading in a 2-story building; however, column costs are from ground floor to 2nd floor only. Joist costs include appropriate bridging. Deflection is limited to 1/360 of the span. Screeds and steel trowel finish.

Design Loads	Min.	Max.
S.S. & Joists	6.3 PSF	15.3 PSF
Slab Form	1.0	1.0
2-1/2″ Concrete	27.0	27.0
Ceiling	3.0	3.0
Misc.	5.7	1.7
	43.0 PSF	48.0 PSF

System Components	QUANTITY	UNIT	COST PER S.F.		
			MAT.	INST.	TOTAL
SYSTEM 3.5-460-2350					
15′X20′BAY 40 PSF S. LOAD, 17″ DEPTH, 83 PSF TOTAL LOAD					
Structural steel	1.974	Lb.	1.14	.50	1.64
Open web joists	3.140	Lb.	1.35	.69	2.04
Slab form, galvanized steel 9/16″ deep, 28 gauge	1.020	S.F.	.50	.32	.82
Welded wire fabric rolls, 6 x 6 - W1.4 x W1.4 (10 x 10), 21 lb/csf	1.000	S.F.	.08	.23	.31
Concrete ready mix, regular weight, 3000 psi	.210	C.F.	.45		.45
Place and vibrate concrete, elevated slab less than 6″, pumped	.210	C.F.		.20	.20
Finishing floor, monolithic steel trowel finish for finish floor	1.000	S.F.		.61	.61
Curing with sprayed membrane curing compound	.010	S.F.	.02	.05	.07
TOTAL			3.54	2.60	6.14

3.5-460		Steel Joists, Beams & Slab on Columns						
	BAY SIZE (FT.)	SUPERIMPOSED LOAD (P.S.F.)	DEPTH (IN.)	TOTAL LOAD (P.S.F.)	COLUMN ADD	COST PER S.F.		
						MAT.	INST.	TOTAL
2350	15x20 R3.5 -100	40	17	83		3.54	2.60	6.14
2400					column	.55	.24	.79
2450	15x20	65	19	108		3.91	2.74	6.65
2500					column	.55	.24	.79
2550	15x20	75	19	119		4.09	2.84	6.93
2600					column	.60	.26	.86
2650	15x20	100	19	144		4.34	2.96	7.30
2700					column	.60	.26	.86
2750	15x20	125	19	170		4.71	3.32	8.03
2800					column	.80	.34	1.14
2850	20x20	40	19	83		3.84	2.71	6.55
2900					column	.45	.20	.65
2950	20x20	65	23	109		4.23	2.90	7.13
3000					column	.60	.26	.86
3100	20x20	75	26	119		4.46	3	7.46
3200					column	.60	.26	.86
3400	20x20	100	23	144		4.63	3.07	7.70
3450					column	.60	.26	.86
3500	20x20	125	23	170		5.15	3.33	8.48
3600					column	.72	.31	1.03
3700	20x25	40	44	83		4.26	3.09	7.35
3800					column	.48	.21	.69

Figure 13.6

3.5-460		Steel Joists, Beams & Slab on Columns						

	BAY SIZE (FT.)	SUPERIMPOSED LOAD (P.S.F.)	DEPTH (IN.)	TOTAL LOAD (P.S.F.)	COLUMN ADD	COST PER S.F.		
						MAT.	INST.	TOTAL
3900	20x25	65	26	110		4.65	3.28	7.93
4000					column	.48	.21	.69
4100	20x25	75	26	120		4.73	3.14	7.87
4200					column	.57	.25	.82
4300	20x25	100	26	145		5	3.28	8.28
4400					column	.57	.25	.82
4500	20x25	125	29	170		5.60	3.55	9.15
4600					column	.67	.29	.96
4700	25x25	40	23	84		4.58	3.22	7.80
4800					column	.46	.20	.66
4900	25x25	65	29	110		4.85	3.35	8.20
5000					column	.46	.20	.66
5100	25x25	75	26	120		5.25	3.36	8.61
5200					column	.54	.24	.78
5300	25x25	100	29	145		5.85	3.65	9.50
5400					column	.54	.24	.78
5500	25x25	125	32	170		6.15	3.80	9.95
5600					column	.59	.25	.84
5700	25x30	40	29	84		4.97	3.35	8.32
5800					column	.45	.20	.65
5900	25x30	65	29	110		5.20	3.49	8.69
6000					column	.45	.20	.65
6050	25x30	75	29	120		5.55	3.22	8.77
6100					column	.49	.21	.70
6150	25x30	100	29	145		6.05	3.41	9.46
6200					column	.49	.21	.70
6250	25x30	125	32	170		6.50	4.12	10.62
6300					column	.57	.25	.82
6350	30x30	40	29	84		5.15	3.06	8.21
6400					column	.41	.17	.58
6500	30x30	65	29	110		5.85	3.36	9.21
6600					column	.41	.17	.58
6700	30x30	75	32	120		6	3.40	9.40
6800					column	.47	.21	.68
6900	30x30	100	35	145		6.65	3.69	10.34
7000					column	.55	.24	.79
7100	30x30	125	35	172		7.30	4.51	11.81
7200					column	.61	.26	.87
7300	30x35	40	29	85		5.85	3.34	9.19
7400					column	.35	.16	.51
7500	30x35	65	29	111		6.55	4.17	10.72
7600					column	.46	.20	.66
7700	30x35	75	32	121		6.55	4.17	10.72
7800					column	.47	.20	.67
7900	30x35	100	35	148		7.10	3.83	10.93
8000					column	.57	.25	.82
8100	30x35	125	38	173		7.90	4.15	12.05
8200					column	.58	.25	.83
8300	35x35	40	32	85		6	3.41	9.41
8400					column	.41	.17	.58
8500	35x35	65	35	111		6.90	4.32	11.22
8600					column	.49	.21	.70
9300	35x35	75	38	121		7.05	4.39	11.44
9400					column	.49	.21	.70
9500	35x35	100	38	148		7.60	4.65	12.25
9600					column	.60	.26	.86

Figure 13.6 (cont.)

Description: Table below lists the cost per S.F. for a roof system with steel columns, beams, and deck, using open web steel joists and 1-1/2″ galvanized metal deck.

Roof deck is 1-1/2″, 22 gauge galvanized steel. Joist cost includes appropriate bridging. Deflection is limited to 1/240 of the span. Fireproofing is not included.

Design and Pricing Assumptions:
Columns are 18′ high.
Building is 4 bays long by 4 bays wide.
Joists are 5′-0″ O.C. and span the long direction of the bay.
Joists at columns have bottom chords extended and are connected to columns.
Column costs are not included but are listed separately per S.F. of floor.

Design Loads	Min.		Max.	
Joists & Beams	3	PSF	5	PSF
Deck	2		2	
Insulation	3		3	
Roofing	6		6	
Misc.	6		6	
Total Dead Load	20	PSF	22	PSF

System Components			COST PER S.F.		
	QUANTITY	UNIT	MAT.	INST.	TOTAL
SYSTEM 3.7-420-1100					
METAL DECK AND JOISTS,15′X20′ BAY,20 PSF S. LOAD					
Structural steel	.954	Lb.	.55	.19	.74
Open web joists	1.260	Lb.	.54	.27	.81
Metal decking, open, galvanized, 1-1/2″ deep, 22 gauge	1.050	S.F.	.76	.37	1.13
TOTAL			1.85	.83	2.68

3.7-420		**Steel Joists, Beams, & Deck on Columns**						
	BAY SIZE (FT.)	SUPERIMPOSED LOAD (P.S.F.)	DEPTH (IN.)	TOTAL LOAD (P.S.F.)	COLUMN ADD	COST PER S.F.		
						MAT.	INST.	TOTAL
1100	15x20	20	16	40		1.85	.83	2.68
1200					columns	.90	.31	1.21
1300		30	16	50		2.03	.90	2.93
1400					columns	.90	.31	1.21
1500		40	18	60		2.07	.94	3.01
1600					columns	.90	.31	1.21
1700	20x20	20	16	40		1.99	.88	2.87
1800					columns	.67	.23	.90
1900		30	18	50		2.18	.95	3.13
2000					columns	.67	.23	.90
2100		40	18	60		2.43	1.05	3.48
2200					columns	.67	.23	.90
2300	20x25	20	18	40		2.10	.93	3.03
2400					columns	.54	.19	.73
2500		30	18	50		2.29	1.08	3.37
2600					columns	.54	.19	.73
2700		40	20	60		2.36	1.05	3.41
2800					columns	.72	.25	.97
2900	25x25	20	18	40		2.43	1.05	3.48
3000					columns	.43	.15	.58
3100		30	22	50		2.57	1.18	3.75
3200					columns	.57	.20	.77
3300		40	20	60		2.81	1.20	4.01
3400					columns	.57	.20	.77

Figure 13.7

Cast-in-Place Flat Plate with Drop Panels: Efficient for heavier loads and longer spans. Requires smaller columns than flat plate. Good for parking areas, storage, or industrial applications. Also good for all types of buildings used in conjunction with suspended ceilings.

Cast-in-Place Flat Plate: Economically efficient for light loads and modest spans and minimum depth of construction. Yields flat ceilings that may be used for finish with paint.

Cast-in-Place Multispan Joist Slab: Effective to reduce dead weight and increased flexibility for long spans.

Cast-in-Place Waffle Slab: Effective to reduce dead weight for long spans and heavy loads. Has uniform depth. May be used with or without suspended ceilings.

Precast Planks: May be used for floors with or without topping. Planks without topping require good alignment of panels, heavy padding, and carpeting for residential or office use. Used in conjunction with concrete beams, steel beams, or bearing walls. Results in fast erection time.

Precast, Double "T" Beams: Used for floors with topping and roofs with no topping. Commonly used with concrete beams or bearing walls, for long spans, and for wall panels.

Precast Beams and Planks, No Topping: Various bay sizes and superimposed loads for precast beams and precast planks, untopped. Ceilings not necessary. Fast erection time.

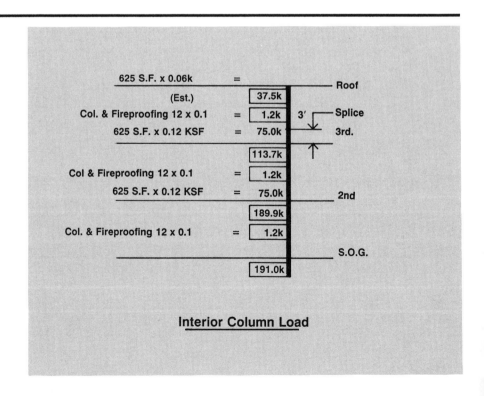

Figure 13.8

(A) Wide Flange		
(B) Pipe		
(C) Pipe, Concrete Filled		
(D) Square Tube		
(E) Square Tube Concrete Filled		
(F) Rectangular Tube		
(G) Rectangular Tube, Concrete Filled		

General: The following pages provide data for seven types of steel columns: wide flange, round pipe, round pipe concrete filled, square tube, square tube concrete filled, rectangular tube and rectangular tube concrete filled.

Design Assumptions: Loads are concentric; wide flange and round pipe bearing capacity is for 36 KSI steel. Square and rectangular tubing bearing capacity is for 46 KSI steel.

The effective length factor K=1.1 is used for determining column values in the tables. K=1.1 is within a frequently used range for pinned connections with cross bracing.

How To Use Tables:
a. Steel columns usually extend through two or more stories to minimize splices. Determine floors with splices.
b. Enter Table No. below with load to column at the splice. Use the unsupported height.
c. Determine the column type desired by price or design.

Cost:
a. Multiply number of columns at the desired level by the total height of the column by the cost/VLF.
b. Repeat the above for all tiers.

Please see the reference section for further design and cost information.

3.1-130						Steel Columns			
	LOAD (KIPS)	UNSUPPORTED HEIGHT (FT.)	WEIGHT (P.L.F.)	SIZE (IN.)	TYPE	COST PER V.L.F.		
						MAT.	INST.	TOTAL
1000	25	10	13	4	A	9.10	6.35	15.45
1020	R3.1 -130		7.58	3	B	8.35	6.35	14.70
1040			15	3-1/2	C	10.15	6.35	16.50
1060			6.87	3	D	7.50	6.35	13.85
1080			15	3	E	9.70	6.35	16.05
1100			8.15	4x3	F	8.60	6.35	14.95
1120			20	4x3	G	11.20	6.35	17.55
1200		16	16	5	A	10.40	4.75	15.15
1220			10.79	4	B	11	4.75	15.75
1240			36	5-1/2	C	15.20	4.75	19.95
1260			11.97	5	D	12.05	4.75	16.80
1280			36	5	E	16.05	4.75	20.80
1300			11.97	6x4	F	11.65	4.75	16.40
1320			64	8x6	G	22.50	4.75	27.25
1400		20	20	6	A	12.30	4.75	17.05
1420			14.62	5	B	14.10	4.75	18.85
1440			49	6-5/8	C	18.75	4.75	23.50
1460			11.97	5	D	11.40	4.75	16.15
1480			49	6	E	18.65	4.75	23.40
1500			14.53	7x5	F	13.40	4.75	18.15
1520			64	8x6	G	21.50	4.75	26.25
1600	50	10	16	5	A	11.25	6.35	17.60
1620			14.62	5	B	16.10	6.35	22.45
1640			24	4-1/2	C	12.10	6.35	18.45
1660			12.21	4	D	13.30	6.35	19.65
1680			25	4	E	13.50	6.35	19.85
1700			11.97	6x4	F	12.60	6.35	18.95
1720			28	6x3	G	14.85	6.35	21.20
1800		16	24	8	A	15.60	4.75	20.35
1820			18.97	6	B	19.35	4.75	24.10
1840			36	5-1/2	C	15.20	4.75	19.95
1860			14.63	6	D	14.75	4.75	19.50
1880			36	5	E	16.05	4.75	20.80
1900			14.53	7x5	F	14.15	4.75	18.90
1920			64	8x6	G	22.50	4.75	27.25
1940								

Figure 13.9

| 3.1-130 | | | | | Steel Columns | | | |

	LOAD (KIPS)	UNSUPPORTED HEIGHT (FT.)	WEIGHT (P.L.F.)	SIZE (IN.)	TYPE	COST PER V.L.F.		
						MAT.	INST.	TOTAL
3600	125	16	40	8	A	26	4.75	30.75
3620			28.55	8	B	29	4.75	33.75
3640			81	8	C	30	4.75	34.75
3660			25.82	8	D	26	4.75	30.75
3680			66	7	E	23.50	4.75	28.25
3700			27.59	8x6	F	27	4.75	31.75
3720			64	8x6	G	22.50	4.75	27.25
3800		20	48	8	A	29.50	4.75	34.25
3820			40.48	10	B	39	4.75	43.75
3840			81	8	C	28.50	4.75	33.25
3860			25.82	8	D	24.50	4.75	29.25
3880			66	7	E	22	4.75	26.75
3900			37.59	10x6	F	34.50	4.75	39.25
3920			60	8x6	G	30.50	4.75	35.25
4000	150	10	35	8	A	24.50	6.35	30.85
4020			40.48	10	B	44.50	6.35	50.85
4040			81	8-5/8	C	32.50	6.35	38.85
4060			25.82	8	D	28	6.35	34.35
4080			66	7	E	25	6.35	31.35
4100			27.48	7x5	F	29	6.35	35.35
4120			64	8x6	G	24.50	6.35	30.85
4200		16	45	10	A	29	4.75	33.75
4220			40.48	10	B	41.50	4.75	46.25
4240			81	8-5/8	C	30	4.75	34.75
4260			31.84	8	D	32	4.75	36.75
4280			66	7	E	23.50	4.75	28.25
4300			37.69	10x6	F	36.50	4.75	41.25
4320			70	8x6	G	32.50	4.75	37.25
4400		20	49	10	A	30	4.75	34.75
4420			40.48	10	B	39	4.75	43.75
4440			123	10-3/4	C	40.50	4.75	45.25
4460			31.84	8	D	30.50	4.75	35.25
4480			82	8	E	25.50	4.75	30.25
4500			37.69	10x6	F	35	4.75	39.75
4520			86	10x6	G	30	4.75	34.75
4600	200	10	45	10	A	31.50	6.35	37.85
4620			40.48	10	B	44.50	6.35	50.85
4640			81	8-5/8	C	32.50	6.35	38.85
4660			31.84	8	D	34.50	6.35	40.85
4680			82	8	E	29	6.35	35.35
4700			37.69	10x6	F	39.50	6.35	45.85
4720			70	8x6	G	35	6.35	41.35
4800		16	49	10	A	32	4.75	36.75
4820			49.56	12	B	50.50	4.75	55.25
4840			123	10-3/4	C	42.50	4.75	47.25
4860			37.60	8	D	38	4.75	42.75
4880			90	8	E	39	4.75	43.75
4900			42.79	12x6	F	41.50	4.75	46.25
4920			85	10x6	G	37.50	4.75	42.25
5000		20	58	12	A	35.50	4.75	40.25
5020			49.56	12	B	48	4.75	52.75
5040			123	10-3/4	C	40.50	4.75	45.25
5060			40.35	10	D	38.50	4.75	43.25
5080			90	8	E	37	4.75	41.75
5100			47.90	12x8	F	44	4.75	48.75
5120			93	10x6	G	46	4.75	50.75

Figure 13.9 (cont.)

Precast Beams and Planks, with Topping: Various bay sizes and superimposed loads for precast beams and precast planks, with 2″ concrete topping. Ceilings not necessary. Fast erection time.

Precast Beams and Precast Double "T's," With Topping: Multiple bay sizes and superimposed loads for precast beams and precast double "T's" with 2″ concrete topping. Effective for long spans and fast erection.

Steel Framing: Multiple bay sizes and beam girder spacing and orientation, and various superimposed loads. System includes sprayed-on fireproofing. For use with cast-in-place slabs; one-way, precast planks; and metal deck with concrete fill.

Light Gauge Steel: Light gauge "C" or punched joists with plywood subfloor. Can be used with bearing walls or steel stud wall systems. Clean, simple erection.

Steel Joists on Walls: Various spans for open web joists, slab form, and 2-1/2″ concrete slab supported by walls or other suitable bearing system. Inexpensive, lightweight; a popular system. Must be fireproofed with fire-rated suspended or gypsum ceiling.

Steel Joists and Beams on Columns with Exterior Bearing Walls: Varied bay sizes and superimposed loads for columns, beams, open-web joists, and 2-1/2″ concrete slabs. Exterior walls are assumed as bearing walls; the cost of these walls is not included. Inexpensive, lightweight; a popular system. Must be fireproofed with fire-rated suspended or gypsum ceiling.

Steel Joists and Beams on Columns: Varied bay sizes and superimposed loads for columns, beams, open-web joists, and 2-1/2″ concrete slabs. Inexpensive, lightweight; a popular system. Must be fireproofed with fire-rated suspended or gypsum ceiling. Columns are included in the cost of the system for use with one suspended floor and one roof.

Composite Beam and Cast-in-Place Slabs: Varied bay sizes and superimposed loads for steel beams with shear connectors and formed reinforced concrete slab. Efficient when loads are heavy and spans are moderately long. Beams are fireproofed with sprayed-on fireproofing.

Steel Beams, Composite Deck, and Concrete Slabs: Varied bay sizes and superimposed loads for steel beams, composite steel deck, and concrete slab. Efficient when loads are heavy and spans are moderately long. Beams and steel deck are fireproofed with sprayed-on fireproofing.

Composite Beams and Deck, Lightweight Concrete Slabs: Varied bay sizes and superimposed loads for composite steel beams with shear connectors, composite steel deck, and lightweight concrete slabs. Popular, stiff, relatively inexpensive steel and concrete framing system. Sprayed-on fireproofing for the beams only is included, as the lightweight concrete deck is considered fireproof by most codes.

Metal Deck and Concrete Fill: Varied spans and superimposed loading for composite steel deck, cellular and non-cellular, and concrete fill of varied thickness. Used with steel framing system.

Wood Joists: Costs are listed per square foot for wood joists and plywood subfloor. Joists are spaced 12″, 16″, and 24″ on-center. Sizes range from 2″ x 6″ to 4″ x 10″.

Wood Beam and Joist: Costs are listed per square foot for wood beams, wood joists, and plywood subfloor for various bay sizes and superimposed loading.

Laminated Floor Beams: Costs are listed per square foot for laminated wood beams, wood joists, and plywood subfloor for various bay sizes and superimposed loading.

Wood Deck: Costs are listed per square foot for wood deck of varied species and thicknesses with allowable load and span.

Roof Framing Assemblies

Once the structural floor assembly has been designed and costs have been established, select a roof assembly. The roof framing assembly normally has the same bay spacing and may or may not be of the same construction.

For single-story, slab-on-grade construction, the choice of roof framing assemblies will follow the same pattern outlined previously under floor assemblies, except that roof loads will be used instead of the heavier floor loads.

Columns

If possible, at each floor level in multi-story concrete structures, use the same size column on every floor to minimize formwork. Usually, this can be done by varying the amount of reinforcing in concrete columns. Steel columns are usually spliced every other floor.

The column cost tables are arranged so that the user can enter and make cost determinations when any one of the following situations exists:

- The exact type, size, weight, and length of the columns being estimated has been predetermined.
- The total load concentrated to the columns as well as the type of column desired are known.
- The bay size has been determined, floor and roof systems (not including columns) have been selected, and the total load concentrated on the columns has been tabulated.

Column Fireproofing

A cost table for the fireproofing of vertical steel columns is shown in Figure 13.9. Using this table, the estimator can choose one of several encasement systems.

Example: 3-Story Office Building

Column fireproofing: The average size of the columns to be fireproofed is 8″. Two-hour protection is required. Drywall is selected as the fireproofing material, as it is compatible with the partitions that will be painted.

Cost:

See Figure 13.10 (#3450): 859 L.F. @ $14.77/L.F. = $12,687

Beam Fireproofing

Costs for fireproofing steel beams using concrete, gypsum, gypsum and perlite plaster, and sprayed fiber (non-asbestos) are shown in Figure 13.11. Multiply the cost per linear foot of the desired system by the total L.F. of beams. If many different-size beams are used in the project, use an average size.

Stairs

Following the assemblies method of estimating to determine the cost of stairs can save valuable time, which can be applied to other aspects of the project that may have a greater cost impact. The following steps will help estimate the cost of stairs:

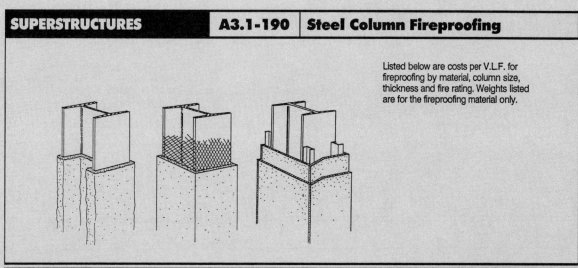

Listed below are costs per V.L.F. for fireproofing by material, column size, thickness and fire rating. Weights listed are for the fireproofing material only.

System Components	QUANTITY	UNIT	COST PER V.L.F.		
			MAT.	INST.	TOTAL
SYSTEM 3.1-190-3000					
CONCRETE FIREPROOFING, 8" STEEL COLUMN, 1" THICK, 1 HR. FIRE RATING					
Forms in place, columns, plywood, 4 uses	3.330	SFCA	3.50	16.65	20.15
Welded wire fabric, 2 x 2 #14 galv. 21 lb./C.S.F., column wrap	2.700	S.F.	.38	3.27	3.65
Concrete ready mix, regular weight, 3000 psi	.621	C.F.	1.32		1.32
Place and vibrate concrete, 12" sq./round columns, pumped	.621	C.F.		1.47	1.47
TOTAL			5.20	21.39	26.59

3.1-190	Steel Column Fireproofing							
	ENCASEMENT SYSTEM	COLUMN SIZE (IN.)	THICKNESS (IN.)	FIRE RATING (HRS.)	WEIGHT (P.L.F.)	COST PER V.L.F.		
						MAT.	INST.	TOTAL
3000	Concrete	8	1	1	110	5.20	21.50	26.70
3050			1-1/2	2	133	5.90	24	29.90
3100			2	3	145	6.60	26	32.60
3150		10	1	1	145	6.55	26.50	33.05
3200			1-1/2	2	168	7.25	28.50	35.75
3250			2	3	196	8	31	39
3300		14	1	1	258	8.10	31.50	39.60
3350			1-1/2	2	294	8.80	33.50	42.30
3400			2	3	325	9.60	35.50	45.10
3450	Gypsum board	8	1/2	2	8	2.32	12.45	14.77
3500	1/2" fire rated	10	1/2	2	11	2.48	13	15.48
3550	1 layer	14	1/2	2	18	2.56	13.30	15.86
3600	Gypsum board	8	1	3	14	3.30	15.90	19.20
3650	1/2" fire rated	10	1	3	17	3.57	16.85	20.42
3700	2 layers	14	1	3	22	3.72	17.35	21.07
3750	Gypsum board	8	1-1/2	3	23	4.47	20	24.47
3800	1/2" fire rated	10	1-1/2	3	27	5.05	22	27.05
3850	3 layers	14	1-1/2	3	35	5.65	24.50	30.15
3900	Sprayed fiber	8	1-1/2	2	6.3	2.78	4.95	7.73
3950	Direct application		2	3	8.3	3.83	6.85	10.68
4000			2-1/2	4	10.4	4.96	8.85	13.81
4050		10	1-1/2	2	7.9	3.36	6	9.36
4100			2	3	10.5	4.60	8.20	12.80
4150			2-1/2	4	13.1	5.90	10.55	16.45
4200		14	1-1/2	2	10.8	4.17	7.45	11.62
4250			2	3	14.5	5.70	10.15	15.85
4300			2-1/2	4	18	7.30	12.95	20.25

Figure 13.10

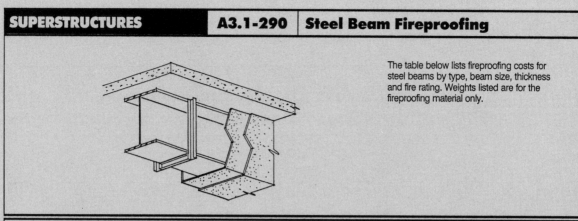

SUPERSTRUCTURES — A3.1-290 — Steel Beam Fireproofing

The table below lists fireproofing costs for steel beams by type, beam size, thickness and fire rating. Weights listed are for the fireproofing material only.

System Components	QUANTITY	UNIT	COST PER L.F. MAT.	COST PER L.F. INST.	COST PER L.F. TOTAL
SYSTEM 3.1-290-1300					
FIREPROOFING, 5/8" F.R. GYP. BOARD, 12"X 4" BEAM, 2" THICK, 2 HR. F.R.					
Corner bead for drywall, 1-1/4" x 1-1/4", galvanized	.020	C.L.F.	.25	1.84	2.09
L bead for drywall, galvanized	.020	C.L.F.	.36	2.14	2.50
Furring, beams & columns, 3/4" galv. channels, 24" O.C.	2.330	S.F.	.37	3.80	4.17
Drywall on beam, no finish, 2 layers at 5/8" thick	3.000	S.F.	1.92	6.45	8.37
Drywall, taping and finishing joints, add	3.000	S.F.	.27	.96	1.23
TOTAL			3.17	15.19	18.36

3.1-290 — Steel Beam Fireproofing

	ENCASEMENT SYSTEM	BEAM SIZE (IN.)	THICKNESS (IN.)	FIRE RATING (HRS.)	WEIGHT (P.L.F.)	MAT.	INST.	TOTAL
0400	Concrete	12x4	1	1	77	4.03	16.60	20.63
0450	3000 PSI		1-1/2	2	93	4.60	18.20	22.80
0500			2	3	121	5.05	19.90	24.95
0550		14x5	1	1	100	5.05	20.50	25.55
0600			1-1/2	2	122	5.70	23	28.70
0650			2	3	142	6.30	25.50	31.80
0700		16x7	1	1	147	6.05	22	28.05
0750			1-1/2	2	169	6.55	23	29.55
0800			2	3	195	7.15	25.50	32.65
0850		18x7-1/2	1	1	172	6.85	25.50	32.35
0900			1-1/2	2	196	7.55	28.50	36.05
0950			2	3	225	8.25	31	39.25
1000		24x9	1	1	264	9.40	31.50	40.90
1050			1-1/2	2	295	10.05	33.50	43.55
1100			2	3	328	10.75	35	45.75
1150		30x10-1/2	1	1	366	12.05	40	52.05
1200			1-1/2	2	404	12.90	42.50	55.40
1250			2	3	449	13.50	45	58.50
1300	5/8" fire rated	12x4	2	2	15	3.17	15.20	18.37
1350	Gypsum board		2-5/8	3	24	4.39	18.80	23.19
1400		14x5	2	2	17	3.24	15.80	19.04
1450			2-5/8	3	27	4.65	19.95	24.60
1500		16x7	2	2	20	3.64	17.70	21.34
1550			2-5/8	3	31	4.69	17.05	21.74
1600	5/8" fire rated	18x7-1/2	2	2	22	3.90	18.95	22.85
1650	Gypsum board		2-5/8	3	34	5.60	24	29.60
1700		24x9	2	2	27	4.84	23.50	28.34
1750			2-5/8	3	42	6.90	29.50	36.40

Figure 13.11

- Decide on the material and select a description from Figure 13.12.
- Determine whether the stairs are to be straight or interrupted with a landing.
- Determine the number of risers from calculations or by using Figure 13.13.
- Select the cost per flight from Figure 13.12. Multiply the cost per flight by the required number of flights.
- Remember that most multi-story buildings with stairs will include a flight that leads up to the roof.

When selecting stair material it is a good idea to keep similar materials together. For example, use concrete stairs in concrete framed buildings and steel pan stairs in steel framed buildings.

> Example: Calculating the S.F. Cost of Stairs for the Sample Building #1
>
> Stairs:
>
> > Use steel with concrete filled metal pans. Landings will be used; see the floor plan in Chapter 16 (Figure 16.3).
> >
> > Number of risers: 20
> >
> > 12'-0" floor-to-floor height (see Figure 13.13).
>
> Cost:
>
> > See Figure 13.12 (line #0760): 5 flights @ $5,425 = $27,125

Analysis of Various Foundation Types with Effects of Superstructures Included

Using the information from Chapters 11 and 12 and this chapter, many comparisons can be made to gain further understanding of the methods presented and the choices offered. For example, the following cost comparisons can be made between:

- Full load and factored load.
- Large-area buildings versus small-area buildings.
- Various foundation costs, using load reduction factors.
- Interior versus exterior footings for a small building and a large building.

Footing Load Reduction Comparisons

Example One: Small Building

See Figure 13.14.

Footings:	1 Interior
	4 Exterior
	4 Corner

Assuming all footings at full bay loading (75 kips load and 3 KSF soil pressure), the footing cost is:

> See Figure 13.15, line #7250:
>
> | 9 ea. @ $286 | = | $2,574 or |
> | $2,574/2,500 S.F. | = | $1.03/S.F. |
>
> 25' x 25' Bays

Now assume interior and exterior footings with load reduction:

Footing Loads:	Interior	=	75 kips
	Exterior	=	45 kips
	Corner	=	34 kips
Soil Pressure = 3 KSF			

General Design: See reference section for code requirements. Maximum height between landings is 12'; usual stair angle is 20° to 50° with 30° to 35° best. Usual relation of riser to treads is:
  Riser + tread = 17.5.
  2x (Riser) + tread = 25.
  Riser x tread = 70 or 75.
Maximum riser height is 7" for commercial, 8-1/4" for residential.
Usual riser height is 6-1/2" to 7-1/4".

Minimum tread width is 11" for commercial and 9" for residential.

For additional information please see reference section.

Cost Per Flight: Table below lists the cost per flight for 4'-0" wide stairs. Side walls are not included. Railings are included.

System Components	QUANTITY	UNIT	COST PER FLIGHT		
			MAT.	INST.	TOTAL
SYSTEM 3.9-100-0560					
STAIRS, C.I.P. CONCRETE WITH LANDING, 12 RISERS					
Concrete in place, free standing stairs not incl. safety treads	48.000	L.F.	273.60	1,152.48	1,426.08
Concrete in place, free standing stair landing	32.000	S.F.	72.96	315.52	388.48
Stair tread C.I. abrasive 4" wide	48.000	L.F.	319.20	193.44	512.64
Industrial railing, welded, 2 rail 3'-6" high 1-1/2" pipe	18.000	L.F.	241.20	125.82	367.02
Wall railing with returns, steel pipe	17.000	L.F.	88.40	118.83	207.23
TOTAL			995.36	1,906.09	2,901.45

3.9-100	Stairs	COST PER FLIGHT		
		MAT.	INST.	TOTAL
0470	Stairs, C.I.P. concrete, w/o landing, 12 risers, w/o nosing R3.9-100	605	1,375	1,980
0480	With nosing	920	1,575	2,495
0550	W/landing, 12 risers, w/o nosing	675	1,725	2,400
0560	With nosing	995	1,925	2,920
0570	16 risers, w/o nosing	840	2,150	2,990
0580	With nosing	1,275	2,400	3,675
0590	20 risers, w/o nosing	1,000	2,600	3,600
0600	With nosing	1,550	2,925	4,475
0610	24 risers, w/o nosing	1,175	3,025	4,200
0620	With nosing	1,800	3,425	5,225
0630	Steel, grate type w/nosing & rails, 12 risers, w/o landing	1,475	500	1,975
0640	With landing	2,300	765	3,065
0660	16 risers, with landing	2,775	930	3,705
0680	20 risers, with landing	3,275	1,100	4,375
0700	24 risers, with landing	3,775	1,250	5,025
0710	Cement fill metal pan & picket rail, 12 risers, w/o landing	1,900	500	2,400
0720	With landing	2,975	855	3,830
0740	16 risers, with landing	3,625	1,025	4,650
0760	20 risers, with landing	4,250	1,175	5,425
0780	24 risers, with landing	4,875	1,350	6,225
0790	Cast iron tread & pipe rail, 12 risers, w/o landing	1,925	500	2,425
0800	With landing	3,025	855	3,880
0820	16 risers, with landing	3,675	1,025	4,700
0840	20 risers, with landing	4,300	1,175	5,475
0860	24 risers, with landing	4,950	1,350	6,300
0870	Pan tread & flat bar rail, pre-assembled, 12 risers, w/o landing	1,375	265	1,640
0880	With landing	2,300	530	2,830
0900	16 risers, with landing	2,550	580	3,130
0920	20 risers, with landing	3,025	670	3,695
0940	24 risers, with landing	3,475	750	4,225
0950	Spiral steel, industrial checkered plate 4'-6" dia., 12 risers	1,100	480	1,580
0960	16 risers	1,475	640	2,115

Figure 13.12

Typical Range of Risers for Various Story Heights

Story Height	Minimum Risers	Maximum Riser Height	Tread Width	Maximum Risers	Minimum Riser Height	Tread Width	Average Risers	Average Riser Height	Tread Width
7'-6"	12	7.50"	10.00"	14	6.43"	11.07"	13	6.92"	10.58"
8'-0"	13	7.38	10.12	15	6.40	11.10	14	6.86	10.64
8'-6"	14	7.29	10.21	16	6.38	11.12	15	6.80	10.70
9'-0"	15	7.20	10.30	17	6.35	11.15	16	6.75	10.75
9'-6"	16	7.13	10.37	18	6.33	11.17	17	6.71	10.79
10'-0"	16	7.50	10.00	19	6.32	11.18	18	6.67	10.83
10'-6"	17	7.41	10.09	20	6.30	11.20	18	7.00	10.50
11'-0"	18	7.33	10.17	21	6.29	11.21	19	6.95	10.55
11'-6"	19	7.26	10.24	22	6.27	11.23	20	6.90	10.60
12'-0"	20	7.20	10.30	23	6.26	11.24	21	6.86	10.64
12'-6"	20	7.50	10.00	24	6.25	11.25	22	6.82	10.68
13'-0"	21	7.43	10.07	25	6.24	11.26	22	7.09	10.41
13'-6"	22	7.36	10.14	25	6.48	11.02	23	7.04	10.46
14'-0"	23	7.30	10.20	26	6.46	11.04	24	7.00	10.50

Figure 13.13

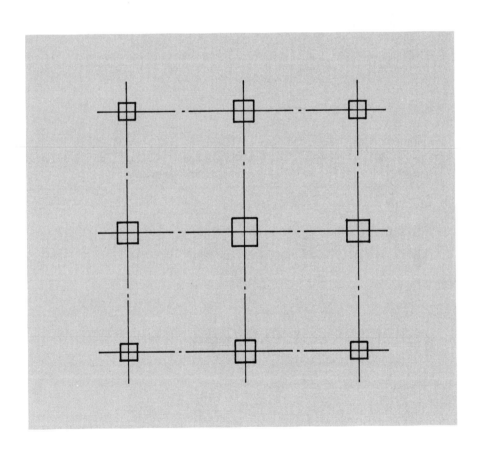

Figure 13.14

The Spread Footing System includes: excavation; backfill; forms (four uses); all reinforcement; 3,000 p.s.i. concrete (chute placed); and screed finish.

Footing systems are priced per individual unit. The Expanded System Listing at the bottom shows footings that range from 3' square x 12" deep, to 18' square x 52" deep. It is assumed that excavation is done by a truck mounted hydraulic excavator with an operator and oiler.

Backfill is with a dozer, and compaction by air tamp. The excavation and backfill equipment is assumed to operate at 30 C.Y. per hour.

Please see the reference section for further design and cost information.

System Components	QUANTITY	UNIT	COST EACH		
			MAT.	INST.	TOTAL
SYSTEM 1.1-120-7100					
SPREAD FOOTINGS, LOAD 25K, SOIL CAPACITY 3 KSF, 3' SQ X 12" DEEP					
Bulk excavation	.590	C.Y.		3.18	3.18
Hand trim	9.000	S.F.		4.77	4.77
Compacted backfill	.260	C.Y.		.55	.55
Formwork, 4 uses	12.000	S.F.	7.20	36.12	43.32
Reinforcing, fy = 60,000 psi	.006	Ton	3.45	4.50	7.95
Dowel or anchor bolt templates	6.000	L.F.	4.02	15	19.02
Concrete, f'c = 3,000 psi	.330	C.Y.	18.98		18.98
Place concrete, direct chute	.330	C.Y.		5	5
Screed finish	9.000	S.F.		2.97	2.97
TOTAL			33.65	72.09	105.74

1.1-120	Spread Footings	COST EACH		
		MAT.	INST.	TOTAL
7090	Spread footings, 3000 psi concrete, chute delivered			
7100	Load 25K, soil capacity 3 KSF, 3'-0" sq. x 12" deep	33.50	72	105.50
7150	Load 50K, soil capacity 3 KSF, 4'-6" sq. x 12" deep	69.50	125	194.50
7200	Load 50K, soil capacity 6 KSF, 3'-0" sq. x 12" deep	33.50	72	105.50
7250	Load 75K, soil capacity 3 KSF, 5'-6" sq. x 13" deep	109	177	286
7300	Load 75K, soil capacity 6 KSF, 4'-0" sq. x 12" deep	57	107	164
7350	Load 100K, soil capacity 3 KSF, 6'-0" sq. x 14" deep	138	212	350
7410	Load 100K, soil capacity 6 KSF, 4'-6" sq. x 15" deep	85.50	147	232.50
7450	Load 125K, soil capacity 3 KSF, 7'-0" sq. x 17" deep	217	305	522
7500	Load 125K, soil capacity 6 KSF, 5'-0" sq. x 16" deep	110	177	287
7550	Load 150K, soil capacity 3 KSF 7'-6" sq. x 18" deep	261	360	621
7610	Load 150K, soil capacity 6 KSF, 5'-6" sq. x 18" deep	145	223	368
7650	Load 200K, soil capacity 3 KSF, 8'-6" sq. x 20" deep	370	475	845
7700	Load 200K, soil capacity 6 KSF, 6'-0" sq. x 20" deep	189	275	464
7750	Load 300K, soil capacity 3 KSF, 10'-6" sq. x 25" deep	675	770	1,445
7810	Load 300K, soil capacity 6 KSF, 7'-6" sq. x 25" deep	355	465	820
7850	Load 400K, soil capacity 3 KSF, 12'-6" sq. x 28" deep	1,075	1,150	2,225
7900	Load 400K, soil capacity 6 KSF, 8'-6" sq. x 27" deep	490	605	1,095
7950	Load 500K, soil capacity 3 KSF, 14'-0" sq. x 31" deep	1,475	1,525	3,000
8010	Load 500K, soil capacity 6 KSF, 9'-6" sq. x 30" deep	675	785	1,460
8050	Load 600K, soil capacity 3 KSF, 16'-0" sq. x 35" deep	2,150	2,075	4,225
8100	Load 600K, soil capacity 6 KSF, 10'-6" sq. x 33" deep	905	1,025	1,930
8150	Load 700K, soil capacity 3 KSF, 17'-0" sq. x 37" deep	2,525	2,375	4,900
8200	Load 700K, soil capacity 6 KSF, 11'-6" sq. x 36" deep	1,150	1,250	2,400

Figure 13.15

Costs:

Interior:	(Figure 13.15 #7250)	1 ea. @ $286	=	$ 286
Exterior:	(Figure 13.15 #7150)	4 ea. @ $194.50	=	778
Corner:	(Figure 13.15)			
	(Interpolated between			
	#7100 & #7150)	4 ea. @ $150	=	600

Total $1,664

Total Cost per Square Foot   = $0.67/S.F.

Compare the previous example with a building 10 bays x 10 bays and 62,500 S.F.

Example Two: Large Building

See Figure 13.16.

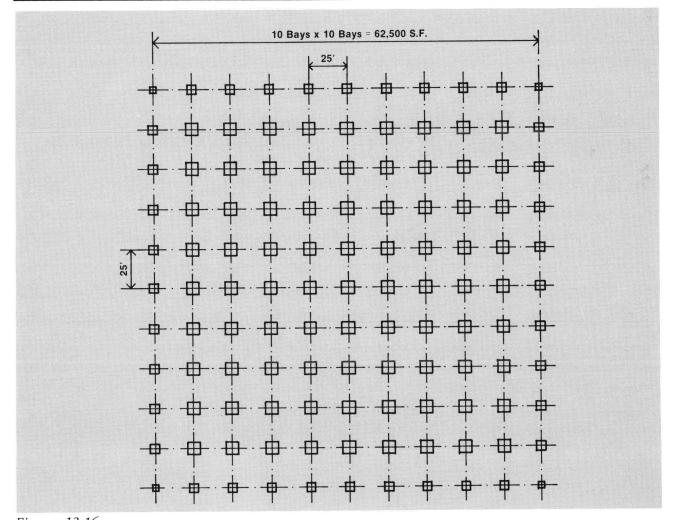

10 Bays x 10 Bays = 62,500 S.F.

25'

25'

Figure 13.16

Footings: 81 Interior
36 Exterior
4 Corner

Assuming all interior footings (75 KSF load and 3 KSF soil pressure), the cost is:

See Figure 13.15, line #7250:

121 ea. @ $286   = $34,606
$34,606/62,500 S.F.   = $0.55/S.F.

25′ x 25′ Bays

Now assume interior and exterior footings with load reduction:

Footing Loads:	Interior	=	75 kips
	Exterior	=	45 kips
	Corner	=	34 kips

Soil Pressure: 3 KSF

Costs:

Interior: (Figure 13.15 #7250)	81 ea. @ $286	=	$23,166
Exterior: (Figure 13.15 #7150)	36 ea. @ $194.50	=	7,002
Corner: (Figure 13.15) (Interpolate between #7100 & #7150)	4 ea. @ $150	=	600
Total			$30,768
Total Cost per Square Foot:		=	$0.49/S.F.

In summary, using full bay loading in lieu of 1/2 and 1/4 bay loading would increase the estimated cost of a 2 bay x 2 bay building by approximately 0.8%. For a 10 bay x 10 bay building, the increase would be 0.12%.

Small versus Large Concrete Building

Example One: Small Building

Follow the same procedure for a Three-Story Concrete Building with 25′ square bays and flat plate construction.

Loads:	Floors	=	242.5 kips
	Roof	=	95.0 kips
			337.5 kips

2 Bay x 2 Bay
2,500 S.F. × 3 floors = 7,500 S.F.

Footing Loads:	Interior	=	338 kips
	Exterior	=	110 kips
	Corner	=	152 kips

Interior	(Figure 13.15 #7850)	1 ea. @ $2,225	=	$2,225
Exterior	(Figure 13.15 #7650)	4 ea. @ $845	=	3,380
Corner	(Figure 13.15 #7550)	4 ea. @ $621	=	2,484
Total				$8,089
Cost per Square Foot			=	$1.08/S.F.

All Footings Full Bay Loading
Footing Load = 338 kips

	(Figure 13.15 #7850)	9 ea. @ $2,225	=	$20,025
Cost per Square Foot			=	$2.67/S.F.

Example Two: Large Building

10 Bay x 10 Bay
62,500 S.F. × 3 floors = 187,500 S.F.
Exterior & Interior Footing Loading

Footing Loads:	Interior	=	338 kips
	Exterior	=	110 kips
	Corner	=	152 kips

Interior	(Figure 13.15 #7850)	81 ea. @ $2,225	=	$180,225
Exterior	(Figure 13.15 #7650)	36 ea. @ $845	=	30,420
Corner	(Figure 13.15 #7550)	4 ea. @ $621	=	2,484
Total Cost			=	$213,129
Cost per Square Foot			=	$1.14/S.F.

All Footings Full Bay Loading
Footing Load = 338 kips

(Figure 13.15, line #7850)	120 ea. @ $2,225	=	267,000
Cost per Square Foot		=	$1.42/S.F.

Conclusions Based on a $40/S.F. Building Cost

On small-perimeter, light steel buildings having a large percentage of exterior footings with all footings total load versus interior, exterior and corner loading, the cost difference is $0.36/S.F. or 0.9%. It's worthwhile to compute the loads to each footing, then cost them out accordingly. The cost savings is worth the time spent.

On large-perimeter, light steel buildings having a large percentage of interior footings with total load versus interior, exterior and corner loading, the cost difference is $0.06/S.F. or 0.2%. The cost saved by sizing each footing is not substantial enough to merit the time spent.

On small-perimeter, heavy concrete buildings having a large percentage of exterior footings with total load versus interior, exterior and corner loading, the cost difference is $1.59/S.F. or 4.0%. The cost saved is worth the time to size each footing and cost each one out accordingly.

On large-perimeter, heavy concrete buildings having a large percentage of interior footings with total load versus interior, exterior and corner loading, the cost difference is $0.28/S.F. or 0.7%. The cost saved may or may not be worth the effort.

Large versus Small Bays, Concrete Building

A larger or deeper steel or wood beam or girder is required to span a 30' clear span than a 15' clear span. Similarly, a deeper concrete slab is required to span a 30' square column spacing versus a 15' square column spacing. Therefore, it can be assumed that with suspended floor systems of the same components, larger column spacings will increase the system cost and increase the ultimate square foot cost.

For example, using Figure 13.17, Cast-in-Place Flat Plates, with a total load of 75 psf and a bay size of 15' x 15', the cost is $7.54 per square foot.

Using the same superimposed load of 75 psf and a bay size of 25' x 25', the cost is now $9.06 per square foot.

25' Square Bay:	$9.06/S.F.
15' Square Bay:	$7.54/S.F.
Difference	$1.52/S.F.

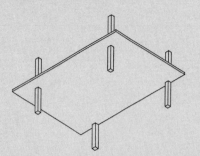

General: Flat Plates: Solid uniform depth concrete two-way slab without drops or interior beams. Primary design limit is shear at columns.

Design and Pricing Assumptions:
Concrete f'c to 4 KSI, placed by concrete pump.
Reinforcement, fy = 60 KSI.
Forms, four use.
Finish, steel trowel.
Curing, spray on membrane.
Based on 4 bay x 4 bay structure.

System Components

System Components	QUANTITY	UNIT	COST PER S.F. MAT.	COST PER S.F. INST.	COST PER S.F. TOTAL
SYSTEM 3.5-150-2000					
15'X15' BAY 40 PSF S. LOAD, 12" MIN. COL.					
Forms in place, flat plate to 15' high, 4 uses	.992	S.F.	.83	3.41	4.24
Edge forms to 6" high on elevated slab, 4 uses	.065	L.F.	.03	.16	.19
Reinforcing in place, elevated slabs #4 to #7	1.706	Lb.	.51	.46	.97
Concrete ready mix, regular weight, 3000 psi	.459	C.F.	.98		.98
Place and vibrate concrete, elevated slab less than 6", pump	.459	C.F.		.44	.44
Finish floor, monolithic steel trowel finish for finish floor	1.000	S.F.		.61	.61
Cure with sprayed membrane curing compound	.010	C.S.F.	.02	.05	.07
		TOTAL	2.37	5.13	7.50

3.5-150		Cast in Place Flat Plate							
	BAY SIZE (FT.)	SUPERIMPOSED LOAD (P.S.F.)	MINIMUM COL. SIZE (IN.)	SLAB THICKNESS (IN.)	TOTAL LOAD (P.S.F.)	COST PER S.F. MAT.	COST PER S.F. INST.	COST PER S.F. TOTAL
---	---	---	---	---	---	---	---	---
2000	15 x 15	40	12	5-1/2	109	2.37	5.15	7.52
2200	R3.5	75	14	5-1/2	144	2.39	5.15	7.54
2400	-010	125	20	5-1/2	194	2.49	5.20	7.69
2600		175	22	5-1/2	244	2.54	5.20	7.74
3000	15 x 20	40	14	7	127	2.71	5.20	7.91
3400	R3.5	75	16	7-1/2	169	2.88	5.30	8.18
3600	-100	125	22	8-1/2	231	3.18	5.45	8.63
3800		175	24	8-1/2	281	3.20	5.45	8.65
4200	20 x 20	40	16	7	127	2.71	5.20	7.91
4400		75	20	7-1/2	175	2.92	5.35	8.27
4600		125	24	8-1/2	231	3.19	5.45	8.64
5000		175	24	8-1/2	281	3.21	5.50	8.71
5600	20 x 25	40	18	8-1/2	146	3.16	5.45	8.61
6000		75	20	9	188	3.27	5.50	8.77
6400		125	26	9-1/2	244	3.53	5.65	9.18
6600		175	30	10	300	3.68	5.70	9.38
7000	25 x 25	40	20	9	152	3.27	5.50	8.77
7400		75	24	9-1/2	194	3.46	5.60	9.06
7600		125	30	10	250	3.68	5.75	9.43
8000								

Figure 13.17

Using a flat plate concrete slab, and increasing the column spacing or bay size from 15' square to 25' square, the cost of the floor system is increased by $1.52 per square foot and apparently increases or affects the final square foot cost of the building. However, for reasons of the architect or owner, it may be necessary or desirable to choose the 25' bay spacing.

This study of column spacing or bay spacing would not be completed without considering the cost of columns or supports for the slab. For the previous Cast-in-Place, Flat Plate System with a superimposed load of 75 psf on a 15' versus 25' square bay, the following analysis may be performed, using Figure 13.17:

Bay Size	Superimposed Load psf	Minimum Col. Size	Slab Thickness	Total Load psf
15' x 15'	75	14"	5-1/2"	144
25' x 25'	75	24"	9-1/2"	194

To support the same 75 psf, but increase the bay spacing from 15' square to 25', the slab thickness would have to increase from 5-1/2" to 9-1/2", adding 50 psf to the dead weight of the structure. Likewise, the minimum allowable concrete column size would increase from 14" square to 24".

To understand the effect of columns on the cost differential, proceed as follows:

Load to columns: 15' Square Bay = 32.4 kips
25' Square Bay = 121.3 kips

Using the "Cast-In-Place Columns — Square Tied" table in Figure 13.5, utilize the "Minimum Reinforcement Table" for the minimum column size shown, and assume the columns are 12' high. The cost for columns can be calculated as follows:

Large Building — Assumed Structural Cost — Summary						
	Footing $/S.F.	Ftg. % Diff.	Floor $/S.F.	Total $/S.F.	Struct. % Diff.	*Bldg. % Diff. Total
2 Bay x 2 Bay						
Full Bay Loads	1.03	65%	9.06	10.09	4%	8%
Full Ext. & Corner Loads	.67		9.06	9.73		
10 Bay x 10 Bay						
Full Bay Loads	.55	11%	9.06	9.61	.6%	.12%
Full, Ext. & Corner Loads	.49		9.06	9.55		
* Assuming the structural cost as 20% of the Total Building Cost, % Differences would be as shown in this column.						

Figure 13.18

14" Square Column
12" Square	=	$44.85/L.F.
Interpolating: 14" Square	=	$51.50/L.F.
16" Square	=	$58.15/L.F.

15' Square Bay, 14" Square Column: $51.50/L.F. × 12' = $618.00 ea.
25' Square Bay, 24" Square Column: $98.00/L.F. × 12' = $1,176.00 ea.

The column cost per square foot of bay area is:

15' Bay: $2.75/S.F.
25' Bay: $1.88/S.F.

Using the costs per square foot found previously, the cost differential is now:

25' Square Bay:	$9.06 + $1.88	=	$10.94/S.F.
15' Square Bay:	$7.54 + $2.75	=	$10.29/S.F.
Difference			$ 0.65/S.F.

Partial Summary: Increasing the bay size from 15' square to 25' square with the same load increases the suspended framing portion of the building cost by $.65/S.F. or approximately 6%.

To further evaluate the bay spacings, compare the foundations or column footings used to support the different bay sizes.

As mentioned, the footing loads are:
| 15' Square Bay | = | 32.4 kips |
| 25' Square Bay | = | 121.3 kips |

Assume a 3 KSF soil pressure design condition.

Costs:

15' Square Bay: (Figure 13.15, line #7150) = $194.50 ea.
Cost per Square Foot = $0.86/S.F.
25' Square Bay: (Figure 13.15, line #7450) = $522.00 ea.
Cost per Square Foot = $0.84/S.F.

To complete the analysis, the cost differential is now:

	Floor	Column	Footing		
25' Square Bay	$9.06 +	$1.88 +	$0.84	=	$11.78/S.F.
15' Square Bay	$7.54 +	$2.75 +	$0.86	=	$11.15/S.F.
Difference					$ 0.63/S.F.

It costs $0.63/S.F. more, a 6% increase, to change the bay size from 15' square to 25' square for this concrete building.

Small versus Large Bays, Steel Building Bays

For comparison purposes, establish a square foot cost for a structural steel framing system using interior columns, beams, open web joists, and concrete slab, using Figure 13.20. Exterior walls are bearing walls. Column costs are included in the systems. Use the same 75 psf.

30' Square Bay (Figure 13.20, line #2070)	=	$7.56/S.F.
20' Square Bay (Figure 13.20, line #1400)	=	$6.07/S.F.
Difference		$1.49/S.F.

| Loading: | 20' Square Bay | = | 47.6 kips |
| | 30' Square Bay | = | 108 kips |

Again, assuming 3 KSF allowable soil pressure, the footing costs are:

20′ Square Bay (Figure 13.15, line #7150)	=	$ 194.50 ea.
Cost per Square Foot	=	$ 0.49/S.F.
30′ Square Bay (Figure 13.15, line #7450)	=	$522 ea.
Cost per Square Foot	=	$ 0.58/S.F.

Total cost for floor and column plus foundation, to increase bay size:

30′ Square Bay:	$7.56 + $0.58	=	$8.14/S.F.
20′ Square Bay:	$6.07 + $0.49	=	$6.56/S.F.
Difference			$1.58/S.F.

It costs approximately 24% more to increase the bay size from 20′ square to 30′ square for a steel and bar joist structural system. See Figure 13.21 for summary of Division 3 costs.

Small Building—Assumed Structural Cost—Summary					
	Footing $/S.F.	Ftg. % Diff.	Floor $/S.F.	Total $/S.F.	Struct. % Diff.
2 Bay x 2 Bay					
Full Bay Loads	2.67	247%	8.07	10.74	17%
Corner Loads	1.08		8.07	9.15	
10 Bay x 10 Bay					
Full Bay Loads	1.42	125%	8.07	9.49	3%
Full, Ext. & Corner Loads	1.14		8.07	9.21	

Figure 13.19

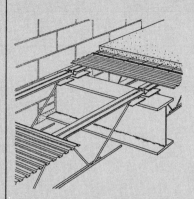

Table below lists costs for a floor system on exterior bearing walls and interior columns and beams using open web steel joists, galvanized steel slab form, 2-1/2″ concrete slab reinforced with welded wire fabric.

Design and Pricing Assumptions:
Structural Steel is A36.
Concrete f'c = 3 KSI placed by pump.
WWF 6 x 6 — W1.4 x W1.4 (10 x 10)
Columns are 12' high.
Building is 4 bays long by 4 bays wide.
Joists are 2' O.C. ± and span the long direction of the bay.

Joists at columns have bottom chords extended and are connected to columns.

Slab form is 28 gauge galvanized. Column costs in table are for columns to support 1 floor plus roof loading in a 2-story building; however, column costs are from ground floor to 2nd floor only. Joist costs include appropriate bridging. Deflection is limited to 1/360 of the span. Screeds and steel trowel finish.

Design Loads	Min.	Max.
S.S & Joists	4.4 PSF	11.5 PSF
Slab Form	1.0	1.0
2-1/2″ Concrete	27.0	27.0
Ceiling	3.0	3.0
Misc.	7.6	5.5
	43.0 PSF	48.0 PSF

System Components			COST PER S.F.		
	QUANTITY	UNIT	MAT.	INST.	TOTAL
SYSTEM 3.5-440-1200					
15'X20' BAY, W.F. STEEL, STEEL JOISTS, SLAB FORM, CONCRETE SLAB					
Structural steel	1.248	Lb.	.54	.24	.78
Open web joists	3.140	Lb.	1.35	.69	2.04
Slab form, galvanized steel 9/16″ deep, 28 gauge	1.020	S.F.	.50	.32	.82
Welded wire fabric 6x6 - W1.4 x W1.4 (10 x 10), 21 lb/CSF roll, 10% lap	1.000	S.F.	.08	.23	.31
Concrete ready mix, regular weight, 3000 psi	.210	C.F.	.45		.45
Place and vibrate concrete, elevated slab less than 6″, pumped	.210	C.F.		.20	.20
Finishing floor, monolithic steel trowel finish for finish floor	1.000	S.F.		.61	.61
Curing with sprayed membrane curing compound	.010	C.S.F.	.02	.05	.07
TOTAL			2.94	2.34	5.28

3.5-440			**Steel Joists on Beam and Wall**					
	BAY SIZE (FT.)	SUPERIMPOSED LOAD (P.S.F.)	DEPTH (IN.)	TOTAL LOAD (P.S.F.)	COLUMN ADD	COST PER S.F.		
						MAT.	INST.	TOTAL
1200	15x20 R3.5 -100	40	17	83		2.94	2.34	5.28
1210					columns	.18	.08	.26
1220	15x20	65	19	108		3.02	2.38	5.40
1230					columns	.18	.08	.26
1250	15x20	75	19	119		3.23	2.48	5.71
1260					columns	.22	.09	.31
1270	15x20	100	19	144		3.56	2.84	6.40
1280					columns	.22	.09	.31
1300	15x20	125	19	170		3.66	2.89	6.55
1310					columns	.29	.12	.41
1350	20x20	40	19	83		3.02	2.37	5.39
1360					columns	.16	.07	.23
1370	20x20	65	23	109		3.36	2.54	5.90
1380					columns	.22	.09	.31
1400	20x20	75	23	119		3.48	2.59	6.07
1410					columns	.22	.09	.31
1420	20x20	100	23	144		3.59	2.63	6.22
1430					columns	.22	.09	.31
1450	20x20	125	23	170		4.19	3.14	7.33
1460					columns	.26	.11	.37
1500	20x25	40	23	84		3.34	2.71	6.05
1510					columns	.17	.08	.25

Figure 13.20

| 3.5-440 | Steel Joists on Beam and Wall | | | | | | | |

	BAY SIZE (FT.)	SUPERIMPOSED LOAD (P.S.F.)	DEPTH (IN.)	TOTAL LOAD (P.S.F.)	COLUMN ADD	COST PER S.F.		
						MAT.	INST.	TOTAL
1520	20x25	65	26	110		3.60	2.85	6.45
1530					columns	.17	.08	.25
1550	20x25	75	26	120		3.85	3	6.85
1560					columns	.21	.09	.30
1570	20x25	100	26	145		3.88	2.80	6.68
1580					columns	.21	.09	.30
1670	20x25	125	29	170		4.43	3.07	7.50
1680					columns	.24	.10	.34
1720	25x25	40	23	84		3.57	2.80	6.37
1730					columns	.17	.07	.24
1750	25x25	65	29	110		3.76	2.90	6.66
1760					columns	.17	.07	.24
1770	25x25	75	26	120		4.05	2.86	6.91
1780					columns	.19	.08	.27
1800	25x25	100	29	145		4.59	3.13	7.72
1810					columns	.19	.08	.27
1820	25x25	125	29	170		4.82	3.24	8.06
1830					columns	.21	.09	.30
1870	25x30	40	29	84		3.88	2.90	6.78
1880					columns	.16	.07	.23
1900	25x30	65	29	110		4.11	2.62	6.73
1910					columns	.16	.07	.23
1920	25x30	75	29	120		4.34	2.71	7.05
1930					columns	.18	.08	.26
1950	25x30	100	29	145		4.72	2.87	7.59
1960					columns	.18	.08	.26
1970	25x30	125	32	170		5.10	3.54	8.64
1980					columns	.20	.09	.29
2020	30x30	40	29	84		4.01	2.58	6.59
2030					columns	.15	.07	.22
2050	30x30	65	29	110		4.56	2.81	7.37
2060					columns	.15	.07	.22
2070	30x30	75	32	120		4.70	2.86	7.56
2080					columns	.17	.08	.25
2100	30x30	100	35	145		5.20	3.07	8.27
2110					columns	.20	.09	.29
2120	30x30	125	35	172		5.70	3.83	9.53
2130					columns	.24	.10	.34
2170	30x35	40	29	85		4.57	2.80	7.37
2180					columns	.14	.06	.20
2200	30x35	65	29	111		5.15	3.58	8.73
2210					columns	.16	.07	.23
2220	30x35	75	32	121		5.15	3.58	8.73
2230					columns	.17	.07	.24
2250	30x35	100	35	148		5.60	3.19	8.79
2260					columns	.21	.09	.30
2270	30x35	125	38	173		6.30	3.48	9.78
2280					columns	.21	.09	.30
2320	35x35	40	32	85		4.70	2.86	7.56
2330					columns	.15	.07	.22
2350	35x35	65	35	111		5.45	3.71	9.16
2360					columns	.18	.08	.26
2370	35x35	75	35	121		5.55	3.76	9.31
2380					columns	.18	.08	.26
2400	35x35	100	38	148		5.80	3.90	9.70
2410					columns	.22	.09	.31

Figure 13.20 (cont.)

Assembly Number	Description	Qty	Unit	Total Cost Unit	Total Cost Total	Cost per S. F.
1.0	**Foundations**					
1.1-120-	Column Footings 6 KSF					
7700	Interior 190K (200K)	9	Ea	464	4176	
7500	Exterior 114K (125K)	12	Ea	287	3444	
7410	Corner 86K (100K)	4	Ea	232.50	930	
1.1-140-	Strip Footings 12" x 24"					
2700	400 LF - (12x5') - (4x4.5')	322	LF	23.50	7567	
1.1-210-	Foundation Wall					
1560	4' x 12" Direct Chute	400	LF	43.80	17520	
3.1-114	Piers : Assume 20" sq.					
9924	16 x 3.5'	56	LF	76.50	4284	
					$37,921	$1.26
2.0	**Substructure**					
2.1-200-	4" non-industrial slab-on-grade					
2240	Reinforced	10000	SF	2.86	28600	
					$28,600	$0.95
3.0	**Superstructure**					
3.7-420-	Roof					
3300	25' Sq. 40 lb.	10000	SF	4.01	40100	
3.5-460-	Floor					
5100	2nd & 3rd 25' Sq. 75 lb.	20000	SF	8.61	172200	
3.1-130-	Columns					
4600	190K (200K) : 25x27.33'	684	LF	37.85	25889	
1600	38K (50K) : 25x9'	175	LF	17.60	3080	
3.1-190-	Column Fireproofing					
3450	8" Average, Gyp. Bd.	859	LF	14.77	12687	
3.9-100-	Stairs					
0760	Steel pan	5	FL	5425	27125	
					$281,082	$9.37
4.0	**Exterior Closure**					

Figure 13.21

Chapter 14

Exterior Closure

Often the owner or architect needs a cost comparison to show the client how the exterior closure assemblies price will vary for the same size building. This relationship can be quickly and accurately determined using illustrated assemblies in *Means Assemblies Cost Data*. Total costs for each assembly are provided, as are adds or deducts for modifications. For curtain wall systems, a selective system worksheet is provided, which makes it possible to quickly estimate or compare costs of a variety of curtain wall materials.

Multi-story buildings with elevators require a penthouse for the elevator shaft, ventilation, and stair towers. Many buildings now have mechanical systems installed in a penthouse on the roof. Exterior closure for these systems is required and may be a different material than that used for the rest of the building. Many times the change of material is overlooked in the estimating process.

The Effects of Energy Requirements

Because energy requirements have such a dramatic impact on the future operating costs of a building, it is critical during the initial design and budget process to consider not only the cost of exterior closure systems, but also their contribution to energy conservation. Simply stated, the least expensive material may be the most costly (in energy) over the life of a building when compared to another material whose energy profile is better (even though it may be more costly to initially purchase and install).

Resistance to heat flow is important in construction. The term *R value* has been adopted as the measure of thermal resistance. The benefit of using R values is realized from the fact that when heat flows through several materials, as in Figure 14.1, the individual R values can be added together. Using R values also allows the calculation of the Overall Coefficient of Thermal Transmittance (U) of a complete assembly. To determine the U values, add up all the individual R values and take the reciprocal of that number ($U = 1/R$).

For example, in Figure 14.1, the first system has a total R value of 4.61. By taking the reciprocal, the U value of 0.22 is found. After adding the insulation, the R value is increased to 14.87 and the U value is reduced to 0.067. The R and U values are inversely proportional: as the R value increases, the U value decreases.

The following example illustrates how the thermal efficiency of a wall can be improved substantially by the addition of low-cost insulation.

In Figure 14.1, the cost per square foot of the wall system before adding rigid insulation is as follows:

Brick Veneer/CMU Backup (Figure 14.2, line #1200)	=	$18.85/S.F.
Plaster (Figure 14.3 #0920)	=	1.93/S.F.
Total Cost per Square Foot		$20.78/S.F.

After adding the insulation, the cost becomes:

Brick Veneer/CMU Backup/Plaster	=	$20.78/S.F.
Insulation (similar to Figure 14.4, line #1835)	=	0.82/S.F.
Total Cost per Square Foot		$21.60/S.F.

If the exterior wall has 11,500 S.F., the cost comparison is as follows:

11,500 S.F. @ $21.60/S.F.	=	$248,400
11,500 S.F. @ $20.78/S.F.	=	238,970
Cost Difference		$ 9,430

It costs 4% more to increase the wall U Value from 0.22 to 0.067, an increase of 328%. Add to this evaluation the cost difference for heating and cooling the building, both with and without the added insulation. The initial added cost may be insignificant when compared to the additional operating costs.

14" Masonry Cavity Wall with 1" Plaster for the Exterior Closure

Construction	Resistance R.
1. Outside surface (15 mph wind)	0.17
2. Face brick (4 in.)	0.44
3. Air space (2 in., 50° mean temp, 10° diff)	1.02
4. Concrete block (8in., lightweight)	2.12
5. Plaster (1 in., sand aggregate)	0.18
6. Inside surface (still air)	0.68
Total resistance	**4.61**
U= 1/R=1/4.61=	**0.22**

Replace item 3 with 2" smooth rigid polystyrene insulation and item 5 with 3/4" furring and 1/2" drywall

Total resistance		4.61
Deduct 3. Airspace	**1.02**	
5. Plaster	**0.18**	
	1.20	
Difference 4.61 - 1.20 =		3.41
Add rigid polystyrene insulation		10.00
3/4" airspace		1.01
1/2" gypsum board		0.45
Total resistance 14.87		
U=1/R=1/14.87=0.067		

Figure 14.1

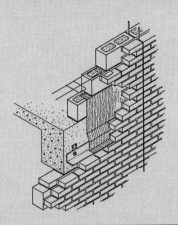

Exterior brick face composite walls are defined in the following terms: type of face brick and backup masonry, thickness of backup masonry and insulation. A special section is included on triple wythe construction at the back. Seven types of face brick are shown with various thicknesses of seven types of backup. All systems include a brick shelf, ties to the backup and necessary dampproofing, flashing, and control joints every 20'.

System Components	QUANTITY	UNIT	COST PER S.F.		
			MAT.	INST.	TOTAL
SYSTEM 4.1-272-1120					
COMPOSITE WALL, STANDARD BRICK FACE, 6" C.M.U. BACKUP, PERLITE FILL					
Face brick veneer, standard, running bond	1.000	S.F.	2.56	6.75	9.31
Wash brick	1.000	S.F.	.11	.58	.69
Concrete block backup, 6" thick	1.000	S.F.	1.12	3.70	4.82
Wall ties	.300	Ea.	.04	.09	.13
Perlite insulation, poured	1.000	S.F.	.29	.19	.48
Flashing, aluminum	.100	S.F.	.09	.25	.34
Shelf angle	1.000	Lb.	.58	.65	1.23
Control joint	.050	L.F.	.11	.03	.14
Backer rod and sealant	.100	L.F.	.02	.25	.27
Collar joint	1.000	S.F.	.39	.33	.72
TOTAL			5.31	12.82	18.13

4.1-272		Brick Face Composite Wall - Double Wythe						
	FACE BRICK	BACKUP MASONRY	BACKUP THICKNESS (IN.)	BACKUP CORE FILL		COST PER S.F.		
						MAT.	INST.	TOTAL
1000	Standard	common brick	4	none		5.90	15.40	21.30
1040		SCR brick	6	none		8.05	13.70	21.75
1080		conc. block	4	none		4.85	12.40	17.25
1120			6	perlite		5.30	12.85	18.15
1160				styrofoam		5.85	12.65	18.50
1200			8	perlite		5.70	13.15	18.85
1240				styrofoam		6.05	12.90	18.95
1280		L.W. block	4	none		4.97	12.30	17.27
1320			6	perlite		5.60	12.75	18.35
1360				styrofoam		6.10	12.55	18.65
1400			8	perlite		6.20	13	19.20
1440				styrofoam		6.55	12.80	19.35
1520		glazed block	4	none		10.65	13.25	23.90
1560			6	perlite		11.45	13.60	25.05
1600				styrofoam		11.95	13.40	25.35
1640			8	perlite		12.25	13.95	26.20
1680				styrofoam		12.60	13.70	26.30

Figure 14.2

6.1-680   Plaster Partition Components

		MAT.	INST.	TOTAL
		COST PER S.F.		
0060	Metal studs, 16" O.C., including track, non load bearing, 25 gage, 1-5/8"	.28	.73	1.01
0080	2-1/2"	.28	.73	1.01
0100	3-1/4"	.32	.75	1.07
0120	3-5/8"	.32	.75	1.07
0140	4"	.36	.77	1.13
0160	6"	.47	.79	1.26
0180	Load bearing, 20 gage, 2-1/2"	.90	1.61	2.51
0200	3-5/8"	1.05	1.69	2.74
0220	4"	1.09	1.79	2.88
0240	6"	1.33	1.89	3.22
0260	16 gage 2-1/2"	1.13	1.79	2.92
0280	3-5/8"	1.19	1.89	3.08
0300	4"	1.38	2.01	3.39
0320	6"	1.68	2.15	3.83
0340	Wood studs, including blocking, shoe and double plate, 2"x4", 12" O.C.	.53	.84	1.37
0360	16" O.C.	.42	.67	1.09
0380	24" O.C.	.32	.54	.86
0400	2"x6", 12" O.C.	.75	.95	1.70
0420	16" O.C.	.61	.74	1.35
0440	24" O.C.	.46	.58	1.04
0460	Furring one face only, steel channels, 3/4", 12" O.C.	.26	1.29	1.55
0480	16" O.C.	.24	1.14	1.38
0500	24" O.C.	.16	.86	1.02
0520	1-1/2", 12" O.C.	.38	1.44	1.82
0540	16" O.C.	.34	1.26	1.60
0560	24"O.C.	.23	.99	1.22
0580	Wood strips 1"x3", on wood., 12" O.C.	.35	.59	.94
0600	16"O.C.	.26	.44	.70
0620	On masonry, 12" O.C.	.35	.65	1
0640	16" O.C.	.26	.49	.75
0660	On concrete, 12" O.C.	.35	1.24	1.59
0680	16" O.C.	.26	.93	1.19
0700	Gypsum lath. plain or perforated, nailed to studs, 3/8" thick	.43	.39	.82
0720	1/2" thick	.45	.42	.87
0740	Clipped to studs, 3/8" thick	.44	.45	.89
0760	1/2" thick	.49	.48	.97
0780	Metal lath, diamond painted, nailed to wood studs, 2.5 lb.	.21	.39	.60
0800	3.4 lb.	.27	.42	.69
0820	Screwed to steel studs, 2.5 lb.	.21	.42	.63
0840	3.4 lb.	.24	.45	.69
0860	Rib painted, wired to steel, 2.75 lb	.20	.45	.65
0880	3.4 lb	.36	.48	.84
0900	4.0 lb	.32	.52	.84
0910				
0920	Gypsum plaster, 2 coats	.40	1.53	1.93
0940	3 coats	.55	1.85	2.40
0960	Perlite or vermiculite plaster, 2 coats	.46	1.74	2.20
0980	3 coats	.73	2.17	2.90
1000	Stucco, 3 coats, 1" thick, on wood framing	.76	3.75	4.51
1020	On masonry	.34	.80	1.14
1100	Metal base galvanized and painted 2-1/2" high	.45	1.26	1.71

Figure 14.3

		A5.7-101	Rigid Insulation

5.7-101	Roof Deck Rigid Insulation	COST PER S.F.		
		MAT.	INST.	TOTAL
0100	Fiberboard low density, 1/2" thick, R1.39			
0150	1" thick R2.78	.34	.39	.73
0300	1 1/2" thick R4.17	.51	.39	.90
0350	2" thick R5.56	.66	.39	1.05
0370	Fiberboard high density, 1/2" thick R1.3	.19	.32	.51
0380	1" thick R2.5	.35	.39	.74
0390	1 1/2" thick R3.8	.61	.39	1
0410	Fiberglass, 3/4" thick R2.78	.42	.32	.74
0450	15/16" thick R3.70	.54	.32	.86
0500	1-1/16" thick R4.17	.68	.32	1
0550	1-5/16" thick R5.26	.95	.32	1.27
0600	2-1/16" thick R8.33	1.01	.39	1.40
0650	2 7/16" thick R10	1.14	.39	1.53
1000	Foamglass, 1 1/2" thick R4.55	1.41	.39	1.80
1100	3" thick R9.09	2.80	.45	3.25
1200	Tapered for drainage	.94	.53	1.47
1260	Perlite, 1/2" thick R1.32	.29	.30	.59
1300	3/4" thick R2.08	.33	.39	.72
1350	1" thick R2.78	.39	.39	.78
1400	1 1/2" thick R4.17	.52	.39	.91
1450	2" thick R5.56	.66	.45	1.11
1510	Polyisocyanurate 2#/CF density, 1" thick R7.14	.37	.23	.60
1550	1 1/2" thick R10.87	.42	.25	.67
1600	2" thick R14.29	.50	.29	.79
1650	2 1/2" thick R16.67	.62	.30	.92
1700	3" thick R21.74	.73	.32	1.05
1750	3 1/2" thick R25	.90	.32	1.22
1800	Tapered for drainage	.44	.23	.67
1810	Expanded polystyrene, 1#/CF density, 3/4" thick R2.89	.19	.21	.40
1820	2" thick R7.69	.33	.25	.58
1825	Extruded Polystyrene			
1830	15 PSI compressive strength, 1" thick R5	.32	.21	.53
1835	2" thick R10	.57	.25	.82
1840	3" thick R15	.80	.32	1.12
1900	25 PSI compressive strength, 1" thick R5	.34	.21	.55
1950	2" thick R10	.63	.25	.88
2000	3" thick R15	.91	.32	1.23
2050	4" thick R20	1.20	.32	1.52
2150	Tapered for drainage	.42	.21	.63
2550	40 PSI compressive strength, 1" thick R5	.40	.21	.61
2600	2" thick R10	.75	.25	1
2650	3" thick R15	1.12	.32	1.44
2700	4" thick R20	1.50	.32	1.82
2750	Tapered for drainage	.53	.23	.76
2810	60 PSI compressive strength, 1" thick R5	.46	.22	.68
2850	2" thick R10	.86	.26	1.12
2900	Tapered for drainage	.64	.23	.87
2910	115 PSI compressive strength, 1" thick R5	.99	.23	1.22
2950	2" thick R10	1.96	.27	2.23
3000	Tapered for drainage	1.07	.23	1.30
4000	Composites with 1-1/2" polyisocyanurate			
4010	1" fiberboard	.77	.39	1.16
4020	1" perlite	.79	.37	1.16
4030	7/16" oriented strand board	.90	.39	1.29

Figure 14.4

Exterior Wall Systems

Choosing exterior wall systems is usually dictated by the client, zoning ordinance, or some corporate policy regarding architectural and/or environmental guidelines requiring that similar architectural elements be present on all buildings in a complex. Whatever the source, exterior building materials have more than just aesthetic importance. Energy, heat loss, codes and regulations, natural light requirements, and, of course, economics are just a few of the issues that must be addressed during the selection process.

The exterior wall systems allow the designer unlimited flexibility and choice of materials or combinations of materials. The exterior of the building is the most visible. It also represents a significant cost when compared to other parts of the building.

Exterior Wall Area

The percentage of exterior wall area in relation to door and window area can change significantly with the size of the building and the window configuration. Consider the example illustrated in Figure 14.5.

In a small building such as Building A (with individual windows), the percentage of windows to wall area is insignificant. The percentage becomes very significant in larger wall areas, such as the wall design used for Building B.

Building "A"	Exterior Wall	92%
	Windows	6%
	Doors	2%
Building "B"	Exterior Wall	58%
	Window Wall	41.5%
	Doors	0.5%

Small Footprint versus Large Footprint

When compared to the building footprint area, the exterior closure can reveal some interesting cost discoveries:

Consider a 50' x 50' building with a 12' high block exterior wall, 12" thick. See Figure 14.6.

Cost: (Figure 14.7, line #1510)

$8.98/S.F. × 2,400 S.F. = $21,552

Cost per Square Foot of Building:

$21,552 / 2,500 S.F. = $8.62/S.F.

Now, compare this building with the larger building shown in Figure 14.8.

Cost: (Figure 14.7, line #1510)

$8.98/S.F. × 12,000 S.F. = $107,760

Cost per Square Foot of Building:

$107,760/62,500 S.F. = $1.72/S.F.

Assuming a building cost of $40/S.F., the exterior wall now looks like:

| 2,500 S.F. Building | $8.62/$40 | = | 22% |
| 62,500 S.F. Building | $1.72/$40 | = | 4% |

As can be determined from the above square foot costs, the exterior closure can have a significant impact on the cost of a building. The choice of

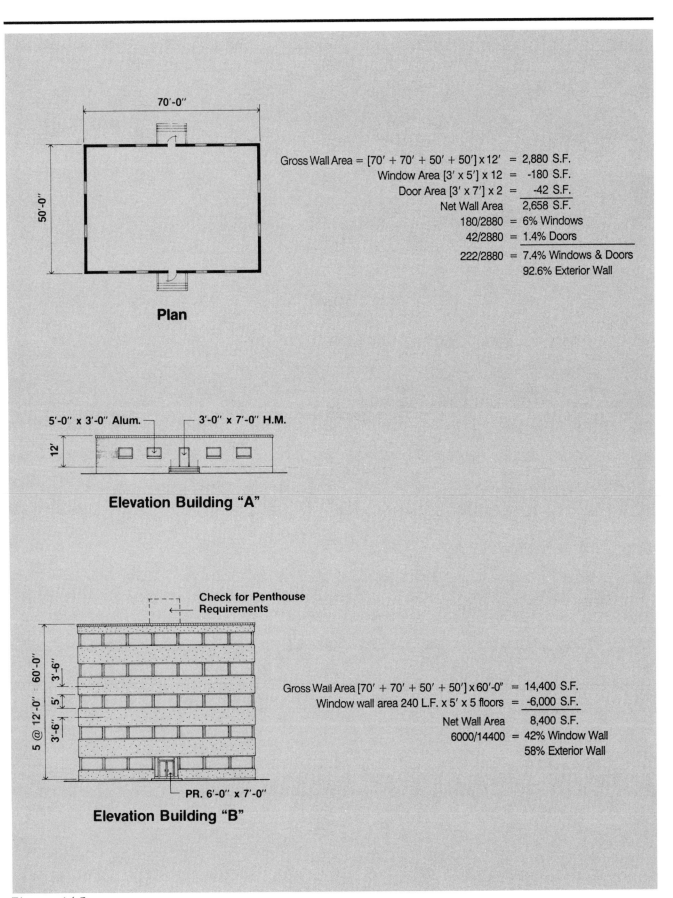

Plan

Gross Wall Area = [70' + 70' + 50' + 50'] x 12' = 2,880 S.F.
Window Area [3' x 5'] x 12 = -180 S.F.
Door Area [3' x 7'] x 2 = -42 S.F.
Net Wall Area 2,658 S.F.
180/2880 = 6% Windows
42/2880 = 1.4% Doors

222/2880 = 7.4% Windows & Doors
92.6% Exterior Wall

5'-0" x 3'-0" Alum.   3'-0" x 7'-0" H.M.

12'

Elevation Building "A"

Check for Penthouse
Requirements

5 @ 12'-0" = 60'-0"
3'-6"
5'
3'-6"

Gross Wall Area [70' + 70' + 50' + 50'] x 60'-0" = 14,400 S.F.
Window wall area 240 L.F. x 5' x 5 floors = -6,000 S.F.

Net Wall Area 8,400 S.F.
6000/14400 = 42% Window Wall
58% Exterior Wall

PR. 6'-0" x 7'-0"

Elevation Building "B"

Figure 14.5

materials should be carefully evaluated to get the most for the dollars spent. Using the exterior closure as bearing walls in low-rise buildings should be considered.

Example: Three-Story Office Building

The office building will have the following included in the exterior closure:

- Insulated concrete masonry units
- Precast concrete wall panels
- Aluminum tube frame windows and door units
- Insulating glass
- Precast terrazzo window sills

Costs:

Insulated Concrete Masonry Units (Penthouse)			
(Figure 14.7, line #1360)	493 S.F. @ $6.24/S.F.	=	$ 3,076
Precast Concrete Wall Panels			
(Figure 14.9, line #5750)	9,338 S.F. @ $14.11/S.F.	=	131,759
Aluminum and Glass Doors			
(Figure 14.10, line #6350)	2 pr. @ $3,050/pr.	=	6,100
Aluminum Flush Tube Windows with Thermal Break			
(Figure 14.11, line #2000)	5,750 S.F. @ $19.50/S.F.	=	112,125
Insulating Glass			
(Figure 14.12, line #1000)	5,750 S.F. @ $11.95/S.F.	=	68,712
Precast Terrazzo Window Sills			
(Figure 14.13, line #1880)	804 L.F. @ $9.87/L.F.	=	7,935
Total Cost			$329,707
Cost per Square Foot		=	$10.99/S.F.

Figures 14.14 and 14.15 are representative for the Three-Story Office Building. All four elevations are similar except that the side elevations have windows instead of doors. The wall section is typical for showing construction components and relationships. Figure 14.16 summarizes the square foot costs for this division for the sample project.

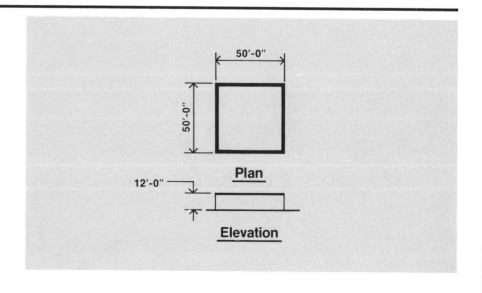

Figure 14.6

EXTERIOR CLOSURE | A4.1-211 | Concrete Block Wall

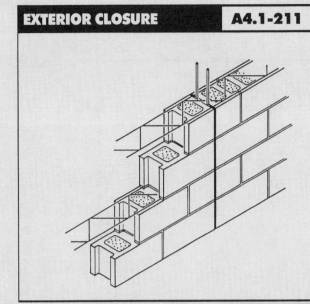

Exterior concrete block walls are defined in the following terms; structural reinforcement, weight, percent solid, size, strength and insulation. Within each of these categories, two to four variations are shown. No costs are included for brick shelf or relieving angles.

System Components

System Components	QUANTITY	UNIT	COST PER S.F. MAT.	COST PER S.F. INST.	COST PER S.F. TOTAL
SYSTEM 4.1-211-1400					
UNREINFORCED CONCRETE BLOCK WALL, 8" X 8" X 16", PERLITE CORE FILL					
Concrete block wall, 8" thick	1.000	S.F.	1.35	3.95	5.30
Perlite insulation	1.000	S.F.	.43	.24	.67
Horizontal joint reinforcing, alternate courses	.800	S.F.	.10	.10	.20
Control joint	.050	L.F.	.11	.03	.14
TOTAL			1.99	4.32	6.31

4.1-211 — Concrete Block Wall - Regular Weight

	TYPE	SIZE (IN.)	STRENGTH (P.S.I.)	CORE FILL		COST PER S.F. MAT.	COST PER S.F. INST.	COST PER S.F. TOTAL
1200	Hollow	4x8x16	2,000	none		1.15	3.56	4.71
1250			4,500	none		1.34	3.56	4.90
1300		6x8x16	2,000	perlite		1.61	4.01	5.62
1310				styrofoam		2.13	3.82	5.95
1340				none		1.32	3.82	5.14
1350			4,500	perlite		1.90	4.01	5.91
1360				styrofoam		2.42	3.82	6.24
1390				none		1.61	3.82	5.43
1400		8x8x16	2,000	perlite		1.99	4.32	6.31
1410				styrofoam		2.37	4.08	6.45
1440				none		1.56	4.08	5.64
1450			4,500	perlite		2.48	4.32	6.80
1460				styrofoam		2.86	4.08	6.94
1490				none		2.05	4.08	6.13
1500		12x8x16	2,000	perlite		3.43	5.75	9.18
1510				styrofoam		3.73	5.25	8.98
1540				none		2.73	5.25	7.98
1550			4,500	perlite		3.50	5.75	9.25
1560				styrofoam		3.80	5.25	9.05
1590				none		2.80	5.25	8.05
2000	75% solid	4x8x16	2,000	none		1.42	3.60	5.02
2050			4,500	none		1.61	3.60	5.21

Figure 14.7

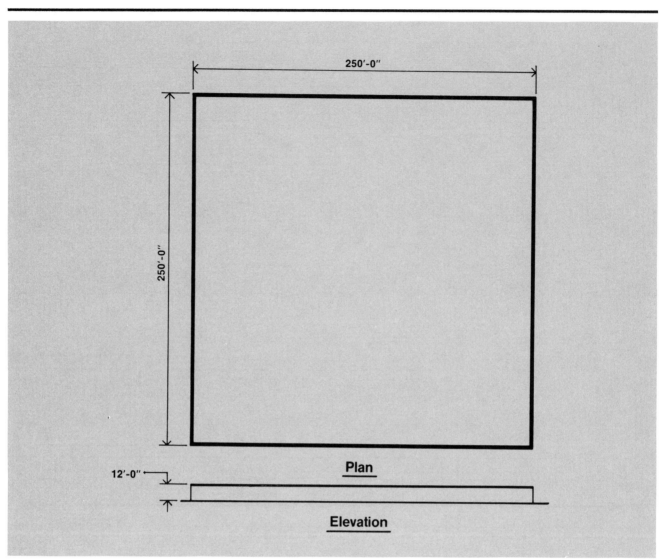

250'-0"

250'-0"

Plan

12'-0"

Elevation

Figure 14.8

182

4.1-140 — Flat Precast Concrete

	THICKNESS (IN.)	PANEL SIZE (FT.)	FINISHES	RIGID INSULATION (IN)	TYPE	COST PER S.F.		
						MAT.	INST.	TOTAL
4800	8	4x8	white face	none	low rise	16.95	3.21	20.16
4850		8x8				13.05	2.47	15.52
4900		10x10				11.75	2.22	13.97
4950		20x10				10.80	2.05	12.85
5000	8	4x8	white face	2	low rise	17.80	3.57	21.37
5050		8x8				13.95	2.83	16.78
5100		10x10				12.60	2.58	15.18
5150		20x10				11.70	2.41	14.11

4.1-140 — Fluted Window or Mullion Precast Concrete

	THICKNESS (IN.)	PANEL SIZE (FT.)	FINISHES	RIGID INSULATION (IN)	TYPE	COST PER S.F.		
						MAT.	INST.	TOTAL
5200	4	4x8	smooth gray	none	high rise	11.60	11.50	23.10
5250		8x8				8.30	8.20	16.50
5300		10x10				11.80	3.01	14.81
5350		20x10				10.35	2.64	12.99
5400	5	4x8	smooth gray	none	high rise	11.80	11.65	23.45
5450		8x8				8.55	8.50	17.05
5500		10x10				12.30	3.14	15.44
5550		20x10				10.90	2.77	13.67
5600	6	4x8	smooth gray	none	high rise	12.10	12	24.10
5650		8x8				8.85	8.75	17.60
5700		10x10				12.65	3.23	15.88
5750		20x10				11.25	2.86	14.11
5800	6	4x8	smooth gray	2	high rise	12.95	12.35	25.30
5850		8x8				9.70	9.15	18.85
5900		10x10				13.50	3.59	17.09
5950		20x10				12.10	3.22	15.32
6000	7	4x8	smooth gray	none	high rise	12.35	12.25	24.60
6050		8x8				9.10	9.05	18.15
6100		10x10				13.20	3.37	16.57
6150		20x10				11.60	2.96	14.56
6200	7	4x8	smooth gray	2	high rise	13.20	12.60	25.80
6250		8x8				9.95	9.40	19.35
6300		10x10				14.05	3.73	17.78
6350		20x10				12.45	3.32	15.77
6400	8	4x8	smooth gray	none	high rise	12.55	12.45	25
6450		8x8				9.35	9.25	18.60
6500		10x10				13.65	3.47	17.12
6550		20x10				12.15	3.10	15.25
6600	8	4x8	smooth gray	2	high rise	13.40	12.80	26.20
6650		8x8				10.20	9.60	19.80
6700		10x10				14.50	3.83	18.33
6750		20x10				13.05	3.46	16.51

Figure 14.9

4.6-100				Wood, Steel & Aluminum				
	MATERIAL	TYPE	DOORS	SPECIFICATION	OPENING	COST PER OPNG.		
						MAT.	INST.	TOTAL
6000	Aluminum	combination	storm & screen	hinged	3'-0" x 6'-8"	262	47.50	309.50
6050					3'-0" x 7'-0"	288	52	340
6100		overhead	rolling grill	manual oper.	12'-0" x 12'-0"	2,450	1,475	3,925
6150				motor oper.	12'-0" x 12'-0"	3,250	1,625	4,875
6200	Alum. & Fbrgls.	overhead	heavy duty	manual oper.	12'-0" x 12'-0"	1,250	430	1,680
6250				electric oper.	12'-0" x 12'-0"	1,650	590	2,240
6300	Alum. & glass	w/o transom	narrow stile	w/panic Hrdwre.	3'-0" x 7'-0"	1,150	615	1,765
6350				dbl. door, Hrdwre.	6'-0" x 7'-0"	2,025	1,025	3,050
6400			wide stile	hdwre.	3'-0" x 7'-0"	1,325	605	1,930
6450				dbl. door, Hdwre.	6'-0" x 7'-0"	2,600	1,200	3,800
6500			full vision	hdwre.	3'-0" x 7'-0"	1,700	960	2,660
6550				dbl. door, Hdwre.	6'-0" x 7'-0"	2,575	1,375	3,950
6600			non-standard	hdwre.	3'-0" x 7'-0"	1,475	605	2,080
6650				dbl. door, Hdwre.	6'-0" x 7'-0"	2,975	1,200	4,175
6700			bronze fin.	hdwre.	3'-0" x 7'-0"	1,500	605	2,105
6750				dbl. door, Hrdwre.	6'-0" x 7'-0"	3,025	1,200	4,225
6800			black fin.	hdwre.	3'-0" x 7'-0"	1,550	605	2,155
6850				dbl. door, Hdwre.	6'-0" x 7'-0"	3,125	1,200	4,325
6900		w/transom	narrow stile	hdwre.	3'-0" x 10'-0"	1,300	700	2,000
6950				dbl. door, Hdwre.	6'-0" x 10'-0"	2,200	1,225	3,425
7000			wide stile	hdwre.	3'-0" x 10'-0"	1,500	840	2,340
7050				dbl. door, Hdwre.	6'-0" x 10'-0"	2,500	1,450	3,950
7100			full vision	hdwre.	3'-0" x 10'-0"	1,675	930	2,605
7150				dbl. door, Hdwre.	6'-0" x 10'-0"	2,750	1,600	4,350
7200			non-standard	hdwre.	3'-0" x 10'-0"	1,525	650	2,175
7250				dbl. door, Hdwre.	6'-0" x 10'-0"	3,025	1,300	4,325
7300			bronze fin.	hdwre.	3'-0" x 10'-0"	1,550	650	2,200
7350				dbl. door, Hdwre.	6'-0" x 10'-0"	3,075	1,300	4,375
7400			black fin.	hdwre.	3'-0" x 10'-0"	1,600	650	2,250
7450				dbl. door, Hdwre.	6'-0" x 10'-0"	3,175	1,300	4,475
7500		revolving	stock design	minimum	6'-10" x 7'-0"	14,900	2,225	17,125
7550				average	6'-0" x 7'-0"	17,100	2,775	19,875
7600				maximum	6'-10" x 7'-0"	20,400	3,700	24,100
7650				min., automatic	6'-10" x 7'-0"	22,100	2,575	24,675
7700				avg., automatic	6'-10" x 7'-0"	24,300	3,125	27,425
7750				max., automatic	6'-10" x 7'-0"	27,600	4,050	31,650
7800		balanced	standard	economy	3'-0" x 7'-0"	3,600	925	4,525
7850				premium	3'-0" x 7'-0"	4,975	1,200	6,175
7900		mall front	sliding panels	alum. fin.	16'-0" x 9'-0"	2,200	470	2,670
7950					24'-0" x 9'-0"	3,200	875	4,075
8000				bronze fin.	16'-0" x 9'-0"	2,575	550	3,125
8050					24'-0" x 9'-0"	3,725	1,025	4,750
8100			fixed panels	alum. fin.	48'-0" x 9'-0"	5,950	680	6,630
8150				bronze fin.	48'-0" x 9'-0"	6,925	790	7,715
8200		sliding entrance	5' x 7' door	electric oper.	12'-0" x 7'-6"	5,600	875	6,475
8250		sliding patio	temp. glass	economy	6'-0" x 7'-0"	950	161	1,111
8300				economy	12'-0" x 7'-0"	1,925	215	2,140
8350				premium	6'-0" x 7'-0"	1,425	242	1,667
8400					12'-0" x 7'-0"	2,900	325	3,225

Figure 14.10

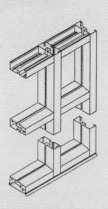

The table below lists costs per S.F of opening for framing with 1-3/4″ x 4-1/2″ clear anodized tubular aluminum framing. This is the type often used for 1/4″ plate glass flush glazing.

For bronze finish, add 18% to material cost. For black finish, add 27% to material cost. For stainless steel, add 75% to material cost. For monumental grade, add 50% to material cost. This tube framing is usually installed by a glazing contractor.

Note: The costs below do not include the glass.

System Components	QUANTITY	UNIT	COST/S.F. OPNG.		
			MAT.	INST.	TOTAL
SYSTEM 4.7-582-1250					
ALUM FLUSH TUBE, FOR 1/4″GLASS, 5′X20′OPENING, 3 INTER. HORIZONTALS					
Flush tube frame, alum mill fin, 1-3/4″x4″ open headr for 1/4″ glass	.450	L.F.	3.49	3.44	6.93
Flush tube frame alum mill fin 1-3/4″x4″ open sill for 1/4″ glass	.050	L.F.	.34	.37	.71
Flush tube frame alum mill fin,1-3/4″x4″closed back sill, 1/4″ glass	.150	L.F.	1.54	1.08	2.62
Aluminum structural shapes, 1″ to 10″ members, under 1 ton	.040	Lb.	.09	.14	.23
Joints for tube frame, 90° clip type	.100	Ea.	1.92		1.92
Caulking/sealants, polysulfide, 1 or 2 part,1/2x1/4″bead 154 lf/gal	.500	L.F.	.17	1.21	1.38
TOTAL			7.55	6.24	13.79

4.7-582	Tubular Aluminum Framing	COST/S.F. OPNG.		
		MAT.	INST.	TOTAL
1100	Alum flush tube frame,for 1/4″glass,1-3/4″x4″,5′x6′opng, no inter horizntls	8.40	7.55	15.95
1150	One intermediate horizontal	11.45	8.85	20.30
1200	Two intermediate horizontals	14.50	10.10	24.60
1250	5′ x 20′ opening, three intermediate horizontals	7.55	6.25	13.80
1400	1-3/4″ x 4-1/2″, 5′ x 6′ opening, no intermediate horizontals	9.55	7.55	17.10
1450	One intermediate horizontal	12.75	8.85	21.60
1500	Two intermediate horizontals	15.90	10.10	26
1550	5′ x 20′ opening, three intermediate horizontals	8.50	6.25	14.75
1700	For insulating glass, 2″x4-1/2″, 5′x6′ opening, no intermediate horizontals	10.75	7.90	18.65
1750	One intermediate horizontal	14.05	9.30	23.35
1800	Two intermediate horizontals	17.35	10.60	27.95
1850	5′ x 20′ opening, three intermediate horizontals	9.40	6.60	16
2000	Thermal break frame, 2-1/4″x4-1/2″, 5′x6′opng, no intermediate horizontals	11.50	8	19.50
2050	One intermediate horizontal	15.45	9.65	25.10
2100	Two intermediate horizontals	19.35	11.25	30.60
2150	5′ x 20′ opening, three intermediate horizontals	10.50	6.90	17.40

Figure 14.11

The table below lists costs of curtain wall and spandrel panels per S.F. Costs do not include structural framing used to hang the panels from.

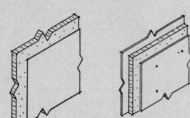

Spandrel Glass Panel   **Sandwich Panel**

4.7-584	Curtain Wall Panels	COST PER S.F.		
		MAT.	INST.	TOTAL
1000	Glazing panel, insulating, 1/2" thick, 2 lites 1/8" float, clear	5.50	6.45	11.95
1100	Tinted	6.10	6.45	12.55
1200	5/8" thick units, 2 lites 3/16" float, clear	6.35	6.80	13.15
1400	1" thick units, 2 lites, 1/4" float, clear	7.50	8.15	15.65
1700	Light and heat reflective glass, tinted	15	7.20	22.20
2000	Plate glass, 1/4" thick, clear	2.92	5.10	8.02
2050	Tempered	4.06	5.10	9.16
2100	Tinted	4	5.10	9.10
2200	3/8" thick, clear	6.20	8.15	14.35
2250	Tempered	9.15	8.15	17.30
2300	Tinted	7.80	8.15	15.95
2400	1/2" thick, clear	11.15	11.15	22.30
2450	Tempered	14.35	11.15	25.50
2500	Tinted	13.35	11.15	24.50
2600	3/4" thick, clear	17.35	17.50	34.85
2650	Tempered	20	17.50	37.50
3000	Spandrel glass, panels, 1/4" plate glass insul w/fiberglass, 1" thick	8.45	5.10	13.55
3100	2" thick	9.85	5.10	14.95
3200	Galvanized steel backing, add	2.92		2.92
3300	3/8" plate glass, 1" thick	14.15	5.10	19.25
3400	2" thick	15.55	5.10	20.65
4000	Polycarbonate, masked, clear or colored, 1/8" thick	4.62	3.60	8.22
4100	3/16" thick	6.05	3.71	9.76
4200	1/4" thick	7.65	3.95	11.60
4300	3/8" thick	11.45	4.09	15.54
5000	Sandwich panel, 1-1/2" fiberglass, 16 Ga. aluminum exterior	6.05	6.45	12.50
5100	16 Ga. porcelainized aluminum exterior	6.85	6.45	13.30
5200	18 Ga. galvanized steel exterior	5.30	6.45	11.75
5300	20 Ga. protected metal exterior	5.80	6.45	12.25
5400	20 Ga stainless steel exterior	7.65	6.45	14.10
5500	22 Ga. galv., both sides 2" insulation, enamel exterior	7.20	3.50	10.70
5600	Polyvinylidene floride exterior finish	7.60	3.50	11.10
5700	26 Ga., galv. both sides, 1" insulation, colored 1 side	3.71	3.32	7.03
5800	Colored 2 sides	4.85	3.32	8.17

Figure 14.12

6.6-100	Tile & Covering	COST PER S.F.		
		MAT.	INST.	TOTAL
1340	Cork tile, minimum	2.53	.95	3.48
1360	Maximum	9.60	.95	10.55
1380	Polyethylene, in rolls, minimum	2.21	1.08	3.29
1400	Maximum	4.42	1.08	5.50
1420	Polyurethane, thermoset, minimum	3.50	2.98	6.48
1440	Maximum	4.17	5.95	10.12
1460	Rubber, sheet goods, minimum	3.16	2.48	5.64
1480	Maximum	5.15	3.31	8.46
1500	Tile, minimum	3.16	.74	3.90
1520	Maximum	6.15	1.08	7.23
1540	Synthetic turf, minimum	2.72	1.42	4.14
1560	Maximum	6.95	1.57	8.52
1580	Vinyl, composition tile, minimum	.71	.60	1.31
1600	Maximum	1.76	.60	2.36
1620	Tile, minimum	1.61	.60	2.21
1640	Maximum	2.52	.60	3.12
1660	Sheet goods, minimum	1.32	1.19	2.51
1680	Maximum	4.95	1.49	6.44
1720	Tile, ceramic natural clay	3.64	2.95	6.59
1730	Marble, synthetic 12"x12"x5/8"	7.30	9	16.30
1740	Porcelain type, minimum	3.94	2.95	6.89
1760	Maximum	4.16	2.84	7
1800	Quarry tile, mud set, minimum	3.19	3.85	7.04
1820	Maximum	4.70	4.91	9.61
1840	Thin set, deduct		.77	.77
1850				
1860	Terrazzo precast, minimum	5.25	3.97	9.22
1880	Maximum	5.90	3.97	9.87
1900	Non-slip, minimum	11.10	19.90	31
1920	Maximum	16.60	27.50	44.10
1960	Wood, block, end grain factory type, creosoted, 2" thick	2.79	1.09	3.88
1980	2-1/2" thick	4.15	2.58	6.73
2000	3" thick	4.15	2.58	6.73
2020	Natural finish, 2" thick	4.51	2.58	7.09
2040	Fir, vertical grain, 1"x4", no finish, minimum	2.43	1.26	3.69
2060	Maximum	2.58	1.26	3.84
2080	Prefinished white oak, prime grade, 2-1/4" wide	6.25	1.89	8.14
2100	3-1/4" wide	7.75	1.74	9.49
2120	Maple strip, sanded and finished, minimum	3.51	2.88	6.39
2140	Maximum	4.06	2.88	6.94
2160	Oak strip, sanded and finished, minimum	4.50	2.88	7.38
2180	Maximum	4.17	2.88	7.05
2200	Parquetry, sanded and finished, minimum	2.30	3	5.30
2220	Maximum	6.60	4.21	10.81
2260	Add for sleepers on concrete, treated, 24" O.C., 1"x2"	1.58	1.70	3.28
2280	1"x3"	1.58	1.34	2.92
2300	2"x4"	.85	.67	1.52
2320	2"x6"	.85	.52	1.37
2340	Underlayment, plywood, 3/8" thick	.48	.44	.92
2350	1/2" thick	.59	.45	1.04
2360	5/8" thick	.71	.47	1.18
2370	3/4" thick	.87	.52	1.39
2380	Particle board, 3/8" thick	.34	.44	.78
2390	1/2" thick	.36	.45	.81
2400	5/8" thick	.39	.47	.86
2410	3/4" thick	.45	.52	.97
2420	Hardboard, 4' x 4', .215" thick	.50	.44	.94

Figure 14.13

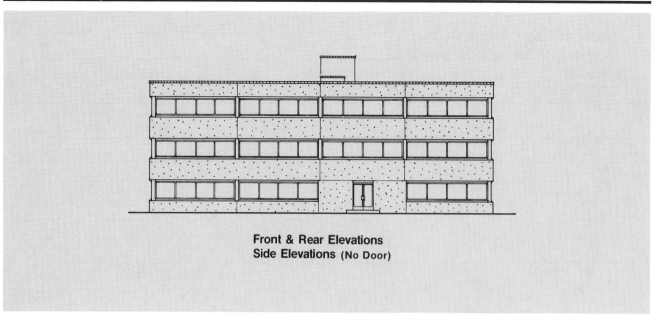

Front & Rear Elevations
Side Elevations (No Door)

Figure 14.14

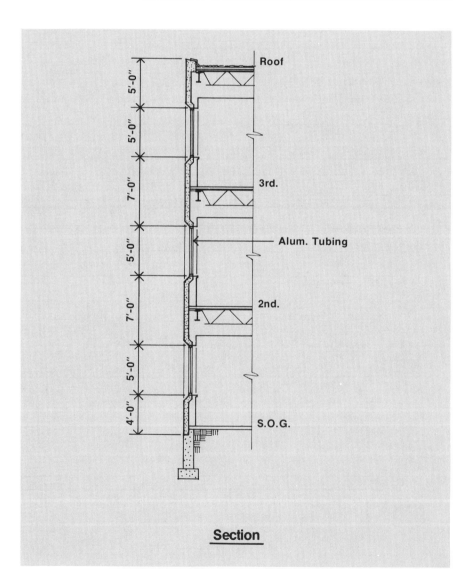

Section

Figure 14.15

Assembly Number	Description	Qty	Unit	Total Cost		Cost per S. F.
				Unit	Total	
1.0	**Foundations**					
1.1-120-	Column Footings 6 KSF					
7700	Interior 190K (200K)	9	Ea	464	4176	
7500	Exterior 114K (125K)	12	Ea	287	3444	
7410	Corner 86K (100K)	4	Ea	232.50	930	
1.1-140-	Strip Footings 12" x 24"					
2700	400 LF - (12x5') - (4x4.5')	322	LF	23.50	7567	
1.1-210-	Foundation Wall					
1560	4' x 12" Direct Chute	400	LF	43.80	17520	
3.1-114	Piers : Assume 20" sq.					
9924	16 x 3.5'	56	LF	76.50	4284	
					$37,921	$1.26
2.0	**Substructure**					
2.1-200-	4" non-industrial slab-on-grade					
2240	Reinforced	10000	SF	2.86	28600	
					$28,600	$0.95
3.0	**Superstructure**					
3.7-420-	Roof					
3300	25' Sq. 40 lb.	10000	SF	4.01	40100	
3.5-460-	Floor					
5100	2nd & 3rd 25' Sq. 75 lb.	20000	SF	8.61	172200	
3.1-130-	Columns					
4600	190K (200K) : 25x27.33'	684	LF	37.85	25889	
1600	38K (50K) : 25x9'	175	LF	17.60	3080	
3.1-190-	Column Fireproofing					
3450	8" Average, Gyp. Bd.	859	LF	14.77	12687	
3.9-100-	Stairs					
0760	Steel pan	5	FL	5425	27125	
					$281,082	$9.37
4.0	**Exterior Closure**					
4.1-211-	Insulated CMU at Penthouse					
1360	6" CMU:(46'x7.35')+(39'x4")	493	SF	6.24	3076	
4.1-140-	Wall Panels					
5750	Precast concrete (175 SF ea.)	9338	SF	14.11	131759	
4.6-100-	Doors					
6350	Aluminum and Glass	2	PR	3050	6100	
4.7-582-	Windows					
2000	Alum. Flush Tube w/ Thermal Break	5750	SF	19.50	112125	
4.7-584-	Glass					
1000	Insulating Type	5750	SF	11.95	68713	
6.6-100-	Window Sills					
1880	Precast Terrazzo	804	LF	9.87	7935	
					$329,708	$10.99

Figure 14.16

Chapter 15

Roofing

Choosing the roofing system is much like choosing the exterior closure system. Many times the owner or client has prescriptive requirements that must be met. These requirements may include: the specific system to be used; required service life; a standard specification; and maintenance agreements with the manufacturer and/or the roofing contractor. The roofing systems in *Means Assemblies Cost Data* are divided into four groups:

- Built-up
- Single-ply
- Metal
- Shingles & tile

Simply select the system desired and multiply the system cost times the number of square feet needed. With the tables provided, it's easy to do a cost comparison between systems.

Example: Three-Story Office Building

The office building (Figure 15.1) will have the following included in the roofing system:

- Built-up roof
- Roof deck insulation
- Aluminum flashing

Costs:
Built-up, 3-ply with gravel on non-nailable deck/insulation
  (Figure 15.2 #1400)   10,000 S.F. @ $1.38/S.F.   = $13,800
Roof Deck Insulation, 3" Urethane
  (Figure 15.3 #2000)   10,000 S.F. @ $1.23/S.F.   =   12,300
Aluminum Flashing
  (Figure 15.4 #0350)   756 S.F. @ $2.22/S.F.   =   1,678
Aluminum Gravel Stop
  (Figure 15.5 #5200)   177 L.F. @ $6.52/L.F.   =   1,154

Total Cost $28,932
Cost per Square Foot   = $0.96/S.F.

Figure 15.6 summarizes the costs for this division.

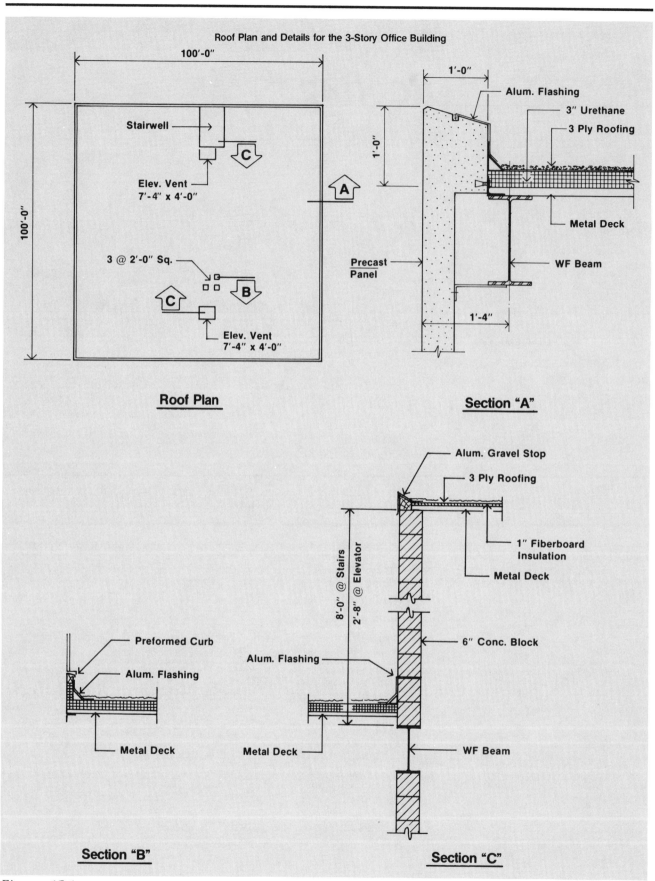

Roof Plan and Details for the 3-Story Office Building

100'-0"

Stairwell

C

Elev. Vent
7'-4" x 4'-0"

100'-0"

3 @ 2'-0" Sq.

C

B

Elev. Vent
7'-4" x 4'-0"

A

Roof Plan

1'-0"

1'-0"

Alum. Flashing

3" Urethane

3 Ply Roofing

Metal Deck

Precast
Panel

WF Beam

1'-4"

Section "A"

Alum. Gravel Stop

3 Ply Roofing

1" Fiberboard
Insulation

Metal Deck

8'-0" @ Stairs
2'-8" @ Elevator

6" Conc. Block

Alum. Flashing

WF Beam

Preformed Curb

Alum. Flashing

Metal Deck

Alum. Flashing

Metal Deck

Section "B"

Section "C"

Figure 15.1

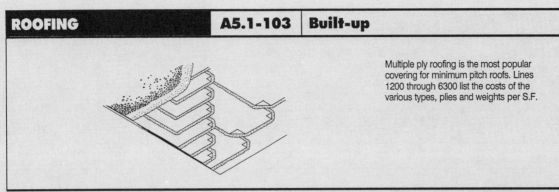

Multiple ply roofing is the most popular covering for minimum pitch roofs. Lines 1200 through 6300 list the costs of the various types, plies and weights per S.F.

System Components	QUANTITY	UNIT	COST PER S.F.		
			MAT.	INST.	TOTAL
SYSTEM 5.1-103-2500					
ASPHALT FLOOD COAT, W/GRAVEL, 4 PLY ORGANIC FELT					
Organic #30 base felt	1.000	S.F.	.05	.05	.10
Organic #15 felt, 3 plies	3.000	S.F.	.08	.16	.24
Asphalt mopping of felts	4.000	S.F.	.12	.45	.57
Asphalt flood coat	1.000	S.F.	.07	.39	.46
Gravel aggregate, washed river stone	4.000	Lb.	.04	.09	.13
TOTAL			.36	1.14	1.50

5.1-103	Built-Up	COST PER S.F.		
		MAT.	INST.	TOTAL
1200	Asphalt flood coat w/gravel; not incl. insul, flash., nailers			
1300				
1400	Asphalt base sheets & 3 plies #15 asphalt felt, mopped	.34	1.04	1.38
1500	On nailable deck	.39	1.08	1.47
1600	4 plies #15 asphalt felt, mopped	.49	1.14	1.63
1700	On nailable deck	.45	1.19	1.64
1800	Coated glass base sheet, 2 plies glass (type IV), mopped	.39	1.04	1.43
1900	For 3 plies	.47	1.14	1.61
2000	On nailable deck	.45	1.19	1.64
2300	4 plies glass fiber felt (type IV), mopped	.57	1.14	1.71
2400	On nailable deck	.52	1.19	1.71
2500	Organic base sheet & 3 plies #15 organic felt, mopped	.36	1.14	1.50
2600	On nailable deck	.39	1.19	1.58
2700	4 plies #15 organic felt, mopped	.47	1.04	1.51
2750				
2800	Asphalt flood coat, smooth surface			
2850				
2900	Asphalt base sheet & 3 plies #15 asphalt felt, mopped	.36	.95	1.31
3000	On nailable deck	.34	.99	1.33
3100	Coated glass fiber base sheet & 2 plies glass fiber felt, mopped	.34	.91	1.25
3200	On nailable deck	.33	.95	1.28
3300	For 3 plies, mopped	.42	.99	1.41
3400	On nailable deck	.40	1.04	1.44
3700	4 plies glass fiber felt (type IV), mopped	.50	.99	1.49
3800	On nailable deck	.48	1.04	1.52
3900	Organic base sheet & 3 plies #15 organic felt, mopped	.36	.95	1.31
4000	On nailable decks	.34	.99	1.33
4100	4 plies #15 organic felt, mopped	.42	1.04	1.46
4200	Coal tar pitch with gravel surfacing			
4300	4 plies #15 tarred felt, mopped	.92	1.08	2
4400	3 plies glass fiber felt (type IV), mopped	.76	1.19	1.95
4500	Coated glass fiber base sheets 2 plies glass fiber felt, mopped	.78	1.19	1.97
4600	On nailable decks	.68	1.26	1.94
4800	3 plies glass fiber felt (type IV), mopped	1.05	1.08	2.13

Figure 15.2

5.7-101	Roof Deck Rigid Insulation	COST PER S.F.		
		MAT.	INST.	TOTAL
0100	Fiberboard low density, 1/2" thick, R1.39			
0150	1" thick R2.78	.34	.39	.73
0300	1 1/2" thick R4.17	.51	.39	.90
0350	2" thick R5.56	.66	.39	1.05
0370	Fiberboard high density, 1/2" thick R1.3	.19	.32	.51
0380	1" thick R2.5	.35	.39	.74
0390	1 1/2" thick R3.8	.61	.39	1
0410	Fiberglass, 3/4" thick R2.78	.42	.32	.74
0450	15/16" thick R3.70	.54	.32	.86
0500	1-1/16" thick R4.17	.68	.32	1
0550	1-5/16" thick R5.26	.95	.32	1.27
0600	2-1/16" thick R8.33	1.01	.39	1.40
0650	2 7/16" thick R10	1.14	.39	1.53
1000	Foamglass, 1 1/2" thick R4.55	1.41	.39	1.80
1100	3" thick R9.09	2.80	.45	3.25
1200	Tapered for drainage	.94	.53	1.47
1260	Perlite, 1/2" thick R1.32	.29	.30	.59
1300	3/4" thick R2.08	.33	.39	.72
1350	1" thick R2.78	.39	.39	.78
1400	1 1/2" thick R4.17	.52	.39	.91
1450	2" thick R5.56	.66	.45	1.11
1510	Polyisocyanurate 2#/CF density, 1" thick R7.14	.37	.23	.60
1550	1 1/2" thick R10.87	.42	.25	.67
1600	2" thick R14.29	.50	.29	.79
1650	2 1/2" thick R16.67	.62	.30	.92
1700	3" thick R21.74	.73	.32	1.05
1750	3 1/2" thick R25	.90	.32	1.22
1800	Tapered for drainage	.44	.23	.67
1810	Expanded polystyrene, 1#/CF density, 3/4" thick R2.89	.19	.21	.40
1820	2" thick R7.69	.33	.25	.58
1825	Extruded Polystyrene			
1830	15 PSI compressive strength, 1" thick R5	.32	.21	.53
1835	2" thick R10	.57	.25	.82
1840	3" thick R15	.80	.32	1.12
1900	25 PSI compressive strength, 1" thick R5	.34	.21	.55
1950	2" thick R10	.63	.25	.88
2000	3" thick R15	.91	.32	1.23
2050	4" thick R20	1.20	.32	1.52
2150	Tapered for drainage	.42	.21	.63
2550	40 PSI compressive strength, 1" thick R5	.40	.21	.61
2600	2" thick R10	.75	.25	1
2650	3" thick R15	1.12	.32	1.44
2700	4" thick R20	1.50	.32	1.82
2750	Tapered for drainage	.53	.23	.76
2810	60 PSI compressive strength, 1" thick R5	.46	.22	.68
2850	2" thick R10	.86	.26	1.12
2900	Tapered for drainage	.64	.23	.87
2910	115 PSI compressive strength, 1" thick R5	.99	.23	1.22
2950	2" thick R10	1.96	.27	2.23
3000	Tapered for drainage	1.07	.23	1.30
4000	Composites with 1-1/2" polyisocyanurate			
4010	1" fiberboard	.77	.39	1.16
4020	1" perlite	.79	.37	1.16
4030	7/16" oriented strand board	.90	.39	1.29

Figure 15.3

5.1-620 Flashing

	MATERIAL	BACKING	SIDES	SPECIFICATION	QUANTITY	COST PER S.F.		
						MAT.	INST.	TOTAL
0040	Aluminum	none		.019″		.85	2.50	3.35
0050				.032″		1.01	2.50	3.51
0100				.040″		1.80	2.50	4.30
0150				.050″		2.10	2.50	4.60
0300		fabric	2	.004″		.96	1.10	2.06
0350				.016″		1.12	1.10	2.22
0400		mastic		.004″		.91	1.10	2.01
0450				.005″		1.08	1.10	2.18
0500				.016″		1.30	1.10	2.40
0510								
0700	Copper	none		16 oz.	<500 lbs.	3.58	3.15	6.73
0710					>2000 lbs.	3.58	2.34	5.92
0750				20 oz.	<500 lbs.	4.47	3.30	7.77
0760					>2000 lbs.	4.47	2.50	6.97
0800				24 oz.	<500 lbs.	5.35	3.46	8.81
0810					>2000 lbs.	5.35	2.69	8.04
0850				32 oz.	<500 lbs.	7.15	3.63	10.78
0860					>2000 lbs.	7.15	2.79	9.94
1000		paper backed	1	2 oz.		1.02	1.10	2.12
1100				3 oz.		1.34	1.10	2.44
1200			2	2 oz.		1.06	1.10	2.16
1300				5 oz.		2.15	1.10	3.25
2000	Copper lead	fabric	1	2 oz.		1.43	1.10	2.53
2100	Coated			5 oz.		2.11	1.10	3.21
2200		mastic	2	2 oz.		1.11	1.10	2.21
2300				5 oz.		1.70	1.10	2.80
2400		paper	1	2 oz.		1.02	1.10	2.12
2500				3 oz.		1.34	1.10	2.44
2600			2	2 oz.		1.06	1.10	2.16
2700				5 oz.		2.15	1.10	3.25
3000	Lead	none		2.5 lb.	to 12″ wide	2.98	2.34	5.32
3100					over 12″	2.98	2.34	5.32
3500	PVC black	none		.010″		.12	1.11	1.23
3600				.020″		.20	1.11	1.31
3700				.030″		.31	1.11	1.42
3800				.056″		.74	1.11	1.85
4000	Rubber butyl	none		1/32″		.71	1.11	1.82
4100				1/16″		1.05	1.11	2.16
4200	Neoprene			1/16″		1.60	1.11	2.71
4300				1/8″		3.25	1.11	4.36
4500	Stainless steel	none		.015″	<500 lbs.	3.29	3.15	6.44
4600	Copper clad				>2000 lbs.	3.25	2.34	5.59
4700				.018″	<500 lbs.	4.46	3.63	8.09
4800					>2000 lbs.	3.19	2.50	5.69
5000	Plain			32 Ga.		2.37	2.34	4.71
5100				28 ga.		2.99	2.34	5.33
5200				26 ga.		3.64	2.34	5.98
5300				24 ga.		4.49	2.34	6.83
5400	Terne coated			28 ga.		4.07	2.34	6.41
5500				26 ga.		4.61	2.34	6.95

Figure 15.4

195

| 5.8-500 | | | | Downspouts | | | | |

	MATERIALS	SECTION	SIZE	FINISH	THICKNESS	COST PER V.L.F.		
						MAT.	INST.	TOTAL
0100	Aluminum	rectangular	2"x3"	embossed mill	.020"	.77	1.91	2.68
0150				enameled	.020"	.86	1.91	2.77
0200				enameled	.024"	1.20	2.02	3.22
0250			3"x4"	enameled	.024"	1.91	2.59	4.50
0300		round corrugated	3"	enameled	.020"	.99	1.91	2.90
0350			4"	enameled	.025"	1.32	2.59	3.91
0500	Copper	rectangular corr.	2"x3"	mill	16 Oz.	4.83	1.91	6.74
0550				lead coated	16 Oz.	7.40	1.91	9.31
0600		smooth		mill	16 Oz.	5.20	1.91	7.11
0650				lead coated	16 Oz.	6.35	1.91	8.26
0700		rectangular corr.	3"x4"	mill	16 Oz.	6.35	2.50	8.85
0750				lead coated	16 Oz.	7	2.17	9.17
0800		smooth		mill	16 Oz.	6.45	2.50	8.95
0850				lead coated	16 Oz.	8.10	2.50	10.60
0900		round corrugated	2"	mill	16 Oz.	4.06	1.97	6.03
0950		smooth		lead coated	16 Oz.	4.95	1.91	6.86
1000		corrugated	3"	mill	16 Oz.	4.63	1.98	6.61
1050		smooth		lead coated	16 Oz.	5.25	1.91	7.16
1100		corrugated	4"	mill	16 Oz.	6.05	2.60	8.65
1150		smooth		lead coated	16 Oz.	6.80	2.50	9.30
1200		corrugated	5"	mill	16 Oz.	7.35	2.92	10.27
1250				lead coated	16 Oz.	9.55	2.79	12.34
1300	Steel	rectangular corr.	2"x3"	galvanized	28 Ga.	.58	1.91	2.49
1350				epoxy coated	24 Ga.	1.06	1.91	2.97
1400		smooth		galvanized	28 Ga.	.58	1.91	2.49
1450		rectangular corr.	3"x4"	galvanized	28 Ga.	1.63	2.50	4.13
1500				epoxy coated	24 Ga.	1.75	2.50	4.25
1550		smooth		galvanized	28 Ga.	1.55	2.50	4.05
1600		round corrugated	2"	galvanized	28 Ga.	.62	1.91	2.53
1650			3"	galvanized	28 Ga.	.62	1.91	2.53
1700			4"	galvanized	28 Ga.	.80	2.50	3.30
1750			5"	galvanized	28 Ga.	.92	2.79	3.71
1800				galvanized	26 Ga.	1.19	2.79	3.98
1850			6"	galvanized	28 Ga.	1.39	3.46	4.85
1900				galvanized	26 Ga.	1.54	3.46	5
1950								
2000	Steel pipe	round	4"	black	X.H.	10.75	18.15	28.90
2050			6"	black	X.H.	17.60	20	37.60
2500	Stainless steel	rectangular	2"x3"	mill		14.10	1.91	16.01
2550	Tubing sch.5		3"x4"	mill		17.95	2.50	20.45
2600			4"x5"	mill		37	2.69	39.69
2650		round	3"	mill		14.10	1.91	16.01
2700			4"	mill		17.95	2.50	20.45
2750			5"	mill		37	2.69	39.69

| 5.8-500 | | | | Gravel Stop | | | | |

	MATERIALS	SECTION	SIZE	FINISH	THICKNESS	COST PER L.F.		
						MAT.	INST.	TOTAL
5100	Aluminum	extruded	4"	mill	.050"	2.26	2.50	4.76
5200			4"	duranodic	.050"	4.02	2.50	6.52
5300			8"	mill	.050"	2.99	2.90	5.89
5400			8"	duranodic	.050"	5.45	2.90	8.35
5500			12"-2 pc.	mill	.050"	3.67	3.63	7.30
6000			12"-2 pc.	duranodic	.050"	6.85	3.63	10.48
6100	Stainless	formed	6"	mill	24 Ga.	7.80	2.69	10.49
6200			12"	mill	24 Ga.	14.95	3.63	18.58

Figure 15.5

Assembly Number	Description	Qty	Unit	Total Cost		Cost per S. F.
				Unit	Total	
5.0	**Roofing**					
	Built-Up					
5.1-103-1400	3 ply w/ gravel, non-nailable, insulated	10000	SF	1.38	13800	
	Insulation					
5.7-101-2000	3" Urethane	10000	SF	1.23	12300	
	Flashing					
5.1-620-0350	Aluminum	756	SF	2.22	1678	
	Gravel Stop					
5.8-520-5200	Aluminum, 4"	117	LF	6.52	763	
					$28,541	$0.95
6.0	**Interior Construction**					
7.0	**Conveying**					

Figure 15.6

Chapter 16

Interior Construction

The Interior Construction division includes many different types of partitions, doors, ceilings, floor treatments, and interior finishes. Many of today's commercial buildings seem to have miles of partitions, unlimited doors, vast square feet of ceilings, and a great variety of floor finishes and wall treatments. Assemblies estimating can dramatically save time in the estimating process.

Fixed Partitions

The types of fixed partitions most commonly found in construction today are:

- Masonry
- Drywall with wood or metal studs
- Plaster on drywall, masonry, or metal lath

The length of interior partitions may be determined from a simple sketch, from the detailed drawings, or by approximation from the "Partition/Door Density" table in Figure 16.1. The table shows the average length of interior partitions based on the particular building square footage for different building types. It also provides the average number of interior doors, based on the building square footage, necessary to service the partitions. Multiply the partition linear footage by the required partition height to find the total square footage of partitions.

Example One: Three-Story Apartment Building and One-Story Department Store

Following is a comparison of a three-story apartment building and a one-story department store in terms of the number of doors and linear feet of partitions required.

Apartment	**3 Stories**		**30,000 S.F.**
Partitions	9 S.F./L.F.	=	3,333 L.F. of partitions
Doors	90 S.F./door	=	333 doors
Department Store	**1 Story**		**50,000 S.F.**
Partitions	60 S.F./L.F.	=	833 L.F. of partitions
Doors	600 S.F./door	=	83 doors

Of course, the quantities found are only approximations and should be used only to point the estimator in a direction. If sketches or drawings are available, you should use them.

Partition/Door Density

Building Type		Stories	Partition/Density	Doors	Description of Partition
Apartments		1 story	9 SF/LF	90 SF/door	Plaster, wood doors & trim
		2 story	8 SF/LF	80 SF/door	Drywall, wood studs, wood doors & trim
		3 story	9 SF/LF	90 SF/door	Plaster, wood studs, wood doors & trim
		5 story	9 SF/LF	90 SF/door	Plaster, wood studs, wood doors & trim
		6-15 story	8 SF/LF	80 SF/door	Drywall, wood studs, wood doors & trim
Bakery		1 story	50 SF/LF	500 SF/door	Conc. block, paint, door & drywall, wood studs
		2 story	50 SF/LF	500 SF/door	Conc. block, paint, door & drywall, wood studs
Bank		1 story	20 SF/LF	200 SF/door	Plaster, wood studs, wood doors & trim
		2-4 story	15 SF/LF	150 SF/door	Plaster, wood studs, wood doors & trim
Bottling Plant		1 story	50 SF/LF	500 SF/door	Conc. block, drywall, wood studs, wood trim
Bowling Alley		1 story	50 SF/LF	500 SF/door	Conc. block, wood & metal doors, wood trim
Bus Terminal		1 story	15 SF/LF	150 SF/door	Conc. block, ceramic tile, wood trim
Cannery		1 story	100 SF/LF	1000 SF/door	Drywall on metal studs
Car Wash		1 story	18 SF/LF	180 SF/door	Concrete block, painted & hollow metal door
Dairy Plant		1 story	30 SF/LF	300 SF/door	Concrete block, glazed tile, insulated cooler doors
Department Store		1 story	60 SF/LF	600 SF/door	Drywall, wood studs, wood doors & trim
		2-5 story	60 SF/LF	600 SF/door	30% concrete block, 70% drywall, wood studs
Dormitory		2 story	9 SF/LF	90 SF/door	Plaster, concrete block, wood doors & trim
		3-5 story	9 SF/LF	90 SF/door	Plaster, concrete block, wood doors & trim
		6-15 story	9 SF/LF	90 SF/door	Plaster, concrete block, wood doors & trim
Funeral Home		1 story	15 SF/LF	150 SF/door	Plaster on concrete block & wood studs, paneling
		2 story	14 SF/LF	140 SF/door	Plaster, wood studs, paneling & wood doors
Garage Sales & Service		1 story	30 SF/LF	300 SF/door	50% conc. block, 50% drywall, wood studs
Hotel		3-8 story	9 SF/LF	90 SF/door	Plaster, conc. block, wood doors & trim
		9-15 story	9 SF/LF	90 SF/door	Plaster, conc. block, wood doors & trim
Laundromat		1 story	25 SF/LF	250 SF/door	Drywall, wood studs, wood doors & trim
Medical Clinic		1 story	6 SF/LF	60 SF/door	Drywall, wood studs, wood doors & trim
		2-4 story	6 SF/LF	60 SF/door	Drywall, wood studs, wood doors & trim
Motel		1 story	7 SF/LF	70 SF/door	Drywall, wood studs, wood doors & trim
		2-3 story	7 SF/LF	70 SF/door	Concrete block, drywall on wood studs, wood paneling
Movie Theater	200-600 seats	1 story	18 SF/LF	180 SF/door	Concrete block, wood, metal, vinyl trim
	601-1400 seats		20 SF/LF	200 SF/door	Concrete block, wood, metal, vinyl trim
	1401-2200 seats		25 SF/LF	250 SF/door	Concrete block, wood, metal, vinyl trim
Nursing Home		1 story	8 SF/LF	80 SF/door	Drywall, wood studs, wood doors & trim
		2-4 story	8 SF/LF	80 SF/door	Drywall, wood studs, wood doors & trim
Office		1 story	20 SF/LF	200-500 SF/door	30% concrete block, 70% drywall on wood studs
		2 story	20 SF/LF	200-500 SF/door	30% concrete block, 70% drywall on wood studs
		3-5 story	20 SF/LF	200-500 SF/door	30% concrete block, 70% movable partitions
		6-10 story	20 SF/LF	200-500 SF/door	30% concrete block, 70% movable partitions
		11-20 story	20 SF/LF	200-500 SF/door	30% concrete block, 70% movable partitions
Parking Ramp (Open)		2-8 story	60 SF/LF	600 SF/door	Stair and elevator enclosures only
Parking garage		2-8 story	60 SF/LF	600 SF/door	Stair and elevator enclosures only
Pre-Engineered	Steel	1 story	0		
	Store	1 story	60 SF/LF	600 SF/door	Drywall on wood studs, wood doors & trim
	Office	1 story	15 SF/LF	150 SF/door	Concrete block, movable wood partitions
	Shop	1 story	15 SF/LF	150 SF/door	Movable wood partitions
	Warehouse	1 story	0		
Radio & TV Broadcasting		1 story	25 SF/LF	250 SF/door	Concrete block, metal, and wood doors
& TV Transmitter		1 story	40 SF/LF	400 SF/door	Concrete block, metal and wood doors
Self Service Restaurant		1 story	15 SF/LF	150 SF/door	Concrete block, wood and aluminum trim
Cafe & Drive-in Restaurant		1 story	18 SF/LF	180 SF/door	Drywall, wood studs, ceramic & plastic trim
Restaurant with seating		1 story	25 SF/LF	250 SF/door	Concrete block, paneling, wood studs & trim
Supper Club		1 story	25 SF/LF	250 SF/door	Concrete block, paneling, wood studs & trim
Bar or Lounge		1 story	24 SF/LF	240 SF/door	Plaster or gypsum lath, wooded studs
Retail Store or Shop		1 story	60 SF/LF	600 SF/door	Drywall wood studs, wood doors & trim
Service Station	Masonry	1 story	15 SF/LF	150 SF/door	Concrete block, paint, door & drywall, wood studs
	Metal panel	1 story	15 SF/LF	150 SF/door	Concrete block, paint, door & drywall, wood studs
	Frame	1 story	15 SF/LF	150 SF/door	Drywall, wood studs, wood doors & trim
Shopping Center	(strip)	1 story	30 SF/LF	300 SF/door	Drywall, wood studs, wood doors & trim
	(group)	1 story	40 SF/LF	400 SF/door	50% concrete block, 50% drywall, wood studs
		2 story	40 SF/LF	400 SF/door	50% concrete block, 50% drywall, wood studs
Small Food Store		1 story	30 SF/LF	300 SF/door	Concrete block drywall, wood studs, wood trim
Store/Apt. above	Masonry	2 story	10 SF/LF	100 SF/door	Plaster, wood studs, wood doors & trim
	Frame	2 story	10 SF/LF	100 SF/door	Plaster, wood studs, wood doors & trim
	Frame	3 story	10 SF/LF	100 SF/door	Plaster, wood studs, wood doors & trim
Supermarkets		1 story	40 SF/LF	400 SF/door	Concrete block, paint, drywall & porcelain panel
Truck Terminal		1 story	0		
Warehouse		1 story	0		

Figure 16.1

Partition Components

The materials used to construct partitions vary from masonry brick and block, structural clay facing tile, gypsum block and glass block, to wood or steel studs with plaster or gypsum board and possibly insulation or acoustical batts. The combinations of these materials are almost limitless and, therefore, require some attention.

Most buildings have a combination of different partition materials, and some partitions are used as structural bearing walls or even fire walls. Exercise caution when choosing partitions because the costs and weights vary considerably.

Example Two: Partition Weight

This example illustrates the cost and weight differences that can be found for various partitions. The partitions shown are representative types and are not necessarily included in the estimate for the Three-Story Office Building.

Partition Component	Cost per Square Foot	Partition Weight
6" Concrete Masonry Unit (CMU)	$ 5.14/S.F.	30 psf to 42 psf
6" CMU with 5/8" gypsum plaster both sides	$ 7.03/S.F.	40 psf to 52 psf
Structural clay facing tile glazed both sides (6")	$20.55/S.F.	47 psf
2" x 4" studs @ 16" O.C. 5/8" fire-rated gypsum board both sides	$ 3.01/S.F.	7 psf
2" x 4" studs @ 16" O.C. Gypsum plaster on gypsum lath both sides	$ 6.76/S.F.	14 psf
3-5/8", 18 gauge steel studs & track @ 24" O.C. 5/8" gypsum board both sides	$ 2.83/S.F.	7 psf
6", 16 gauge steel studs & track @ 16" O.C. 3 coats gypsum plaster on metal lath	$ 6.23/S.F.	18 psf

Toilet Partitions

Quality and ease of installation are the biggest factors used to select and price toilet partitions. Floor-mounted painted metal and plastic laminate units are the most economical, whereas ceiling-hung units of the same materials are more costly to install. Wall-hung units of the same materials cost more to fabricate, thus the material price is higher, but the installation cost is about the same as for floor-mounted units. For buildings requiring handicapped accessibility, be certain not to forget the additional partition accessories and the blocking necessary in the walls to receive them.

Wall and Floor Finishes

In assemblies estimating, it will be necessary to determine a percentage of total area if different wall and floor finishes are used in the building. Don't forget to include the interior side of exterior walls and the ceiling finish, if required. For simplicity when computing the wall area to be finished, include only the wall area up to the finished ceiling.

The table in Figure 16.2 is useful to determine interior wall finish area when compared to floor area. Ratio numbers are based on a 10' floor-to-ceiling height.

Example Three: 500 Square Foot Room

A room of approximately 500 S.F. requires approximately 895 S.F. of finish treatment.

$$500 \times 1.79 = 895 \text{ S.F.}$$

If the room had a L:W ratio of 2:1, the finish requirement would be:

$$500 \times 1.90 = 950 \text{ S.F.}$$

Example Four: Three-Story Office Building

For the office building, the following items are included in the Interior Construction division:

- Concrete masonry unit partitions
- Drywall partitions

Area	S.F. of Interior Wall Finish Required as Compared to Floor Area (based on 10' floor to ceiling height)			
	Room Length-To-Width Ratio		Approximate Room Size	
	1:1	2:1	1:1	1:2
100	4.00	4.26	10 x 10	8 x 14
150	3.27	3.48	12 x 12	9 x 17
200	2.82	3.00	15 x 15	10 x 10
250	2.53	2.69	16 x 16	11 x 22
300	2.31	2.45	17 x 17	12 x 25
350	2.14	2.27	19 x 19	13 x 27
400	2.00	2.12	20 x 20	14 x 28
450	1.89	2.00	21 x 21	15 x 30
500	1.79	1.90	22 x 22	16 x 32
550	1.71	1.81	23 x 23	17 x 33
600	1.63	1.73	24 x 24	17 x 35
650	1.57	1.66	26 x 26	28 x 36
700	1.51	1.60	26 x 26	19 x 37
750	1.46	1.55	27 x 27	19 x 39
800	1.41	1.50	28 x 28	20 x 40
850	1.37	1.45	29 x 29	21 x 41
900	1.33	1.41	30 x 30	21 x 42
950	1.30	1.37	31 x 31	22 x 44
1000	1.27	1.34	32 x 32	22 x 45
1500	1.03	1.10	39 x 39	27 x 55
2000	0.89	0.95	45 x 45	32 x 63

Note: For obtaining total S.F. of Interior Partitions, divide ratio multiplier by 2.

Figure 16.2

- Interior doors
- Toilet partitions
- Paint, ceramic tile, & vinyl wall finishes
- Carpet, ceramic, & quarry tile floor finishes
- Suspended acoustical ceiling

Figure 16.3 shows the approximate location of partitions in the proposed Three-Story Office Building. The second and third floors are identical, except for the exit doors in the lobbies by the stairways. It is assumed that future tenants will add other partitions to divide the large areas into smaller spaces. Of course, those costs will be borne by the future tenants and are not included in the initial budget estimate.

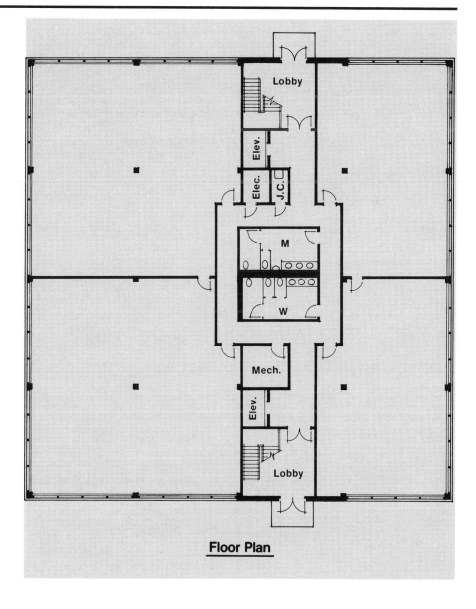

Floor Plan

Figure 16.3

3-Story Office Building   100′ x 100′

Quantities:

Partitions:	8″ Block	534 L.F.
	Steel stud & gyp. board	966 L.F.
Finishes:		
Walls	Paint	85%
	Ceramic Tile	5%
	Vinyl Wall Covering	10%
Floors	Carpet	93%
	Ceramic Tile	4%
	Quarry Tile	3%
Ceiling:	Suspended Acoustic, 9′-0″	

Costs:

Concrete Masonry Unit Partition			
	(Figure 16.4 #6000)		
	6,410 S.F. @ $6.04/S.F.	=	$ 38,716
Drywall Partition			
	(Figure 16.5 #5400)		
	11,600 S.F. @ $2.83/S.F.	=	32,828
Exterior Wall			
Furring	(Figure 16.6 #0649)		
	6,000 S.F. @ $1.22/S.F.	=	7,320
Gyp. Bd.	(Figure 16.6 #0700)		
	6,000 S.F. @ $0.55/S.F.	=	3,300
Insulation	(Figure 16.6 #0920)		
	8,400 S.F. @ $0.67/S.F.	=	5,628
Tape & Fin.	(Figure 16.6 #0960)		
	6,000 S.F. @ $0.41/S.F.	=	2,460
Interior Doors	(Figure 16.7 #0140)		
	60 ea. @ $460 ea.	=	27,600
Toilet Partitions			
Paint. Mtl.	(Figure 16.8 #0680)		
	15 ea. @ $457 ea.	=	6,855
Handicapped	(Figure 16.8 #0760)		
	6 ea. @ $255 ea.	=	1,530
Urinal Scr.	(Figure 16.8 #1340)		
	3 ea. @ $254.50 ea.	=	763
Wall Finish			
Paint G.B.	(Figure 16.9 #0140)		
	16,200 S.F. @ $0.48/S.F.	=	7,776
Paint Blk.	(Figure 16.9 #0320)		
	5,600 S.F. @ $1.01/S.F.	=	5,656
V.W.C.	(Figure 16.9 #1800)		
	1,900 S.F. @ $1.45/S.F.	=	2,755
Cer. Tile	(Figure 16.9 #1940)		
	950 S.F. @ $4.95/S.F.	=	4,703
Floor Finish			
Carpet	(Figure 16.10 #0160)		
	27,900 S.F. @ $4.55/S.F.	=	126,945
Cer. Tile	(Figure 16.10 #1720)		
	1,200 S.F. @ $6.59/S.F.	=	7,908
Quarry Tile	(Figure 16.10 #1820)		
	900 S.F. @ $9.61/S.F.	=	8,649
Ceiling			
Susp. Acous.	(Figure 16.11 #6000)		
	30,000 S.F. @ $1.90/S.F.	=	57,000
Total Cost			$348,392
Cost per Square Foot			$ 11.61/S.F.

Figure 16.12 summarizes the costs for this division.

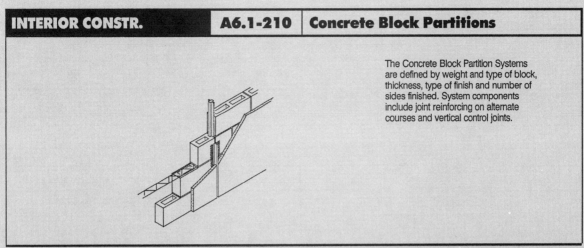

INTERIOR CONSTR.   A6.1-210   Concrete Block Partitions

The Concrete Block Partition Systems are defined by weight and type of block, thickness, type of finish and number of sides finished. System components include joint reinforcing on alternate courses and vertical control joints.

System Components	QUANTITY	UNIT	COST PER S.F.		
			MAT.	INST.	TOTAL
SYSTEM 6.1-210-1020					
CONC. BLOCK PARTITION, 8" X 16", 4" TK., 2 CT. GYP. PLASTER 2 SIDES					
Conc. block partition, 4" thick	1.000	S.F.	.95	3.44	4.39
Control joint	.050	L.F.	.09	.09	.18
Horizontal joint reinforcing	.800	L.F.	.11	.03	.14
Gypsum plaster, 2 coat, on masonry	2.000	S.F.	.62	3.21	3.83
TOTAL			1.77	6.77	8.54

6.1-210   Concrete Block Partitions - Regular Weight

	TYPE	THICKNESS (IN.)	TYPE FINISH	SIDES FINISHED		COST PER S.F.		
						MAT.	INST.	TOTAL
1000	Hollow	4	none	0		1.15	3.56	4.71
1010			gyp. plaster 2 coat	1		1.46	5.15	6.61
1020				2		1.77	6.75	8.52
1100			lime plaster - 2 coat	1		1.33	5.15	6.48
1150			lime portland - 2 coat	1		1.34	5.15	6.49
1200			portland - 3 coat	1		1.36	5.40	6.76
1400			5/8" drywall	1		1.54	4.74	6.28
1410				2		1.93	5.90	7.83
1500		6	none	0		1.32	3.82	5.14
1510			gyp. plaster 2 coat	1		1.63	5.40	7.03
1520				2		1.94	7.05	8.99
1600			lime plaster - 2 coat	1		1.50	5.40	6.90
1650			lime portland - 2 coat	1		1.51	5.40	6.91
1700			portland - 3 coat	1		1.53	5.65	7.18
1900			5/8" drywall	1		1.71	5	6.71
1910				2		2.10	6.20	8.30
2000		8	none	0		1.56	4.08	5.64
2010			gyp. plaster 2 coat	1		1.87	5.70	7.57
2020			gyp. plaster 2 coat	2		2.18	7.30	9.48
2100		8	lime plaster - 2 coat	1		1.74	5.70	7.44
2150			lime portland - 2 coat	1		1.75	5.70	7.45
2200			portland - 3 coat	1		1.77	5.90	7.67
2400			5/8" drywall	1		1.95	5.25	7.20
2410				2		2.34	6.45	8.79
2500		10	none	0		2.45	4.27	6.72
2510			gyp. plaster 2 coat	1		2.76	5.85	8.61
2520				2		3.07	7.50	10.57
2600			lime plaster - 2 coat	1		2.63	5.85	8.48

Figure 16.4

6.1-210 Concrete Block Partitions - Regular Weight

	TYPE	THICKNESS (IN.)	TYPE FINISH	SIDES FINISHED		COST PER S.F. MAT.	INST.	TOTAL
2650	Hollow	10	lime portland - 2 coat	1		2.64	5.85	8.49
2700			portland - 3 coat	1		2.66	6.10	8.76
2900			5/8" drywall	1		2.84	5.45	8.29
2910				2		3.23	6.65	9.88
3000	Solid	2	none	0		1.20	3.52	4.72
3010			gyp. plaster	1		1.51	5.10	6.61
3020				2		1.82	6.75	8.57
3100			lime plaster - 2 coat	1		1.38	5.10	6.48
3150			lime portland - 2 coat	1		1.39	5.10	6.49
3200			portland - 3 coat	1		1.41	5.35	6.76
3400			5/8" drywall	1		1.59	4.70	6.29
3410				2		1.98	5.90	7.88
3500		4	none	0		1.50	3.69	5.19
3510			gyp. plaster	1		1.88	5.30	7.18
3520				2		2.12	6.90	9.02
3600			lime plaster - 2 coat	1		1.68	5.30	6.98
3650			lime portland - 2 coat	1		1.69	5.30	6.99
3700			portland - 3 coat	1		1.71	5.55	7.26
3900			5/8" drywall	1		1.89	4.87	6.76
3910				2		2.28	6.05	8.33
4000		6	none	0		1.90	3.96	5.86
4010			gyp. plaster	1		2.21	5.55	7.76
4020				2		2.52	7.15	9.67
4100			lime plaster - 2 coat	1		2.08	5.55	7.63
4150			lime portland - 2 coat	1		2.09	5.55	7.64
4200			portland - 3 coat	1		2.11	5.80	7.91
4400			5/8" drywall	1		2.29	5.15	7.44
4410				2		2.68	6.30	8.98

6.1-210 Concrete Block Partitions - Lightweight

	TYPE	THICKNESS (IN.)	TYPE FINISH	SIDES FINISHED		COST PER S.F. MAT.	INST.	TOTAL
5000	Hollow	4	none	0		1.27	3.48	4.75
5010			gyp. plaster	1		1.58	5.10	6.68
5020				2		1.89	6.70	8.59
5100			lime plaster - 2 coat	1		1.45	5.10	6.55
5150			lime portland - 2 coat	1		1.46	5.10	6.56
5200			portland - 3 coat	1		1.48	5.30	6.78
5400			5/8" drywall	1		1.66	4.66	6.32
5410				2		2.05	5.85	7.90
5500		6	none	0		1.60	3.73	5.33
5510			gyp. plaster	1		1.91	5.35	7.26
5520			gyp. plaster	2		2.22	6.95	9.17
5600			lime plaster - 2 coat	1		1.78	5.35	7.13
5650			lime portland - 2 coat	1		1.79	5.35	7.14
5700			portland - 3 coat	1		1.81	5.55	7.36
5900			5/8" drywall	1		1.99	4.91	6.90
5910				2		2.38	6.10	8.48
6000		8	none	0		2.07	3.97	6.04
6010			gyp. plaster	1		2.38	5.55	7.93
6020				2		2.69	7.20	9.89
6100			lime plaster - 2 coat	1		2.25	5.55	7.80
6150			lime portland - 2 coat	1		2.26	5.55	7.81
6200			portland - 3 coat	1		2.28	5.80	8.08
6400			5/8" drywall	1		2.46	5.15	7.61
6410				2		2.85	6.35	9.20

Figure 16.4 (cont.)

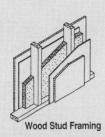
Wood Stud Framing

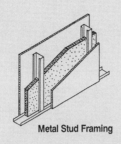
Metal Stud Framing

The Drywall Partitions/Stud Framing Systems are defined by type of drywall and number of layers, type and spacing of stud framing, and treatment on the opposite face. Components include taping and finishing.

Cost differences between regular and fire resistant drywall are negligible, and terminology is interchangeable. In some cases fiberglass insulation is included for additional sound deadening.

System Components	QUANTITY	UNIT	COST PER S.F.		
			MAT.	INST.	TOTAL
SYSTEM 6.1-510-1250					
DRYWALL PARTITION,5/8" F.R.1 SIDE,5/8" REG.1 SIDE,2"X4"STUDS,16" O.C.					
Gypsum plasterboard, nailed/screwed to studs, 5/8"F.R. fire resistant	1.000	S.F.	.23	.32	.55
Gypsum plasterboard, nailed/screwed to studs, 5/8" regular	1.000	S.F.	.23	.32	.55
Taping and finishing joints	2.000	S.F.	.18	.64	.82
Framing, 2 x 4 studs @ 16" O.C., 10' high	1.000	S.F.	.42	.67	1.09
TOTAL			1.06	1.95	3.01

6.1-510 Drywall Partitions/Wood Stud Framing

	FACE LAYER	BASE LAYER	FRAMING	OPPOSITE FACE	INSULATION	COST PER S.F.		
						MAT.	INST.	TOTAL
1200	5/8" FR drywall	none	2 x 4, @ 16" O.C.	same	0	1.06	1.95	3.01
1250				5/8" reg. drywall	0	1.06	1.95	3.01
1300				nothing	0	.74	1.31	2.05
1400		1/4" SD gypsum	2 x 4 @ 16" O.C.	same	1-1/2" fiberglass	1.72	2.99	4.71
1450				5/8" FR drywall	1-1/2" fiberglass	1.55	2.63	4.18
1500				nothing	1-1/2" fiberglass	1.23	1.99	3.22
1600		resil. channels	2 x 4 @ 16", O.C.	same	1-1/2" fiberglass	1.61	3.78	5.39
1650				5/8" FR drywall	1-1/2" fiberglass	1.50	3.03	4.53
1700				nothing	1-1/2" fiberglass	1.18	2.39	3.57
1800		5/8" FR drywall	2 x 4 @ 24" O.C.	same	0	1.42	2.46	3.88
1850				5/8" FR drywall	0	1.19	2.14	3.33
1900				nothing	0	.87	1.50	2.37
1950		5/8" FR drywall	2 x 4, 16" O.C.	same	0	1.52	2.59	4.11
1955				5/8" FR drywall	0	1.29	2.27	3.56
2000				nothing	0	.97	1.63	2.60
2010		5/8" FR drywall	staggered, 6" plate	same	0	1.96	3.26	5.22
2015				5/8" FR drywall	0	1.73	2.94	4.67
2020				nothing	0	1.41	2.30	3.71
2200		5/8" FR drywall	2 rows-2 x 4	same	2" fiberglass	2.32	3.57	5.89
2250			16"O.C.	5/8" FR drywall	2" fiberglass	2.09	3.25	5.34
2300				nothing	2" fiberglass	1.77	2.61	4.38
2400	5/8" WR drywall	none	2 x 4, @ 16" O.C.	same	0	1.16	1.95	3.11
2450				5/8" FR drywall	0	1.11	1.95	3.06
2500				nothing	0	.79	1.31	2.10
2600		5/8" FR drywall	2 x 4, @ 24" O.C.	same	0	1.52	2.46	3.98
2650				5/8" FR drywall	0	1.24	2.14	3.38
2700				nothing	0	.92	1.50	2.42
2800	5/8 VF drywall	none	2 x 4, @ 16" O.C.	same	0	1.80	2.11	3.91
2850				5/8" FR drywall	0	1.43	2.03	3.46
2900				nothing	0	1.11	1.39	2.50

Figure 16.5

INTERIOR CONSTR. — A6.1-510 — Drywall Partitions w/Studs

6.1-510 — Drywall Partitions/Wood Stud Framing

	FACE LAYER	BASE LAYER	FRAMING	OPPOSITE FACE	INSULATION	COST PER S.F. MAT.	INST.	TOTAL
3000	5/8 VF drywall	5/8" FR drywall	2 x 4 , 24" O.C.	same	0	2.16	2.62	4.78
3050				5/8" FR drywall	0	1.56	2.22	3.78
3100				nothing	0	1.24	1.58	2.82
3200	1/2" reg drywall	3/8" reg drywall	2 x 4, @ 16" O.C.	same	0	1.28	2.59	3.87
3250				5/8" FR drywall	0	1.17	2.27	3.44
3300				nothing	0	.85	1.63	2.48

6.1-510 — Drywall Partitions/Metal Stud Framing

	FACE LAYER	BASE LAYER	FRAMING	OPPOSITE FACE	INSULATION	COST PER S.F. MAT.	INST.	TOTAL
5200	5/8" FR drywall	none	1-5/8" @ 24" O.C.	same	0	.84	1.90	2.74
5250				5/8" reg. drywall	0	.84	1.90	2.74
5300				nothing	0	.52	1.26	1.78
5400			3-5/8" @ 24" O.C.	same	0	.91	1.92	2.83
5450				5/8" reg. drywall	0	.91	1.92	2.83
5500				nothing	0	.59	1.28	1.87
5600		1/4" SD gypsum	1-5/8" @ 24" O.C.	same	0	1.18	2.62	3.80
5650				5/8" FR drywall	0	1.01	2.26	3.27
5700				nothing	0	.69	1.62	2.31
5800			2-1/2" @ 24" O.C.	same	0	1.20	2.63	3.83
5850				5/8" FR drywall	0	1.03	2.27	3.30
5900				nothing	0	.71	1.63	2.34
6000		5/8" FR drywall	2-1/2" @ 16" O.C.	same	0	1.60	2.65	4.25
6050				5/8" FR drywall	0	1.37	2.33	3.70
6100				nothing	0	1.05	1.69	2.74
6200			3-5/8" @ 24" O.C.	same	0	1.37	2.56	3.93
6250				5/8"FR drywall	3-1/2" fiberglass	1.40	2.44	3.84
6300				nothing	0	.82	1.60	2.42
6400	5/8" WR drywall	none	1-5/8" @ 24" O.C.	same	0	.94	1.90	2.84
6450				5/8" FR drywall	0	.89	1.90	2.79
6500				nothing	0	.57	1.26	1.83
6600			3-5/8" @ 24" O.C.	same	0	1.01	1.92	2.93
6650				5/8" FR drywall	0	.96	1.92	2.88
6700				nothing	0	.64	1.28	1.92
6800		5/8" FR drywall	2-1/2" @ 16" O.C.	same	0	1.70	2.65	4.35
6850				5/8" FR drywall	0	1.42	2.33	3.75
6900				nothing	0	1.10	1.69	2.79
7000			3-5/8" @ 24" O.C.	same	0	1.47	2.56	4.03
7050				5/8"FR drywall	3-1/2" fiberglass	1.45	2.44	3.89
7100				nothing	0	.87	1.60	2.47
7200	5/8" VF drywall	none	1-5/8" @ 24" O.C.	same	0	1.58	2.06	3.64
7250				5/8" FR drywall	0	1.21	1.98	3.19
7300				nothing	0	.89	1.34	2.23
7400			3-5/8" @ 24" O.C.	same	0	1.65	2.08	3.73
7450				5/8" FR drywall	0	1.28	2	3.28
7500				nothing	0	.96	1.36	2.32
7600		5/8" FR drywall	2-1/2" @ 16" O.C.	same	0	2.34	2.81	5.15
7650				5/8" FR drywall	0	1.74	2.41	4.15
7700				nothing	0	1.42	1.77	3.19
7800			3-5/8" @ 24" O.C.	same	0	2.11	2.72	4.83
7850				5/8"FR drywall	3-1/2" fiberglass	1.77	2.52	4.29
7900				nothing	0	1.19	1.68	2.87

Figure 16.5 (cont.)

6.1-580	Drywall Components	COST PER S.F.		
		MAT.	INST.	TOTAL
0140	Metal studs, 24" O.C. including track, load bearing, 18 gage, 2-1/2"	.72	1.34	2.06
0160	3-5/8"	.84	1.40	2.24
0180	4"	.87	1.46	2.33
0200	6"	1.06	1.53	2.59
0220	16 gage, 2-1/2"	.91	1.46	2.37
0240	3-5/8"	1.01	1.53	2.54
0260	4"	1.10	1.61	2.71
0280	6"	1.35	1.69	3.04
0300	Non load bearing, 25 gage, 1-5/8"	.20	.62	.82
0340	3-5/8"	.27	.64	.91
0360	4"	.29	.66	.95
0380	6"	.38	.67	1.05
0400	20 gage, 2-1/2"	.40	.63	1.03
0420	3-5/8"	.47	.64	1.11
0440	4"	.49	.66	1.15
0460	6"	.62	.67	1.29
0540	Wood studs including blocking, shoe and double top plate, 2"x4", 12"O.C.	.53	.84	1.37
0560	16" O.C.	.42	.67	1.09
0580	24" O.C.	.32	.54	.86
0600	2"x6", 12" O.C.	.75	.95	1.70
0620	16" O.C.	.61	.74	1.35
0640	24" O.C.	.46	.58	1.04
0642	Furring one side only, steel channels, 3/4", 12" O.C.	.26	1.29	1.55
0644	16" O.C.	.24	1.14	1.38
0646	24" O.C.	.16	.86	1.02
0647	1-1/2", 12" O.C.	.38	1.44	1.82
0648	16" O.C.	.34	1.26	1.60
0649	24" O.C.	.23	.99	1.22
0650	Wood strips, 1" x 3", on wood, 12" O.C.	.35	.59	.94
0651	16" O.C.	.26	.44	.70
0652	On masonry, 12" O.C.	.35	.65	1
0653	16" O.C.	.26	.49	.75
0654	On concrete, 12" O.C.	.35	1.24	1.59
0655	16" O.C.	.26	.93	1.19
0665	Gypsum board, one face only, exterior sheathing, 1/2"	.39	.59	.98
0680	Fire resistant, 1/2"	.23	.32	.55
0700	5/8"	.23	.32	.55
0720	Sound deadening board 1/4"	.17	.36	.53
0740	Standard drywall 3/8"	.17	.32	.49
0760	1/2"	.17	.32	.49
0780	5/8"	.23	.32	.55
0800	Tongue & groove coreboard 1"	.53	1.34	1.87
0820	Water resistant, 1/2"	.25	.32	.57
0840	5/8"	.28	.32	.60
0860	Add for the following:, foil backing	.08		.08
0880	Fiberglass insulation, 3-1/2"	.26	.20	.46
0900	6"	.33	.24	.57
0920	Rigid insulation 1"	.35	.32	.67
0940	Resilient furring @ 16" O.C.	.15	1.01	1.16
0960	Taping and finishing	.09	.32	.41
0980	Texture spray	.13	.38	.51
1000	Thin coat plaster	.12	.40	.52
1040	2"x4" staggered studs 2"x6" plates & blocking	.61	.72	1.33

Figure 16.6

6.4-260 — Wood Door/Metal Frame

	TYPE	FACE	SIZE	FRAME	DEPTH	COST EACH MAT.	COST EACH INST.	COST EACH TOTAL
6400	1 hr/flush	walnut	2'-8" x 6'-8"	drywall K.D.	4-7/8"	380	135	515
6420				butt welded	8-3/4"	425	140	565
6560			6'-0" x 7'-0"	drywall K.D.	4-7/8"	740	248	988
6580				butt welded	8-3/4"	800	254	1,054
7600	1-1/2 hr/flush	birch	2'-8" x 6'-8"	drywall K.D.	4-7/8"	299	135	434
7620				butt welded	8-3/4"	340	140	480
7760			6'-0" x 7'-0"	drywall K.D.	4-7/8"	525	248	773
7780				butt welded	8-3/4"	585	254	839
7800		oak	2'-8" x 6'-8"	drywall K.D.	4-7/8"	295	135	430
7820				butt welded	8-3/4"	335	140	475
7960			6'-0" x 7'-0"	drywall K.D.	4-7/8"	525	248	773
7980				butt welded	8-3/4"	585	254	839
8000		walnut	2'-8" x 6'-8"	drywall K.D.	4-7/8"	395	135	530
8020				butt welded	8-3/4"	435	140	575
8160			6'-0" x 7'-0"	drywall K. D.	4 7/8"	760	248	1,008
8180				butt welded	8-3/4"	820	254	1,074

6.4-290 — Doors/Metal Frames

	MATERIAL	GAUGE	THICKNESS (IN.)	TYPE	SIZE W(FT.)XH(FT.)	COST EACH MAT.	COST EACH INST.	COST EACH TOTAL
0100								
0120	Hollow metal	20	1-3/4"	flush	2/8 6/8	273	168	441
0140					3/0 7/0	289	171	460
0160				half glass	2/8 6/8	345	188	533
0180					3/0 7/0	365	192	557
0200		18	1-3/4"	flush	2/8 6/8	295	171	466
0220					3/0 7/0	310	174	484
0240				half glass	2/8 6/8	370	192	562
0260					3/0 7/0	385	195	580
0280	Wood		1-3/4	flush-lauan	2/8 6/8	158	132	290
0320	5 ply			birch	2/8 6/8	157	132	289
0340					3/0 7/0	167	142	309
0360				oak	2/8 6/8	173	132	305
0380					3/0 7/0	189	142	331
0400				walnut	2/8 6/8	217	135	352
0420					3/0 7/0	1,025	540	1,565

6.4-290 — Metal Fire Doors/Metal Frames

	FIRE RATING (HOURS)	GAUGE	THICKNESS (IN.)	TYPE	SIZE W(FT)XH(FT)	COST EACH MAT.	COST EACH INST.	COST EACH TOTAL
0520	1-1/2	20	1-3/4	flush	2/8 6/8	310	168	478
0540					3/0 7/0	320	171	491
0560				composite	2/8 6/8	325	171	496
0580					3/0 7/0	345	174	519
0600		18	1-3/4	flush	3/0 6/8	340	174	514
0620					3/0 7/0	345	174	519
0720	3	18		composite	2/8 6/8	375	176	551
0740					3/0 7/0	380	176	556

Figure 16.7

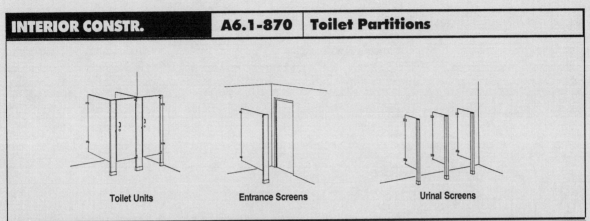

| Toilet Units | Entrance Screens | Urinal Screens |

6.1-870	Toilet Partitions	COST PER UNIT		
		MAT.	INST.	TOTAL
0380	Toilet partitions, cubicles, ceiling hung, marble	1,325	320	1,645
0400	Painted metal	400	161	561
0420	Plastic laminate	530	161	691
0440	Porcelain enamel	810	161	971
0460	Stainless steel	980	161	1,141
0480	Handicap addition	255		255
0520	Floor and ceiling anchored, marble	1,425	258	1,683
0540	Painted metal	405	129	534
0560	Plastic laminate	535	129	664
0580	Porcelain enamel	835	129	964
0600	Stainless steel	1,125	129	1,254
0620	Handicap addition	255		255
0660	Floor mounted marble	850	215	1,065
0680	Painted metal	365	92	457
0700	Plastic laminate	495	92	587
0720	Porcelain enamel	810	92	902
0740	Stainless steel	1,100	92	1,192
0760	Handicap addition	255		255
0780	Juvenile deduction	36.50		36.50
0820	Floor mounted with handrail marble	985	215	1,200
0840	Painted metal	375	107	482
0860	Plastic laminate	490	107	597
0880	Porcelain enamel	820	107	927
0900	Stainless steel	1,100	107	1,207
0920	Handicap addition	255		255
0960	Wall hung, painted metal	460	92	552
1000	Porcelain enamel	825	92	917
1020	Stainless steel	1,100	92	1,192
1040	Handicap addition	255		255
1080	Entrance screens, floor mounted, 54" high, marble	540	71.50	611.50
1100	Painted metal	167	43	210
1120	Porcelain enamel	330	43	373
1140	Stainless steel	605	43	648
1300	Urinal screens, floor mounted, 24" wide, laminated plastic	226	80.50	306.50
1320	Marble	505	96	601
1340	Painted metal	174	80.50	254.50
1360	Porcelain enamel	330	80.50	410.50
1380	Stainless steel	505	80.50	585.50
1428	Wall mounted wedge type, painted metal	196	64.50	260.50
1440	Porcelain enamel	283	64.50	347.50
1460	Stainless steel	405	64.50	469.50

Figure 16.8

6.5-100	Paint & Covering	COST PER S.F.		
		MAT.	INST.	TOTAL
0060	Painting, interior on plaster and drywall, brushwork, primer & 1 coat	.08	.41	.49
0080	Primer & 2 coats	.12	.57	.69
0100	Primer & 3 coats	.16	.75	.91
0120	Walls & ceilings, roller work, primer & 1 coat	.08	.24	.32
0140	Primer & 2 coats	.12	.36	.48
0160	Woodwork incl. puttying, brushwork, primer & 1 coat	.08	.63	.71
0180	Primer & 2 coats	.12	.84	.96
0200	Primer & 3 coats	.16	1.14	1.30
0260	Cabinets and casework, enamel, primer & 1 coat	.08	.66	.74
0280	Primer & 2 coats	.12	.84	.96
0300	Masonry or concrete, latex, brushwork, primer & 1 coat	.15	.57	.72
0320	Primer & 2 coats	.20	.81	1.01
0340	Addition for block filler	.10	.15	.25
0380	Fireproof paints, intumescent, 1/8" thick 3/4 hour	1.55	.58	2.13
0400	3/16" thick 1 hour	3.47	.87	4.34
0420	7/16" thick 2 hour	4.47	2.03	6.50
0440	1-1/16" thick 3 hour	7.30	4.06	11.36
0480	Miscellaneous metal brushwork, exposed metal, primer & 1 coat	.05	.57	.62
0500	Gratings, primer & 1 coat	.14	.84	.98
0600	Pipes over 12" diameter	.38	2.47	2.85
0700	Structural steel, brushwork, light framing 300-500 S.F./Ton	.05	.48	.53
0720	Heavy framing 50-100 S.F./Ton	.05	.38	.43
0740	Spraywork, light framing 300-500 S.F./Ton	.05	.24	.29
0760	Heavy framing 50-100 S.F./Ton	.05	.12	.17
0800	Varnish, interior wood trim, no sanding sealer & 1 coat	.06	.71	.77
0820	Hardwood floor, no sanding 2 coats	.11	.15	.26
0840	Wall coatings, acrylic glazed coatings, minimum	.22	.54	.76
0860	Maximum	.50	.93	1.43
0880	Epoxy coatings, minimum	.30	.54	.84
0900	Maximum	.91	1.67	2.58
0940	Exposed epoxy aggregate, troweled on, 1/16" to 1/4" aggregate, minimum	.47	1.21	1.68
0960	Maximum	.98	2.18	3.16
0980	1/2" to 5/8" aggregate, minimum	.91	2.18	3.09
1000	Maximum	1.55	3.55	5.10
1020	1" aggregate, minimum	1.60	3.16	4.76
1040	Maximum	2.37	5.15	7.52
1060	Sprayed on, minimum	.43	.96	1.39
1080	Maximum	.79	1.96	2.75
1100	High build epoxy 50 mil, minimum	.48	.73	1.21
1120	Maximum	.84	2.99	3.83
1140	Laminated epoxy with fiberglass minimum	.55	.96	1.51
1160	Maximum	.97	1.96	2.93
1180	Sprayed perlite or vermiculite 1/16" thick, minimum	.19	.10	.29
1200	Maximum	.55	.44	.99
1260	Wall coatings, vinyl plastic, minimum	.24	.39	.63
1280	Maximum	.62	1.18	1.80
1300	Urethane on smooth surface, 2 coats, minimum	.18	.25	.43
1320	Maximum	.42	.43	.85
1340	3 coats, minimum	.24	.34	.58
1360	Maximum	.55	.60	1.15
1380	Ceramic-like glazed coating, cementitious, minimum	.36	.65	1.01
1400	Maximum	.60	.82	1.42
1420	Resin base, minimum	.24	.44	.68
1440	Maximum	.41	.86	1.27
1460	Wall coverings, aluminum foil	.78	1.05	1.83
1480	Copper sheets, .025" thick, phenolic backing	5.35	1.20	6.55
1500	Vinyl backing	4.13	1.20	5.33
1510				

Figure 16.9

6.5-100	Paint & Covering	COST PER S.F.		
		MAT.	INST.	TOTAL
1520	Cork tiles, 12"x12", light or dark, 3/16" thick	2.64	1.20	3.84
1540	5/16" thick	2.75	1.23	3.98
1560	Basketweave, 1/4" thick	4.23	1.20	5.43
1580	Natural, non-directional, 1/2" thick	4.16	1.20	5.36
1600	12"x36", granular, 3/16" thick	.91	.75	1.66
1620	1" thick	1.19	.78	1.97
1640	12"x12", polyurethane coated, 3/16" thick	2.86	1.20	4.06
1660	5/16" thick	4.05	1.23	5.28
1661	Paneling, prefinished plywood, birch	1.10	1.58	2.68
1662	Mahogany, African	2.04	1.66	3.70
1663	Philippine (lauan)	.88	1.33	2.21
1664	Oak or cherry	2.86	1.66	4.52
1665	Rosewood	4.07	2.07	6.14
1666	Teak	2.86	1.66	4.52
1667	Chestnut	4.23	1.77	6
1668	Pecan	1.82	1.66	3.48
1669	Walnut	4.62	1.66	6.28
1670	Wood board, knotty pine, finished	1.54	2.76	4.30
1671	Rough sawn cedar	1.92	2.76	4.68
1672	Redwood	4.35	2.76	7.11
1673	Aromatic cedar	3.25	2.95	6.20
1680	Cork wallpaper, paper backed, natural	1.62	.60	2.22
1700	Color	2.01	.60	2.61
1720	Gypsum based, fabric backed, minimum	.59	.36	.95
1740	Average	.89	.40	1.29
1760	Maximum	.99	.45	1.44
1780	Vinyl wall covering, fabric back, light weight	.68	.45	1.13
1800	Medium weight	.85	.60	1.45
1820	Heavy weight	1.15	.66	1.81
1840	Wall paper, @ $9.00 per double roll	.28	.45	.73
1860	@ $15.00 per double roll	.47	.54	1.01
1880	@ $25.00 per double roll	.91	.66	1.57
1900	Grass cloths with lining paper, minimum	.58	.72	1.30
1920	Maximum	1.87	.82	2.69
1940	Ceramic tile, thin set, 4-1/4" x 4-1/4"	2.11	2.84	4.95
1941				

6.5-100	Paint Trim	COST PER L.F.		
		MAT.	INST.	TOTAL
2040	Painting, wood trim, to 6" wide, enamel, primer & 1 coat	.08	.26	.34
2060	Primer & 2 coats	.12	.36	.48
2080	Misc. metal brushwork, ladders	.28	2.58	2.86
2100	Pipes, to 4" dia.	.05	.67	.72
2120	6" to 8" dia.	.07	1.23	1.30
2140	10" to 12" dia.	.30	1.89	2.19
2160	Railings, 2" pipe	.10	.67	.77
2180	Handrail, single	.08	1.01	1.09

Figure 16.9 (cont.)

6.6-100	Tile & Covering	COST PER S.F.		
		MAT.	INST.	TOTAL
0060	Carpet tile, nylon, fusion bonded, 18″ x 18″ or 24″ x 24″, 24 oz.	2.66	.41	3.07
0080	35 oz.	3.16	.41	3.57
0100	42 oz.	4.05	.41	4.46
0120				
0140	Carpet, tufted, nylon, roll goods, 12′ wide, 26 oz.	2.55	.66	3.21
0160	36 oz.	3.89	.66	4.55
0180	Woven, wool, 36 oz.	4.33	.68	5.01
0200	42 oz.	5.50	.68	6.18
0220	Padding, add to above, minimum	.36	.22	.58
0240	Maximum	.62	.22	.84
0260	Composition flooring, acrylic, 1/4″ thick	1.10	3.20	4.30
0280	3/8″ thick	1.49	3.71	5.20
0300	Epoxy, minimum	1.98	2.48	4.46
0320	Maximum	2.42	3.40	5.82
0340	Epoxy terrazzo, minimum	4.13	3.26	7.39
0360	Maximum	6.80	4.35	11.15
0380	Mastic, hot laid, 1-1/2″ thick, minimum	2.86	2.41	5.27
0400	Maximum	3.68	3.20	6.88
0420	Neoprene 1/4″ thick, minimum	2.75	3.06	5.81
0440	Maximum	3.85	3.88	7.73
0460	Polyacrylate with ground granite 1/4″, minimum	2.15	2.26	4.41
0480	Maximum	4.40	3.48	7.88
0500	Polyester with colored quart 2 chips 1/16″, minimum	2	1.57	3.57
0520	Maximum	3.49	2.48	5.97
0540	Polyurethane with vinyl chips, minimum	5.50	1.57	7.07
0560	Maximum	7.80	1.94	9.74
0600	Concrete topping, granolithic concrete, 1/2″ thick	.16	1.57	1.73
0620	1″ thick	.32	1.61	1.93
0640	2″ thick	.65	1.86	2.51
0650				
0660	Heavy duty 3/4″ thick, minimum	.23	2.44	2.67
0680	Maximum	.23	2.90	3.13
0700	For colors, add to above, minimum	.45	.56	1.01
0720	Maximum	2.01	.62	2.63
0740	Exposed aggregate finish, minimum	.39	.47	.86
0760	Maximum	1.23	.63	1.86
0780	Abrasives, .25 P.S.F. add to above, minimum	.10	.39	.49
0800	Maximum	.15	.39	.54
0820	Dust on coloring, add, minimum	.55	.26	.81
0840	Maximum	1.93	.54	2.47
0860	1/2″ integral, minimum	.17	1.57	1.74
0880	Maximum	.17	1.57	1.74
0900	Dustproofing, add, minimum	.06	.15	.21
0920	Maximum	.06	.23	.29
0930	Paint	.20	.81	1.01
0940	Hardeners, metallic add, minimum	.23	.39	.62
0960	Maximum	.52	.59	1.11
0980	Non-metallic, minimum	.06	.39	.45
1000	Maximum	.15	.59	.74
1020	Integral topping, 3/16″ thick	.13	.93	1.06
1040	1/2″ thick	.17	.97	1.14
1060	3/4″ thick	.25	1.09	1.34
1080	1″ thick	.34	1.24	1.58
1100	Terrazzo, minimum	2.15	5	7.15
1120	Maximum	3.96	10.90	14.86
1280	Resilient, asphalt tile, 1/8″ thick on concrete, minimum	.99	.74	1.73
1300	Maximum	1.06	.74	1.80
1320	On wood, add for felt underlay	.19		.19

Figure 16.10

6.6-100	Tile & Covering	COST PER S.F.		
		MAT.	INST.	TOTAL
1340	Cork tile, minimum	2.53	.95	3.48
1360	Maximum	9.60	.95	10.55
1380	Polyethylene, in rolls, minimum	2.21	1.08	3.29
1400	Maximum	4.42	1.08	5.50
1420	Polyurethane, thermoset, minimum	3.50	2.98	6.48
1440	Maximum	4.17	5.95	10.12
1460	Rubber, sheet goods, minimum	3.16	2.48	5.64
1480	Maximum	5.15	3.31	8.46
1500	Tile, minimum	3.16	.74	3.90
1520	Maximum	6.15	1.08	7.23
1540	Synthetic turf, minimum	2.72	1.42	4.14
1560	Maximum	6.95	1.57	8.52
1580	Vinyl, composition tile, minimum	.71	.60	1.31
1600	Maximum	1.76	.60	2.36
1620	Tile, minimum	1.61	.60	2.21
1640	Maximum	2.52	.60	3.12
1660	Sheet goods, minimum	1.32	1.19	2.51
1680	Maximum	4.95	1.49	6.44
1720	Tile, ceramic natural clay	3.64	2.95	6.59
1730	Marble, synthetic 12"x12"x5/8"	7.30	9	16.30
1740	Porcelain type, minimum	3.94	2.95	6.89
1760	Maximum	4.16	2.84	7
1800	Quarry tile, mud set, minimum	3.19	3.85	7.04
1820	Maximum	4.70	4.91	9.61
1840	Thin set, deduct		.77	.77
1850				
1860	Terrazzo precast, minimum	5.25	3.97	9.22
1880	Maximum	5.90	3.97	9.87
1900	Non-slip, minimum	11.10	19.90	31
1920	Maximum	16.60	27.50	44.10
1960	Wood, block, end grain factory type, creosoted, 2" thick	2.79	1.09	3.88
1980	2-1/2" thick	4.15	2.58	6.73
2000	3" thick	4.15	2.58	6.73
2020	Natural finish, 2" thick	4.51	2.58	7.09
2040	Fir, vertical grain, 1"x4", no finish, minimum	2.43	1.26	3.69
2060	Maximum	2.58	1.26	3.84
2080	Prefinished white oak, prime grade, 2-1/4" wide	6.25	1.89	8.14
2100	3-1/4" wide	7.75	1.74	9.49
2120	Maple strip, sanded and finished, minimum	3.51	2.88	6.39
2140	Maximum	4.06	2.88	6.94
2160	Oak strip, sanded and finished, minimum	4.50	2.88	7.38
2180	Maximum	4.17	2.88	7.05
2200	Parquetry, sanded and finished, minimum	2.30	3	5.30
2220	Maximum	6.60	4.21	10.81
2260	Add for sleepers on concrete, treated, 24" O.C., 1"x2"	1.58	1.70	3.28
2280	1"x3"	1.58	1.34	2.92
2300	2"x4"	.85	.67	1.52
2320	2"x6"	.85	.52	1.37
2340	Underlayment, plywood, 3/8" thick	.48	.44	.92
2350	1/2" thick	.59	.45	1.04
2360	5/8" thick	.71	.47	1.18
2370	3/4" thick	.87	.52	1.39
2380	Particle board, 3/8" thick	.34	.44	.78
2390	1/2" thick	.36	.45	.81
2400	5/8" thick	.39	.47	.86
2410	3/4" thick	.45	.52	.97
2420	Hardboard, 4' x 4', .215" thick	.50	.44	.94

Figure 16.10 (cont.)

6.7-100 Drywall Ceilings

	TYPE	FINISH	FURRING	SUPPORT		COST PER S.F. MAT.	COST PER S.F. INST.	COST PER S.F. TOTAL
4800	1/2" F.R. drywall	painted and textured	1"x3" wood, 16" O.C.	wood		.79	1.98	2.77
4900				masonry		.79	2.05	2.84
5000				concrete		.79	2.44	3.23
5100	5/8" F.R. drywall	painted and textured	1"x3" wood, 16" O.C.	wood		.79	1.98	2.77
5200				masonry		.79	2.05	2.84
5300				concrete		.79	2.44	3.23
5400	1/2" F.R. drywall	painted and textured	7/8"resil. channels	24" O.C.		.63	1.92	2.55
5500			1"x2" wood	stud clips		.76	1.86	2.62
5600			1-5/8"metal studs	24" O.C.		.73	1.91	2.64
5700	5/8" F.R. drywall	painted and textured	1-5/8"metal studs	24" O.C.		.73	1.91	2.64
5702								

6.7-100 Acoustical Ceilings

	TYPE	TILE	GRID	SUPPORT		COST PER S.F. MAT.	COST PER S.F. INST.	COST PER S.F. TOTAL
5800	5/8" fiberglass board	24" x 48"	tee	suspended		.87	.93	1.80
5900		24" x 24"	tee	suspended		.95	1.03	1.98
6000	3/4" fiberglass board	24" x 48"	tee	suspended		.95	.95	1.90
6100		24" x 24"	tee	suspended		1.03	1.05	2.08
6500	5/8" mineral fiber	12" x 12"	1"x3" wood, 12" O.C.	wood		.87	1.24	2.11
6600				masonry		.87	1.33	2.20
6700				concrete		.87	1.85	2.72
6800	3/4" mineral fiber	12" x 12"	1"x3" wood, 12" O.C.	wood		1.13	1.24	2.37
6900				masonry		1.13	1.24	2.37
7000				concrete		1.13	1.24	2.37
7100	3/4"mineral fiber on	12" x 12"	25 ga. channels	runners		1.27	1.70	2.97
7102	5/8" F.R. drywall							
7200	5/8" plastic coated	12" x 12"		adhesive backed		.81	.32	1.13
7201	Mineral fiber							
7202								
7300	3/4" plastic coated	12" x 12"		adhesive backed		1.07	.32	1.39
7301	Mineral fiber							
7302								
7400	3/4" mineral fiber	12" x 12"	conceal 2" bar &	suspended		1.59	1.64	3.23
7401			channels					
7402								

Figure 16.11

Assembly Number	Description	Qty	Unit	Total Cost Unit	Total Cost Total	Cost per S. F.
5.0	**Roofing**					
	Built-Up					
5.1-103-1400	3 ply w/ gravel, non-nailable, insulated	10000	SF	1.38	13800	
	Insulation					
5.7-101-2000	3" Urethane	10000	SF	1.23	12300	
	Flashing					
5.1-620-0350	Aluminum	756	SF	2.22	1678	
	Gravel Stop					
5.8-500-5200	Aluminum, 4"	117	LF	6.52	763	
					$28,541	$0.95
6.0	**Interior Construction**					
	Partitions					
6.1-210-6000	8" CMU at stairs & elevator	6410	SF	6.04	38716	
6.1-510-5400	Drywall, 5/8" G.B., 3 5/8" Stl. Studs	11600	SF	2.83	32828	
	Exterior Wall					
6.1-580-0649	Furring 1 1/2"@24"	6000	SF	1.22	7320	
6.1-580-0920	Insulation, Rigid 1"	8400	SF	0.67	5628	
6.1-580-0700	Gyp. Bd. 5/8"	6000	SF	0.55	3300	
6.1-580-0960	Tape and Finish	6000	SF	0.41	2460	
6.4-290-0140	Interior Doors	60	EA	460	27600	
	Toilet Partitions					
6.1-870-0680	Painted Metal	15	EA	457	6855	
6.1-870-0760	Handicapped Accessories	6	EA	255	1530	
6.1-870-1340	Urinal Screens	3	EA	254.50	764	
	Wall Finish					
6.5-100-0140	Painted Gyp. Bd.	16200	SF	0.48	7776	
6.5-100-0320	Painted CMU	5600	SF	1.01	5656	
6.5-100-1940	Ceramic Tile	950	SF	4.95	4703	
6.5-100-1800	Vinyl Wall Covering	1900	SF	1.45	2755	
	Floor Finish					
6.6-100-0160	Carpet	27900	SF	4.55	126945	
6.6-100-1720	Ceramic Tile	1200	SF	6.59	7908	
6.6-100-1820	Quarry Tile	900	SF	9.61	8649	
6.7-100-6000	Ceiling	30000	SF	1.90	57000	
					$348,392	$11.61
7.0	**Conveying**					

Figure 16.12

Chapter 17

Conveying

In construction, conveying assemblies are very specialized and can have a dramatic impact on the final cost of a building. Manufacturers and special consultants should be contacted for specific design and cost data, particularly prior to bidding a job. However, it is possible to put together a reasonable budget estimate during preliminary discussions and design.

There are many types of conveying assemblies that can be installed during construction. The ones most frequently used are:

- Conveyors
- Dumbwaiters
- Elevators
- Escalators
- Hoists & cranes
- Lifts
- Moving walks
- Pneumatic tubes

The discussion here will be limited to elevators, since these are used more frequently than the other systems. The points made, however, do apply to the other conveying systems.

Elevators

Before selection and pricing can even begin, there are specific questions that must be answered:

- Is the system to be hydraulic or electric, geared or gearless?
- What is the required capacity and cab size?
- How many floors or stops must the system have?
- How many elevators will be needed?
- What speed is necessary to service the building efficiently?
- Where will the associated machinery be located?
- What type of door, finish, and/or signals are necessary?
- Are there any special requirements that must be met?

Installed Costs

In almost all situations, elevators are provided and installed by the manufacturer. Typically, the costs quoted from the installer do not include such items as:

- The pit
- Any necessary sump pumps
- Shaft wall
- Temporary doors or shaft front
- Steel framing other than the rails that are considered part of the elevator equipment

Drilling into common earth is typically included in the installer's price, as is the piston casing for hydraulic elevators. Be sure to verify exactly what is and is not included in the installing contractor's price.

During the preliminary stages of a project, specific costs may not be possible simply because sufficient design detail is not available. Wherever possible, current prices should be obtained from the manufacturer or the manufacturer's certified installer. Verify who is responsible for inspections, tests, and permits. Some manufacturers include in their price a monthly maintenance program for the first year of operation with provisions to extend the program each year afterward for a specified price. Most, if not all, municipalities require some sort of annual inspection and certification of elevator systems.

Example: Elevator Costs for Three-Story Office Building

From the floor plan in Figure 16.3, it is apparent that two elevators are required, one at each entry lobby. Since the building is only three stories tall, hydraulic elevators with a 3000-pound capacity are most practical.

Figure 17.1 provides several choices of hydraulic elevators; however, the table gives prices only for two- and five-story systems. It is possible to accurately interpolate between the two systems to arrive at a reasonable price for a three-story system.

```
Hydraulic, 3000 lb.
2 Floors, 100 FPM (Figure 17.1 #2200) = $ 53,400 ea.
Hydraulic, 3000 lb.
5 Floors, 100 FPM (Figure 17.1 #2300) = $100,000 ea.
Difference                                $ 46,600
```

Taking the difference between the two systems and dividing by three, add the product to the two-floor system cost. The result will be a reasonably accurate system price for the three-floor elevator.

$46,600/3 = $15,533 + $53,400 = $68,933, rounded off to $69,000 ea.

Since two elevators are needed, the total cost for conveying assemblies is:

```
$69,000 ea. × 2 ea.   = $138,000
Cost per Square Foot = $4.60/S.F.
```

If there is subsurface rock at the building site or if it is anticipated, expect the drilling and excavation costs to be substantial. "Rock Shafts," as they are known, are very expensive and must be dealt with in the preliminary stages of the estimate. The presence and depth of rock are easily determined from analysis of nearby soil borings. Soil boring information may be obtained from local architects, engineers, or building or city inspectors. It is absolutely necessary that costs for drilling or excavation of any subsurface rock be provided for in the elevator estimate.

Figure 17.2 summarizes the costs for this division.

CONVEYING — A7.1-100 Hydraulic

The hydraulic elevator obtains its motion from the movement of liquid under pressure in the piston connected to the car bottom. These pistons can provide travel to a maximum rise of 70' and are sized for the intended load. As the rise reaches the upper limits the cost tends to exceed that of a geared electric unit.

System Components	QUANTITY	UNIT	MAT.	INST.	TOTAL
SYSTEM 7.1-100-2000					
PASS. ELEV. HYDRAULIC 2500 LB. 5 FLOORS, 100 FPM					
Passenger elevator, hydraulic, 1500 lb capacity, 2 stop, 100 FPM	1.000	Ea.	25,871.03	21,515.86	47,386.89
Over 10' travel height, passenger elevator, hydraulic, add	40.000	V.L.F.	12,600	13,440	26,040
Passenger elevator, hydraulic, 2500 lb capacity over standard, add	1.000	Ea.	1,250		1,250
Over 2 stops, passenger elevator, hydraulic, add	3.000	Stop	9,750	9,447	19,197
Hall lantern	5.000	Ea.	2,050		2,050
Maintenance agreement for pass. elev. 9 months	1.000	Ea.	2,450		2,450
Position indicator at lobby	1.000	Ea.	85	85.50	170.50
TOTAL			54,056.03	44,488.36	98,544.39

7.1-100	Hydraulic		MAT.	INST.	TOTAL
1300	Pass. elev., 1500 lb., 2 Floors, 100 FPM	R7.1 -010	29,100	21,600	50,700
1400	5 Floors, 100 FPM		53,000	44,500	97,500
1600	2000 lb., 2 Floors, 100 FPM		30,000	21,600	51,600
1700	5 floors, 100 FPM		53,500	44,500	98,000
1900	2500 lb., 2 Floors, 100 FPM		30,400	21,600	52,000
2000	5 floors, 100 FPM		54,000	44,500	98,500
2200	3000 lb., 2 Floors, 100 FPM		31,800	21,600	53,400
2300	5 floors, 100 FPM		55,500	44,500	100,000
2500	3500 lb., 2 Floors, 100 FPM		34,000	21,600	55,600
2600	5 floors, 100 FPM		57,500	44,500	102,000
2800	4000 lb., 2 Floors, 100 FPM		34,900	21,600	56,500
2900	5 floors, 100 FPM		58,500	44,500	103,000
3100	4500 lb., 2 Floors, 100 FPM		38,400	21,600	60,000
3200	5 floors, 100 FPM		64,000	44,400	108,400
4000	Hospital elevators, 3500 lb., 2 Floors, 100 FPM		34,700	23,600	58,300
4100	5 floors, 100 FPM		76,000	49,500	125,500
4300	4000 lb., 2 Floors, 100 FPM		34,700	23,600	58,300
4400	5 floors, 100 FPM		76,000	49,500	125,500
4600	4500 lb., 2 Floors, 100 FPM		40,000	23,600	63,600
4800	5 floors, 100 FPM		81,500	49,500	131,000
4900	5000 lb., 2 Floors, 100 FPM		42,500	23,600	66,100
5000	5 floors, 100 FPM		84,000	49,500	133,500
6700	Freight elevators (Class "B"), 3000 lb., 2 Floors, 50 FPM		40,800	18,900	59,700
6800	5 floors, 100 FPM		77,000	47,200	124,200
7000	4000 lb., 2 Floors, 50 FPM		47,300	21,100	68,400
7100	5 floors, 100 FPM		76,000	45,000	121,000
7500	10,000 lb., 2 Floors, 50 FPM		54,000	18,900	72,900
7600	5 floors, 100 FPM		90,000	47,200	137,200
8100	20,000 lb., 2 Floors, 50 FPM		65,500	18,900	84,400
8200	5 Floors, 100 FPM		101,500	47,200	148,700

Figure 17.1

Assembly Number	Description	Qty	Unit	Total Cost Unit	Total Cost Total	Cost per S. F.
5.0	**Roofing**					
	Built-Up					
5.1-103-1400	3 ply w/ gravel, non-nailable, insulated	10000	SF	1.38	13800	
	Insulation					
5.7-101-2000	3" Urethane	10000	SF	1.23	12300	
	Flashing					
5.1-620-0350	Aluminum	756	SF	2.22	1678	
	Gravel Stop					
5.8-500-5200	Aluminum, 4"	117	LF	6.52	763	
					$28,541	$0.95
6.0	**Interior Construction**					
	Partitions					
6.1-210-6000	8" CMU at stairs & elevator	6410	SF	6.04	38716	
6.1-510-5400	Drywall, 5/8" G.B., 3 5/8" Stl. Studs	11600	SF	2.83	32828	
	Exterior Wall					
6.1-580-0649	Furring 1 1/2"@24"	6000	SF	1.22	7320	
6.1-580-0920	Insulation, Rigid 1"	8400	SF	0.67	5628	
6.1-580-0700	Gyp. Bd. 5/8"	6000	SF	0.55	3300	
6.1-580-0960	Tape and Finish	6000	SF	0.41	2460	
6.4-290-0140	Interior Doors	60	EA	460	27600	
	Toilet Partitions					
6.1-870-0680	Painted Metal	15	EA	457	6855	
6.1-870-0760	Handicapped Accessories	6	EA	255	1530	
6.1-870-1340	Urinal Screens	3	EA	254.50	764	
	Wall Finish					
6.5-100-0140	Painted Gyp. Bd.	16200	SF	0.48	7776	
6.5-100-0320	Painted CMU	5600	SF	1.01	5656	
6.5-100-1940	Ceramic Tile	950	SF	4.95	4703	
6.5-100-1800	Vinyl Wall Covering	1900	SF	1.45	2755	
	Floor Finish					
6.6-100-0160	Carpet	27900	SF	4.55	126945	
6.6-100-1720	Ceramic Tile	1200	SF	6.59	7908	
6.6-100-1820	Quarry Tile	900	SF	9.61	8649	
6.7-100-6000	Ceiling	30000	SF	1.90	57000	
					$348,392	$11.61
7.0	**Conveying**					
	Hydraulic 3000lb.					
7.1-100-2200	2 Floors 53400					
7.1-100-2300	5 Floors 100000					
	Difference 46600					
	Per Floor 46000/3 = 15533					
	+53400					
	68933	2	EA	69000	$138,000	$4.60

Figure 17.2

Chapter 18

Mechanical

Developing the assemblies estimate and extending it to a square foot estimate is a reasonably easy exercise for the mechanical portion of a building when certain parameters of the building have been defined. These include:

- Building shape
- Number of floors
- Floor area
- Volume
- Expected occupancy (number of persons and male/female ratio)
- Some idea of the systems desired

With a fair amount of knowledge and care, an estimate for the mechanical portion can be prepared with confidence using the assemblies format.

Scope

The scope of the mechanical estimate includes work within the building that is usually performed by the mechanical contractors, but does not include the utility work beyond the 5' building perimeter. Division 12, Site Work, provides the necessary data for the utility work required outside the building's 5' perimeter. These costs are included in the Site Work portion of the estimate. However, when mechanical equipment such as cooling towers, boilers, and chillers are physically located outside the building, the equipment and its piping are assumed to be part of the mechanical system.

Systems

A very convenient feature of mechanical work is its ability to be broken down into four identifiable systems:

- Plumbing
- Fire Protection
- Heating
- Air Conditioning

Why four subdivisions? Often on the construction site the mechanical work is done by two, three, or four separate and different subcontractors. This is possible because each subdivision is independent of the others. The only exception would be the heating and air conditioning when these systems are combined or must use the same or similar equipment. The lack of dependence of systems on one another adds another advantage to assemblies square foot estimating—the estimates for each system can be accomplished without regard to the others.

The mechanical estimate for the Three-Story Office Building is divided into the four subdivisions noted above. Each part is fully discussed in this chapter and followed by an Assemblies Estimate Worksheet similar to one shown in Figure 18.1.

Plumbing Assemblies

There is a definite method to follow when developing the estimate for plumbing systems:

1. Determine the type and number of fixtures.
2. Select the appropriate price for each fixture system.
3. Develop the necessary piping costs.
4. Size and price the water heating components.
5. Work up the costs for the storm and roof drainage system.

The Estimate Worksheet
The Plumbing Estimate Worksheet in Figure 18.1 is useful for working up the estimate for the plumbing systems. It has sufficient space for development of the plumbing assemblies costs, with ample room for cost source, quantities, extensions, and notes. Using it prior to transferring the total figures to the estimate form can help streamline the estimating process.

Plumbing Fixtures
In Chapter 8, the minimum plumbing fixtures required on each floor of the Three-Story Office Building were determined and are reproduced again in Figure 18.2.

Once the fixtures are selected, it is rather easy to price each one by finding the appropriate system price from the tables in Figure 18.3 through 18.7 or from some other source. Note that these systems from *Means Assemblies Cost Data* include approximately 10′ of the supply and drain piping, as well as the associated stops, traps, tail pieces, etc. This allowance assumes that the fixtures are connected to branches from a main.

Piping

If drawings of the proposed building are available and the estimator has enough knowledge and experience with plumbing, then it is possible to put together an estimate for whatever piping, controls, valves, etc., are necessary for the various fixtures to work successfully. Usually the mechanical contractor and plumber have this required knowledge and experience.

The "Plumbing Approximations" table in Figure 18.8 can be used to estimate the costs for water control, pipe and fittings, and project quality/complexity. The factors are a percentage of the total costs of all the plumbing fixtures in the project. They serve only as allowances and are not sufficient for special or process piping applications. Since the percentages are only guides, the estimator must decide whether the percentages are too low, too high, or just right for the particular project. Use the lower percentages for compact buildings, or for structures with the plumbing in one area or with fixtures vertically stacked one upon the other. The larger percentages should be used when the plumbing is spread out in the building. The client's requirements and preliminary drawings determine which of the quality/complexity multipliers to use. The percentages can be used separately for both material and installation or applied just once against the total cost.

Plumbing Estimate Worksheet

Fixture	Table Number	Quantity	Unit Cost	Total Costs	Calculations
Bathtubs		$	$		
Drinking Fount		$	$		
Kitchen Sink		$	$		
Laundry Sink		S	S		
Lavatory		$	$		
Service Sink		S	S		
Shower		$	$		
Urinal		$	S		
Water Cooler		S	$		
Water Closet		S	S		
W/C Group		S	S		
Wash Fount. Group		S	S		
Bathrooms		$	$		
		$	$		
		$	$		
		$	$		
Water Heater		$	$		
		$	$		
		$	$		
		$	$		
Roof Drains		$	$		
SUBTOTAL		$	$		**Comments**
Water Control		% ST.	$		
Pipe & Fittings		% ST.	$		
Quality Complexity		% ST.	$		
Other		% ST.	$		
TOTAL			$		

Figure 18.1

Example One: Plumbing Costs for the Three-Story Office Building

Plumbing Costs:

Lavatory, Vanity Top (Figure 18.3 #1600)	18 ea. @ $ 661 ea.	=	$11,898
Service Sink (Figure 18.4 #4380)	3 ea. @ $1,415 ea.	=	4,245
Urinal (Figure 18.5 #2000)	3 ea. @ $1,005 ea.	=	3,015
Water Cooler (Figure 18.6 #1880)	3 ea. @ $1,085 ea.	=	3,255
Water Closet (Figure 18.7 #1840)	15 ea. @ $ 1,370 ea.	=	20,550
Subtotal			$42,963
Water Control (Figure 18.8)	10% of Subtotal	=	4,296
Pipe and Fittings (Figure 18.8)	30% of Subtotal	=	12,889
Quality/Complexity (Figure 18.8)	10% of Subtotal	=	4,296
Total Cost			$64,444
Cost per Square Foot		=	$2.15/S.F.

Water Heaters

In Chapter 8, it was determined that the Three-Story Office Building needed 30 gallons/floor for domestic hot water. Figure 18.9 (line #1820) shows an electric, 50-gallon tank with 37 gallons per hour heating capacity.

Cost:

(Figure 18.9 #1820) 3 ea. @ $3,235 ea. = $9,705
Cost per Square Foot = $0.32/S.F.

Storm Drainage Roof Drains

The roof drainage system was sized in Chapter 8. The key to pricing is to remember the drainage pipe. Most estimators remember to include the roof drains, but fail to catch the vertical and horizontal drain pipes.

	Fixtures Per Floor			
Fixture	Minimum Required	Women	Men	Handicapped
Water Closets	4	2	1	2
Urinals	—	—	1	—
Lavatories	4	2	2	2
Drinking Fountain	1	—	—	—
Service Sink	1	—	—	—

Figure 18.2

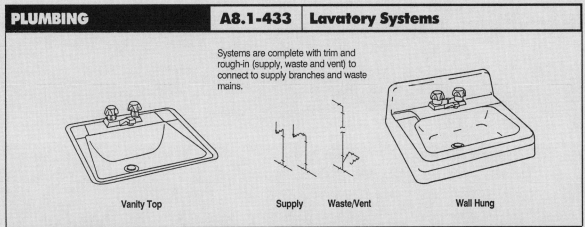

Systems are complete with trim and rough-in (supply, waste and vent) to connect to supply branches and waste mains.

Vanity Top   **Supply**   **Waste/Vent**   **Wall Hung**

System Components	QUANTITY	UNIT	COST EACH		
			MAT.	INST.	TOTAL
SYSTEM 8.1-433-1560					
LAVATORY W/TRIM, VANITY TOP, P.E. ON C.I., 20″ X 18″					
Lavatory w/trim, PE on CI, white, vanity top, 20″ x 18″ oval	1.000	Ea.	176	104	280
Pipe, steel, galvanized, schedule 40, threaded, 1-1/4″ diam	4.000	L.F.	19.28	29.80	49.08
Copper tubing type DWV, solder joint, hanger 10'OC 1-1/4″ diam	4.000	L.F.	14.20	24.60	38.80
Wrought copper DWV, Tee, sanitary, 1-1/4″ diam	1.000	Ea.	6.50	41	47.50
P trap w/cleanout, 20 ga, 1-1/4″ diam	1.000	Ea.	31	20.50	51.50
Copper tubing type L, solder joint, hanger 10' OC 1/2″ diam	10.000	L.F.	16.30	45.40	61.70
Wrought copper 90° elbow for solder joints 1/2″ diam	2.000	Ea.	.82	36.80	37.62
Wrought copper Tee for solder joints, 1/2″ diam	2.000	Ea.	1.32	57	58.32
Stop, chrome, angle supply, 1/2″ diam	2.000	Ea.	12.50	33.50	46
TOTAL			277.92	392.60	670.52

8.1-433	**Lavatory Systems**		COST EACH		
			MAT.	INST.	TOTAL
1560	Lavatory w/trim, vanity top, PE on CI, 20″ x 18″, Vanity top by others.		278	395	673
1600	19″ x 16″ oval		266	395	661
1640	18″ round	R8.1 -400	271	395	666
1680	Cultured marble, 19″ x 17″		194	395	589
1720	25″ x 19″		210	395	605
1760	Stainless, self-rimming, 25″ x 22″		249	395	644
1800	17″ x 22″		243	395	638
1840	Steel enameled, 20″ x 17″		193	405	598
1880	19″ round		191	405	596
1920	Vitreous china, 20″ x 16″		283	410	693
1960	19″ x 16″		283	410	693
2000	22″ x 13″		293	410	703
2040	Wall hung, PE on CI, 18″ x 15″		515	435	950
2080	19″ x 17″		500	435	935
2120	20″ x 18″		375	435	810
2160	Vitreous china, 18″ x 15″		375	445	820
2200	19″ x 17″		320	445	765
2240	24″ x 20″		450	445	895

Figure 18.3

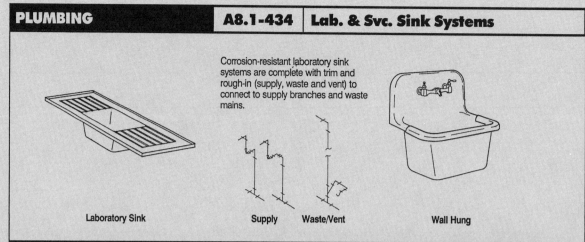

Corrosion-resistant laboratory sink systems are complete with trim and rough-in (supply, waste and vent) to connect to supply branches and waste mains.

Laboratory Sink   Supply   Waste/Vent   Wall Hung

System Components	QUANTITY	UNIT	COST EACH MAT.	COST EACH INST.	COST EACH TOTAL
SYSTEM 8.1-434-1600					
LABORATORY SINK W/TRIM, POLYETHYLENE, SINGLE BOWL					
DOUBLE DRAINBOARD, 54" X 24" OD					
Sink w/trim, polyethylene, 1 bowl, 2 drainboards 54" x 24" OD	1.000	Ea.	820	221	1,041
Pipe, polypropylene, schedule 40, acid resistant 1-1/2" diam	10.000	L.F.	13.40	97.50	110.90
Tee, sanitary, polypropylene, acid resistant, 1-1/2" diam	1.000	Ea.	20	37	57
P trap, polypropylene, acid resistant, 1-1/2" diam	1.000	Ea.	23	21.50	44.50
Copper tubing type L, solder joint, hanger 10' OC 1/2" diam	10.000	L.F.	16.30	45.40	61.70
Wrought copper 90° elbow for solder joints 1/2" diam	2.000	Ea.	.82	36.80	37.62
Wrought copper Tee for solder joints, 1/2" diam	2.000	Ea.	1.32	57	58.32
Stop, angle supply, chrome, 1/2" diam	2.000	Ea.	12.50	33.50	46
TOTAL			907.34	549.70	1,457.04

8.1-434	Laboratory & Service Sink Systems		COST EACH MAT.	COST EACH INST.	COST EACH TOTAL
1580	Laboratory sink w/trim, polyethylene, single bowl,				
1600	Double drainboard, 54" x 24" O.D.		905	550	1,455
1640	Single drainboard, 47" x 24"O.D.	R8.1 -400	945	550	1,495
1680	70" x 24" O.D.		960	550	1,510
1760	Flanged, 14-1/2" x 14-1/2" O.D.		350	495	845
1800	18-1/2" x 18-1/2" O.D.		315	495	810
1840	23-1/2" x 20-1/2" O.D.		365	495	860
1920	Polypropylene, cup sink, oval, 7" x 4" O.D.		161	435	596
1960	10" x 4-1/2" O.D.		172	435	607
2000					
4260	Service sink w/trim, PE on CI, corner floor, 28" x 28", w/rim guard		835	555	1,390
4300	Wall hung w/rim guard, 22" x 18"		720	640	1,360
4340	24" x 20"		760	640	1,400
4380	Vitreous china, wall hung 22" x 20"		775	640	1,415

Figure 18.4

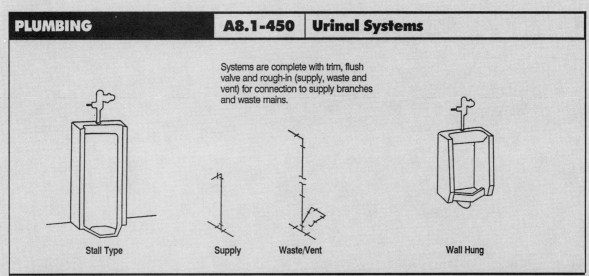

Systems are complete with trim, flush valve and rough-in (supply, waste and vent) for connection to supply branches and waste mains.

Stall Type   Supply   Waste/Vent   Wall Hung

System Components	QUANTITY	UNIT	COST EACH		
			MAT.	INST.	TOTAL
SYSTEM 8.1-450-2000					
URINAL, VITREOUS CHINA, WALL HUNG					
Urinal, wall hung, vitreous china, incl. hanger	1.000	Ea.	420	221	641
Pipe, steel, galvanized, schedule 40, threaded, 1-1/2" diam	5.000	L.F.	27	41.50	68.50
Copper tubing type DWV, solder joint, hangers 10'OC, 2" diam	3.000	L.F.	16.95	25.05	42
Combination Y & 1/8 bend for CI soil pipe, no hub, 3" diam	1.000	Ea.	8.15		8.15
Pipe, CI, no hub, cplg 10' OC, hanger 5' OC, 3" diam	4.000	L.F.	28.60	41.40	70
Pipe coupling standard, CI soil, no hub, 3" diam	3.000	Ea.	9.44	34.90	44.34
Copper tubing type L, solder joint, hanger 10' OC 3/4" diam	5.000	L.F.	11.35	24.20	35.55
Wrought copper 90° elbow for solder joints 3/4" diam	1.000	Ea.	.86	19.35	20.21
Wrought copper Tee for solder joints, 3/4" diam	1.000	Ea.	1.60	30.50	32.10
TOTAL			530.80	474.90	1,005.70

8.1-450	Urinal Systems		COST EACH		
			MAT.	INST.	TOTAL
2000	Urinal, vitreous china, wall hung	R8.1 -400	530	475	1,005
2040	Stall type		895	525	1,420

Figure 18.5

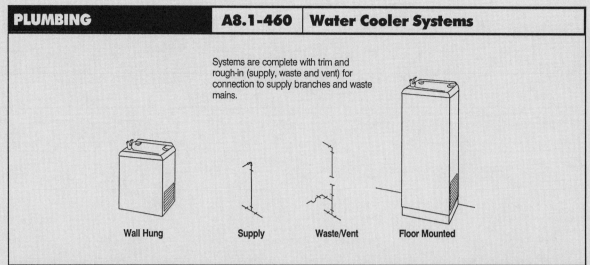

PLUMBING	A8.1-460	Water Cooler Systems

Systems are complete with trim and rough-in (supply, waste and vent) for connection to supply branches and waste mains.

Wall Hung   **Supply**   **Waste/Vent**   **Floor Mounted**

System Components			COST EACH		
	QUANTITY	UNIT	MAT.	INST.	TOTAL
SYSTEM 8.1-460-1840					
WATER COOLER, ELECTRIC, SELF CONTAINED, WALL HUNG, 8.2 GPH					
Water cooler, wall mounted, 8.2 GPH	1.000	Ea.	475	166	641
Copper tubing type DWV, solder joint, hanger 10'OC 1-1/4" diam	4.000	L.F.	14.20	24.60	38.80
Wrought copper DWV, Tee, sanitary 1-1/4" diam	1.000	Ea.	6.50	41	47.50
P trap, copper drainage, 1-1/4" diam	1.000	Ea.	31	20.50	51.50
Copper tubing type L, solder joint, hanger 10' OC 3/8" diam	5.000	L.F.	7.50	21.90	29.40
Wrought copper 90° elbow for solder joints 3/8" diam	1.000	Ea.	1.09	16.75	17.84
Wrought copper Tee for solder joints, 3/8" diam	1.000	Ea.	1.83	26.50	28.33
Stop and waste, straightway, bronze, solder, 3/8" diam	1.000	Ea.	4.17	15.35	19.52
TOTAL			541.29	332.60	873.89

8.1-460	Water Cooler Systems	COST EACH		
		MAT.	INST.	TOTAL
1840	Water cooler, electric, wall hung, 8.2 GPH	540	335	875
1880	Dual height, 14.3 GPH	745	340	1,085
1920	Wheelchair type, 7.5 G.P.H.	1,200	335	1,535
1960	Semi recessed, 8.1 G.P.H.	720	335	1,055
2000	Full recessed, 8 G.P.H.	1,125	355	1,480
2040	Floor mounted, 14.3 G.P.H.	570	290	860
2080	Dual height, 14.3 G.P.H.	790	350	1,140
2120	Refrigerated compartment type, 1.5 G.P.H.	1,000	290	1,290

R8.1 -400

Figure 18.6

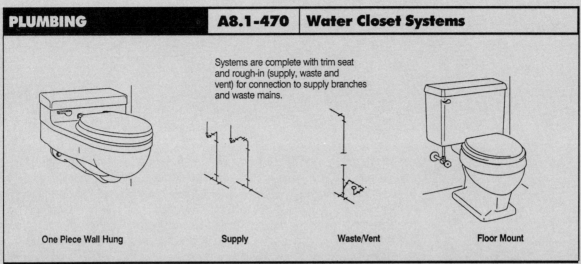

PLUMBING	A8.1-470	Water Closet Systems

Systems are complete with trim seat and rough-in (supply, waste and vent) for connection to supply branches and waste mains.

One Piece Wall Hung   **Supply**   **Waste/Vent**   **Floor Mount**

System Components	QUANTITY	UNIT	COST EACH		
			MAT.	INST.	TOTAL
SYSTEM 8.1-470-1840 WATER CLOSET, VITREOUS CHINA, ELONGATED TANK TYPE, WALL HUNG, ONE PIECE					
Wtr closet tank type vit china wall hung 1 pc w/seat supply & stop	1.000	Ea.	790	125	915
Pipe steel galvanized, schedule 40, threaded, 2" diam	4.000	L.F.	29.80	41.40	71.20
Pipe, CI soil, no hub, cplg 10' OC, hanger 5' OC, 4" diam	2.000	L.F.	18.10	22.80	40.90
Pipe, coupling, standard coupling, CI soil, no hub, 4" diam	2.000	Ea.	11.20	40	51.20
Copper tubing type L, solder joint, hanger 10'OC, 1/2" diam	6.000	L.F.	9.78	27.24	37.02
Wrought copper 90° elbow for solder joints 1/2" diam	2.000	Ea.	.82	36.80	37.62
Wrought copper Tee for solder joints 1/2" diam	1.000	Ea.	.66	28.50	29.16
Support/carrier, for water closet, siphon jet, horiz, single, 4" waste	1.000	Ea.	126	61.50	187.50
TOTAL			986.36	383.24	1,369.60

8.1-470	Water Closet Systems		COST EACH		
			MAT.	INST.	TOTAL
1800	Water closet, vitreous china, elongated				
1840	Tank type, wall hung, one piece		985	385	1,370
1880	Close coupled two piece	R8.1 -400	620	385	1,005
1920	Floor mount, one piece		740	415	1,155
1960	One piece low profile		1,000	415	1,415
2000	Two piece close coupled		310	415	725
2040	Bowl only with flush valve				
2080	Wall hung		565	435	1,000
2120	Floor mount		470	420	890

Figure 18.7

Costs:
4″ Drain with 10′ of pipe (Cast Iron)
    (Figure 18.10 #4200)   4 ea. @ $775 ea.   = $3,100
5″ Drain Pipe (Cast Iron)
    (Figure 18.10 #4320)   60 L.F. @ $24.20/L.F.   =   1,452
6″ Drain Pipe (Cast Iron)
    (Figure 18.10 #4400)   200 L.F. @ $27.10/L.F. =   5,420
Total Cost $9,972
Cost per Square Foot   = $0.33/S.F.

Fire Protection

There are many different types, or "families," of fire protection and suppression systems available to the construction industry today. The choice depends to a great extent on the building in question, the materials used, and the type of occupancy. In most cases, the code agency determines the fire hazard classification: Light, Ordinary, or Extra Hazard. Figure 18.11 explains the three different classifications. Once the classification is determined and the type of system is selected, the cost estimate becomes very easy.

Once the fire hazard classification is determined and the type of system is selected, find the appropriate table for pricing. The pricing tables are set up by system type, fire hazard classification, and square footage that the system will cover. Standpipes and the other components necessary for system operation can be priced out from the tables that follow.

Plumbing Approximations for Quick Estimating

Water Control

Water Meter; Backflow Preventer; .... 10 to 15% of Fixtures
Shock Absorbers; Vacuum Breakers;
Mixer.

Pipe And Fittings .... 30 to 60% of Fixtures

Note: Lower percentage for compact buildings or larger buildings with plumbing in one area.
Larger percentage for large buildings with plumbing spread out.
In extreme cases pipe may be more than 100% of fixtures.
Percentages **do not** include special purpose or process piping.

Plumbing Labor

1 & 2 Story Residential .... Rough-in Labor = 80% of Materials
Apartment Buildings .... Rough-in Labor = 90 to 100% of Materials
Labor for handling and placing fixtures is approximately 25 to 30% of fixtures.

Quality/Complexity Multiplier (for all installations)

Economy installation, add .... 0 to 5%
Good quality, medium complexity, add .... 5 to 15%
Above average quality and complexity, add .... 15 to 25%

Figure 18.8

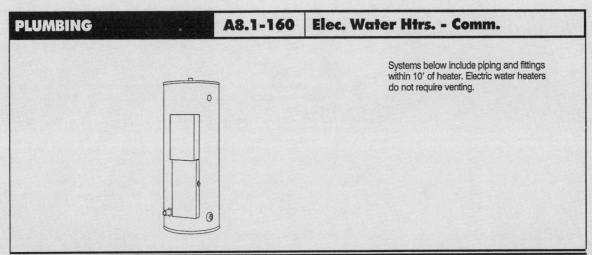

Systems below include piping and fittings within 10' of heater. Electric water heaters do not require venting.

System Components	QUANTITY	UNIT	COST EACH		
			MAT.	INST.	TOTAL
SYSTEM 8.1-160-1820					
ELECTRIC WATER HEATER, COMMERCIAL, 100° F RISE					
50 GALLON TANK, 9 KW, 37 GPH					
Water heater, commercial, electric, 50 Gal, 9 KW, 37 GPH	1.000	Ea.	2,425	204	2,629
Copper tubing, type L, solder joint, hanger 10' OC, 3/4" diam	34.000	L.F.	77.18	164.56	241.74
Wrought copper 90° elbow for solder joints 3/4" diam	5.000	Ea.	4.30	96.75	101.05
Wrought copper Tee for solder joints, 3/4" diam	2.000	Ea.	3.20	61	64.20
Wrought copper union for soldered joints, 3/4" diam	2.000	Ea.	9.50	41	50.50
Valve, gate, bronze, 125 lb, NRS, soldered 3/4" diam	2.000	Ea.	30.90	36.80	67.70
Relief valve, bronze, press & temp, self-close, 3/4" IPS	1.000	Ea.	56.50	13.15	69.65
Wrought copper adapter, copper tubing to male, 3/4" IPS	1.000	Ea.	1.34	17.50	18.84
TOTAL			2,607.92	634.76	3,242.68

8.1-160	Electric Water Heaters - Commercial Systems		COST EACH		
			MAT.	INST.	TOTAL
1800	Electric water heater, commercial, 100° F rise	R8.1 -100			
1820	50 gallon tank, 9 KW 37 GPH		2,600	635	3,235
1860	80 gal, 12 KW 49 GPH		3,400	785	4,185
1900	36 KW 147 GPH		4,650	850	5,500
1940	120 gal, 36 KW 147 GPH		5,025	915	5,940
1980	150 gal, 120 KW 490 GPH		17,300	985	18,285
2020	200 gal, 120 KW 490 GPH		18,200	1,000	19,200
2060	250 gal, 150 KW 615 GPH		20,200	1,175	21,375
2100	300 gal, 180 KW 738 GPH		22,100	1,250	23,350
2140	350 gal, 30 KW 123 GPH		15,500	1,350	16,850
2180	180 KW 738 GPH		22,600	1,350	23,950
2220	500 gal, 30 KW 123 GPH		19,900	1,575	21,475
2260	240 KW 984 GPH		31,500	1,575	33,075
2300	700 gal, 30 KW 123 GPH		24,100	1,800	25,900
2340	300 KW 1230 GPH		38,300	1,800	40,100
2380	1000 gal, 60 KW 245 GPH		29,300	2,525	31,825
2420	480 KW 1970 GPH		52,000	2,525	54,525
2460	1500 gal, 60 KW 245 GPH		43,300	3,100	46,400
2500	480 KW 1970 GPH		61,500	3,100	64,600

Figure 18.9

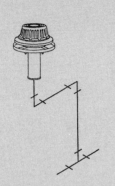

Design Assumptions: Vertical conductor size is based on a maximum rate of rainfall of 4″ per hour. To convert roof area to other rates multiply "Max. S.F. Roof Area" shown by four and divide the result by desired local rate. The answer is the local roof area that may be handled by the indicated pipe diameter.

Basic cost is for roof drain, 10′ of vertical leader and 10′ of horizontal, plus connection to the main.

Pipe Dia.	Max. S.F. Roof Area	Gallons per Min.
2″	544	23
3″	1610	67
4″	3460	144
5″	6280	261
6″	10,200	424
8″	22,000	913

System Components	QUANTITY	UNIT	COST EACH		
			MAT.	INST.	TOTAL
SYSTEM 8.1-310-1880					
ROOF DRAIN, DWV PVC PIPE, 2″ DIAM., 10′ HIGH					
Drain, roof, main, PVC, dome type 2″ pipe size	1.000	Ea.	37.50	47.50	85
Clamp, roof drain, underdeck	1.000	Ea.	18.70	27.50	46.20
Pipe, Tee, PVC DWV, schedule 40, 2″ pipe size	1.000	Ea.	1.53	39	40.53
Pipe, PVC, DWV, schedule 40, 2″ diam.	20.000	L.F.	49.20	225	274.20
Pipe, elbow, PVC schedule 40, 2″ diam.	2.000	Ea.	3.20	47	50.20
TOTAL			110.13	386	496.13

8.1-310	Roof Drain Systems	COST PER SYSTEM		
		MAT.	INST.	TOTAL
1880	Roof drain, DWV PVC, 2″ diam., piping, 10′ high	110	385	495
1920	For each additional foot add	2.46	11.25	13.71
1960	3″ diam., 10′ high	150	465	615
2000	For each additional foot add	3.92	12.50	16.42
2040	4″ diam., 10′ high	186	520	706
2080	For each additional foot add	5.05	13.80	18.85
2120	5″ diam., 10′ high	375	560	935
2160	For each additional foot add	7.70	15.40	23.10
2200	6″ diam., 10′ high	415	665	1,080
2240	For each additional foot add	8.70	17	25.70
2280	8″ diam., 10′ high	900	1,050	1,950
2320	For each additional foot add	14.35	21.50	35.85
3940	C.I., soil, single hub, service wt., 2″ diam. piping, 10′ high	184	405	589
3980	For each additional foot add	4.53	10.50	15.03
4120	3″ diam., 10′ high	237	435	672
4160	For each additional foot add	6.20	11.05	17.25
4200	4″ diam., 10′ high	300	475	775
4240	For each additional foot add	8.50	12.05	20.55
4280	5″ diam., 10′ high	425	525	950
4320	For each additional foot add	10.65	13.55	24.20
4360	6″ diam., 10′ high	500	565	1,065
4400	For each additional foot add	13	14.10	27.10
4440	8″ diam., 10′ high	1,025	1,150	2,175
4480	For each additional foot add	20.50	24	44.50
6040	Steel galv. sch 40 threaded, 2″ diam. piping, 10′ high	270	390	660
6080	For each additional foot add	7.45	10.35	17.80
6120	3″ diam., 10′ high	475	560	1,035
6160	For each additional foot add	13.70	15.40	29.10
6200	4″ diam., 10′ high	690	720	1,410
6240	For each additional foot add	19.90	18.40	38.30

Figure 18.10

Example Two: Fire Protection Costs for the Three-Story Office Building

The office building is in the Light Hazard classification. The owner wants a wet pipe sprinkler system because of the potential for varied occupancies in the building.

Costs:
Wet Pipe Sprinkler System

(Figure 18.12 #0620)	10,000 S.F. @ $1.58/S.F.	=	$15,800
(Figure 18.12 #0740)	20,000 S.F. @ $1.27/S.F.	=	25,400

Fire Protection Classification

System Classification

Rules for installation of sprinkler systems vary depending on the classification of occupancy falling into one of three categories as follows:

Light Hazard Occupancy

The protection area allotted per sprinkler should not exceed 200 S.F. with the maximum distance between lines and sprinklers on lines being 15'. The sprinklers do not need to be staggered. Branch lines should not exceed eight sprinklers on either side of a cross main. Each large area requiring more than 100 sprinklers and without a sub-dividing partition should be supplied by feed mains or risers sized for ordinary hazard occupancy.

Included in this group are:

Auditoriums	Museums
Churches	Nursing Homes
Clubs	Offices
Educational	Residential
Hospitals	Restaurants
Institutional	Schools
Libraries	Theaters
(except large stack rooms)	

Ordinary Hazard Occupancy

The protection area allotted per sprinkler shall not exceed 130 S.F. of noncombustible ceiling and 120 S.F. of combustible ceiling. The maximum allowable distance between sprinkler lines and sprinklers on line is 15'. Sprinklers shall be staggered if the distance between heads exceeds 12'. Branch lines should not exceed eight sprinklers on either side of a cross main.

Included in this group are:

Automotive garages	Electric generating stations
Bakeries	Feed mills
Beverage manufacturing	Grain elevators
Bleacheries	Ice manufacturing
Boiler houses	Laundries
Canneries	Machine shops
Cement plants	Mercantiles
Clothing factories	Paper mills
Cold storage warehouses	Printing and Publishing
Dairy products manufacturing	Shoe factories
Distilleries	Warehouses
Dry cleaning	Wood product assembly

Extra Hazard Occupancy

The protection area allotted per sprinkler shall not exceed 90 S.F. of noncombustible ceiling and 80 S.F. of combustible ceiling. The maximum allowable distance between lines and between sprinklers on lines is 12'. Sprinklers on alternate lines shall be staggered if the distance between sprinklers on lines exceeds 8'. Branch lines should not exceed six sprinklers on either side of a cross main.

Included in this group are:

Aircraft hangars	Paint shops
Chemical works	Shade cloth manufacturing
Explosives manufacturing	Solvent extracting
Linoleum manufacturing	Varnish works
Linseed oil mills	Volatile flammable
Oil refineries	liquid manufacturing & use

Figure 18.11

Standpipe				
(Figure 18.13 #1540)	(12'/10') @ $3,900/floor	=	$ 4,680
(Figure 18.13 #1560)	(12'/10') @ $965/floor	=	1,158

Components:
Cabinets				
(Figure 18.14 #0750)	3 ea. @ $257 ea.	=	771
Hose			
(Figure 18.14 #5200)	300 L.F. @ $2.11/L.F.	=	633
Nozzle			
(Figure 18.14 #5500)	3 ea. @ $28 ea.	=	84
Rack			
(Figure 18.14 #6000)	3 ea. @ $67 ea.	=	201
Siamese			
(Figure 18.14 #7200)	1 ea. @ $468 ea.	=	468
Alarm			
(Figure 18.14 #1650)	1 ea. @ $131.75 ea.	=	132
Valves			
(Figure 18.14 #8050)	3 ea. @ $115.50 ea.	=	346
Hydrolator			
(Figure 18.14 #8300)	3 ea. @ $65.50 ea.	=	197
Adapter			
(Figure 18.14 #0100)	3 ea. @ $32 ea.	=	96
Total Cost			$49,966
Cost per Square Foot		=	$1.67/S.F.

Heating, Ventilation, and Air Conditioning

Mechanical costs in most buildings represent as much as 20% of the total building cost for typical buildings. In special building types such as hospitals, laboratories, and some manufacturing plants, the mechanical costs can be even higher. Since the mechanical work has the potential to be costly, designers must select the needed systems in consideration of not only initial construction costs, but also future operating costs.

In some cases, the type of HVAC system selected may be dictated by the owner's preference or specific direction. The best possible selection comes from the engineer who, based on some initial design assumptions and calculations, can recommend one or two system types that would be most suitable.

The key is to provide the most correct information possible. For example, in the conceptual stages of a future project it may be decided that a commercial building for general lease space will be constructed. The HVAC system selected is a roof-top, multizone type. A little further into the design process, the client decides that the building will be used by a specific industry as a regional headquarters. The new functions in the proposed building require that the HVAC system be much more responsive to changing occupancies and uses on a daily basis. In addition, several areas in the building will generate an inordinant amount of waste heat, while other areas will require very specific and critical environmental control. Consequently, the original costs must be quickly reassessed to determine if there is enough money in the HVAC budget for the new building use.

Heating Assemblies

The most appropriate heating assembly depends on the desired efficiency, required capacity (which is a function of the building envelope's heat loss characteristics), amount of space available for equipment, type of heating source (steam, hot water), and so on. Several code agencies have specific

A8.2-110 | Wet Pipe Sprinkler Systems

Wet Pipe System. A system employing automatic sprinklers attached to a piping system containing water and connected to a water supply so that water discharges immediately from sprinklers opened by heat from a fire.

All areas are assumed to be open.

System Components	QUANTITY	UNIT	COST EACH		
			MAT.	INST.	TOTAL
SYSTEM 8.2-110-0580					
WET PIPE SPRINKLER, STEEL, BLACK, SCH. 40 PIPE					
LIGHT HAZARD, ONE FLOOR, 2000 S.F.					
Valve, gate, iron body, 125 lb, OS&Y, flanged, 4" diam	1.000	Ea.	174.75	165.75	340.50
Valve, swing check, bronze, 125 lb, regrinding disc, 2-1/2" pipe size	1.000	Ea.	173.25	33	206.25
Valve, angle, bronze, 150 lb, rising stem, threaded, 2" diam	1.000	Ea.	198	25.13	223.13
*Alarm valve, 2-1/2" pipe size	1.000	Ea.	502.50	172.50	675
Alarm, water motor, complete with gong	1.000	Ea.	105.75	72	177.75
Valve, swing check, w/balldrip CI with brass trim 4" pipe size	1.000	Ea.	73.13	172.50	245.63
Pipe, steel, black, schedule 40, 4" diam	10.000	L.F.	87.38	146.03	233.41
*Flow control valve, trim & gauges, 4" pipe size	1.000	Set	900	393.75	1,293.75
Fire alarm horn, electric	1.000	Ea.	30	39.75	69.75
Pipe, steel, black, schedule 40, threaded, cplg & hngr 10'OC, 2-1/2" diam	20.000	L.F.	117.75	198.75	316.50
Pipe, steel, black, schedule 40, threaded, cplg & hngr 10'OC, 2" diam	12.500	L.F.	46.31	97.03	143.34
Pipe, steel, black, schedule 40, threaded, cplg & hngr 10'OC, 1-1/4" diam	37.500	L.F.	91.97	209.53	301.50
Pipe steel, black, schedule 40, threaded cplg & hngr 10'OC, 1" diam	112.000	L.F.	231	583.80	814.80
Pipe Tee, malleable iron black, 150 lb threaded, 4" pipe size	2.000	Ea.	123	249	372
Pipe Tee, malleable iron black, 150 lb threaded, 2-1/2" pipe size	2.000	Ea.	38.25	110.25	148.50
Pipe Tee, malleable iron black, 150 lb threaded, 2" pipe size	1.000	Ea.	9.15	45	54.15
Pipe Tee, malleable iron black, 150 lb threaded, 1-1/4" pipe size	5.000	Ea.	21.75	178.13	199.88
Pipe Tee, malleable iron black, 150 lb threaded, 1" pipe size	4.000	Ea.	10.74	138	148.74
Pipe 90° elbow, malleable iron black, 150 lb threaded, 1" pipe size	6.000	Ea.	10.40	128.25	138.65
Sprinkler head, standard spray, brass 135°-286°F 1/2" NPT, 3/8" orifice	12.000	Ea.	59.76	288	347.76
Valve, gate, bronze, NRS, class 150, threaded, 1" pipe size	1.000	Ea.	21.38	14.51	35.89
*Standpipe connection, wall, single, flush w/plug & chain 2-1/2"x2-1/2"	1.000	Ea.	67.50	103.50	171
TOTAL			3,093.72	3,564.16	6,657.88
COST PER S.F.			1.55	1.79	3.34

*Not included in systems under 2000 S.F.

8.2-110	Wet Pipe Sprinkler Systems		COST PER S.F.		
			MAT.	INST.	TOTAL
0520	Wet pipe sprinkler systems, steel, black, sch. 40 pipe				
0530	Light hazard, one floor, 500 S.F.		1.16	1.70	2.86
0560	1000 S.F.	R8.2 -100	1.45	1.77	3.22
0580	2000 S.F.		1.55	1.79	3.34
0600	5000 S.F.	R8.2 -300	.76	1.25	2.01
0620	10,000 S.F.		.55	1.03	1.58
0640	50,000 S.F.		.46	.95	1.41
0660	Each additional floor, 500 S.F.		.62	1.44	2.06

Figure 18.12

8.2-110	Wet Pipe Sprinkler Systems	COST PER S.F.		
		MAT.	INST.	TOTAL
0680	1000 S.F.	.51	1.34	1.85
0700	2000 S.F.	.48	1.21	1.69
0720	5000 S.F.	.35	1.04	1.39
0740	10,000 S.F.	.34	.93	1.27
0760	50,000 S.F.	.31	.74	1.05
1000	Ordinary hazard, one floor, 500 S.F.	1.28	1.82	3.10
1020	1000 S.F.	1.44	1.73	3.17
1040	2000 S.F.	1.61	1.88	3.49
1060	5000 S.F.	.87	1.33	2.20
1080	10,000 S.F.	.69	1.40	2.09
1100	50,000 S.F.	.65	1.36	2.01
1140	Each additional floor, 500 S.F.	.78	1.63	2.41
1160	1000 S.F.	.50	1.32	1.82
1180	2000 S.F.	.55	1.33	1.88
1200	5000 S.F.	.56	1.26	1.82
1220	10,000 S.F.	.48	1.30	1.78
1240	50,000 S.F.	.49	1.18	1.67
1500	Extra hazard, one floor, 500 S.F.	2.65	2.83	5.48
1520	1000 S.F.	1.78	2.45	4.23
1540	2000 S.F.	1.71	2.53	4.24
1560	5000 S.F.	1.20	2.16	3.36
1580	10,000 S.F.	1.14	2.06	3.20
1600	50,000 S.F.	1.20	1.94	3.14
1660	Each additional floor, 500 S.F.	.77	2.01	2.78
1680	1000 S.F.	.74	1.91	2.65
1700	2000 S.F.	.69	1.93	2.62
1720	5000 S.F.	.60	1.68	2.28
1740	10,000 S.F.	.70	1.53	2.23
1760	50,000 S.F.	.70	1.44	2.14
2020	Grooved steel, black sch. 40 pipe, light hazard, one floor, 2000 S.F.	1.51	1.54	3.05
2060	10,000 S.F.	.62	.93	1.55
2100	Each additional floor, 2000 S.F.	.46	.99	1.45
2150	10,000 S.F.	.32	.79	1.11
2200	Ordinary hazard, one floor, 2000 S.F.	1.56	1.63	3.19
2250	10,000 S.F.	.65	1.15	1.80
2300	Each additional floor, 2000 S.F.	.50	1.08	1.58
2350	10,000 S.F.	.44	1.05	1.49
2400	Extra hazard, one floor, 2000 S.F.	1.69	2.09	3.78
2450	10,000 S.F.	.87	1.54	2.41
2500	Each additional floor, 2000 S.F.	.69	1.59	2.28
2550	10,000 S.F.	.60	1.36	1.96
3050	Grooved steel black sch. 10 pipe, light hazard, one floor, 2000 S.F.	1.47	1.52	2.99
3100	10,000 S.F.	.49	.88	1.37
3150	Each additional floor, 2000 S.F.	.42	.97	1.39
3200	10,000 S.F.	.28	.78	1.06
3250	Ordinary hazard, one floor, 2000 S.F.	1.52	1.61	3.13
3300	10,000 S.F.	.59	1.14	1.73
3350	Each additional floor, 2000 S.F.	.46	1.06	1.52
3400	10,000 S.F.	.38	1.04	1.42
3450	Extra hazard, one floor, 2000 S.F.	1.65	2.07	3.72
3500	10,000 S.F.	.79	1.51	2.30
3550	Each additional floor, 2000 S.F.	.65	1.57	2.22
3600	10,000 S.F.	.54	1.33	1.87
4050	Copper tubing, type M, light hazard, one floor, 2000 S.F.	1.58	1.51	3.09
4100	10,000 S.F.	.64	.91	1.55
4150	Each additional floor, 2000 S.F.	.53	.98	1.51
4200	10,000 S.F.	.43	.81	1.24
4250	Ordinary hazard, one floor, 2000 S.F.	1.65	1.68	3.33

Figure 18.12 (cont.)

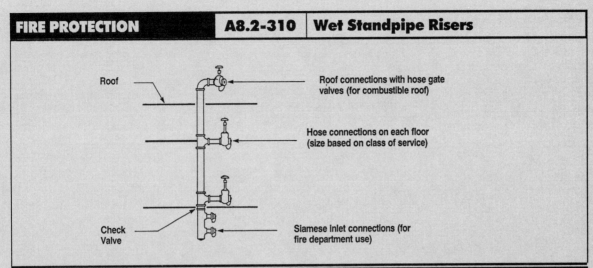

System Components	QUANTITY	UNIT	COST PER FLOOR		
			MAT.	INST.	TOTAL
SYSTEM 8.2-310-0560					
WET STANDPIPE RISER, CLASS I, STEEL, BLACK, SCH. 40 PIPE, 10' HEIGHT					
4" DIAMETER PIPE, ONE FLOOR					
Pipe, steel, black, schedule 40, threaded, 4" diam	20.000	L.F.	291	368	659
Pipe, Tee, malleable iron, black, 150 lb threaded, 4" pipe size	2.000	Ea.	164	332	496
Pipe, 90° elbow, malleable iron, black, 150 lb threaded 4" pipe size	1.000	Ea.	58.50	110	168.50
Pipe, nipple, steel, black, schedule 40, 2-1/2" pipe size x 3" long	2.000	Ea.	17.30	83	100.30
Fire valve, gate, 300 lb, brass w/handwheel, 2-1/2" pipe size	1.000	Ea.	114	55	169
Fire valve, pressure restricting, adj, rgh brs, 2-1/2" pipe size	1.000	Ea.	210	110	320
Valve, swing check, w/ball drip, CI w/brs ftngs, 4" pipe size	1.000	Ea.	97.50	230	327.50
Standpipe conn wall dble flush brs w/plugs & chains 2-1/2"x2-1/2"x4"	1.000	Ea.	330	138	468
Valve, swing check, bronze, 125 lb, regrinding disc, 2-1/2" pipe size	1.000	Ea.	231	44	275
Roof manifold, fire, w/valves & caps, horiz/vert brs 2-1/2"x2-1/2"x4"	1.000	Ea.	107	144	251
Fire, hydrolator, vent & drain, 2-1/2" pipe size	1.000	Ea.	93.50	55	148.50
Valve, gate, iron body 125 lb, OS&Y, threaded, 4" pipe size	1.000	Ea.	233	221	454
TOTAL			1,946.80	1,890	3,836.80

8.2-310	Wet Standpipe Risers, Class I		COST PER FLOOR		
			MAT.	INST.	TOTAL
0550	Wet standpipe risers, Class I, steel black sch. 40, 10' height				
0560	4" diameter pipe, one floor		1,950	1,900	3,850
0580	Additional floors		555	600	1,155
0600	6" diameter pipe, one floor		3,275	3,300	6,575
0620	Additional floors	R8.2 -300	920	935	1,855
0640	8" diameter pipe, one floor		4,825	3,950	8,775
0660	Additional floors		1,250	1,125	2,375
0680					

8.2-310	Wet Standpipe Risers, Class II	COST PER FLOOR		
		MAT.	INST.	TOTAL
1030	Wet standpipe risers, Class II, steel black sch. 40, 10' height			
1040	2" diameter pipe, one floor	765	670	1,435
1060	Additional floors	269	260	529
1080	2-1/2" diameter pipe, one floor	1,050	980	2,030
1100	Additional floors	310	300	610
1120				

Figure 18.13

recommendations with respect to heat loss, fuels used, and system efficiency. The heating system selected should provide the maximum heating for the minimum expense.

Air Conditioning Assemblies

The heating and air conditioning assemblies are usually linked together in one system. One exception is perimeter fin tube radiation heating with a separate forced-air cooling system.

Air conditioning system prices in *Means Assemblies Cost Data* are listed by the cost per square foot for several building types, as shown in Figure 18.16. The buildings include: banks and libraries, bars and taverns, bowling alleys, department stores, drug stores, factories, food supermarkets, medical centers, offices, restaurants, and schools and colleges. For each system presented, there are several building sizes to choose from. If a particular building type or use is not listed, then it is a matter of finding another that is as similar as possible. The costs determined from interpolation and extrapolation are still reasonable enough for the accuracy found in assemblies estimating. Large systems can, for estimating purposes, be multiples of the systems shown at approximately the same price per square foot.

Listed below are the most common air conditioning systems. Heating may be included directly or through supplemental re-heat.

- Chilled water with air-cooled condenser
- Chilled water with cooling tower
- Roof-top single-zone units
- Roof-top multi-zone units
- Self-contained water-cooled units
- Self-contained air-cooled units
- Split systems with air-cooled condensing

Assemblies square foot costs for these are easily assembled, as long as the user remembers to add to the basic costs for items such as the following:

- Ductwork, if not included
- Boilers and the necessary piping, if not included
- Coils, if not included

FIRE PROTECTION		A8.2-310	Wet Standpipe Risers			
8.2-310	Wet Standpipe Risers, Class III			COST PER FLOOR		
				MAT.	INST.	TOTAL
1530	Wet standpipe risers, Class III, steel black sch. 40, 10' height					
1540	4" diameter pipe, one floor			2,000	1,900	3,900
1560	Additional floors			465	500	965
1580	6" diameter pipe, one floor			3,325	3,300	6,625
1600	Additional floors			955	935	1,890
1620	8" diameter pipe, one floor			4,900	3,950	8,850
1640	Additional floors			1,300	1,125	2,425

Figure 18.13 (cont.)

8.2-390	Standpipe Equipment	COST EACH		
		MAT.	INST.	TOTAL
0100	Adapters, reducing, 1 piece, FxM, hexagon, cast brass, 2-1/2" x 1-1/2"	32		32
0200	Pin lug, 1-1/2" x 1"	11		11
0250	3" x 2-1/2"	39.50		39.50
0300	For polished chrome, add 75% mat.			
0400	Cabinets, D.S. glass in door, recessed, steel box, not equipped			
0500	Single extinguisher, steel door & frame	67	86.50	153.50
0550	Stainless steel door & frame	164	86.50	250.50
0600	Valve, 2-1/2" angle, steel door & frame	79	57.50	136.50
0650	Aluminum door & frame	144	57.50	201.50
0700	Stainless steel door & frame	182	57.50	239.50
0750	Hose rack assy, 2-1/2" x 1-1/2" valve & 100' hose, steel door & frame	142	115	257
0800	Aluminum door & frame	259	115	374
0850	Stainless steel door & frame	375	115	490
0900	Hose rack assy.,& extinguisher,2-1/2"x1-1/2" valve & hose,steel door & frame	163	138	301
0950	Aluminum	330	138	468
1000	Stainless steel	435	138	573
1550	Compressor, air, dry pipe system, automatic, 200 gal., 1/3 H.P.	740	295	1,035
1600	520 gal., 1 H.P.	780	295	1,075
1650	Alarm, electric pressure switch (circuit closer)	117	14.75	131.75
2500	Couplings, hose, rocker lug, cast brass, 1-1/2"	25		25
2550	2-1/2"	51		51
3000	Escutcheon plate, for angle valves, polished brass, 1-1/2"	9.90		9.90
3050	2-1/2"	27		27
3500	Fire pump, electric, w/controller, fittings, relief valve			
3550	4" pump, 30 H.P., 500 G.P.M.	12,600	2,150	14,750
3600	5" pump, 40 H.P., 1000 G.P.M.	18,500	2,450	20,950
3650	5" pump, 100 H.P., 1000 G.P.M.	20,700	2,700	23,400
3700	For jockey pump system, add	2,275	345	2,620
5000	Hose, per linear foot, synthetic jacket, lined,			
5100	300 lb. test, 1-1/2" diameter	2.02		2.02
5150	2-1/2" diameter	3.32		3.32
5200	500 lb. test, 1-1/2" diameter	2.11		2.11
5250	2-1/2" diameter	3.82		3.82
5500	Nozzle, plain stream, polished brass, 1-1/2" x 10"	28		28
5550	2-1/2" x 15" x 13/16" or 1-1/2"	101		101
5600	Heavy duty combination adjustable fog and straight stream w/handle 1-1/2"	246		246
5650	2-1/2" direct connection	350		350
6000	Rack, for 1-1/2" diameter hose 100 ft. long, steel	32.50	34.50	67
6050	Brass	51.50	34.50	86
6500	Reel, steel, for 50 ft. long 1-1/2" diameter hose	76.50	49.50	126
6550	For 75 ft. long 2-1/2" diameter hose	124	49.50	173.50
7050	Siamese, w/plugs & chains, polished brass, sidewalk, 4" x 2-1/2" x 2-1/2"	299	276	575
7100	6" x 2-1/2" x 2-1/2"	440	345	785
7200	Wall type, flush, 4" x 2-1/2" x 2-1/2"	330	138	468
7250	6" x 2-1/2" x 2-1/2"	415	150	565
7300	Projecting, 4" x 2-1/2" x 2-1/2"	289	138	427
7350	6" x 2-1/2" x 2-1/2"	480	150	630
7400	For chrome plate, add 15% mat.			
8000	Valves, angle, wheel handle, 300 Lb., rough brass, 1-1/2"	31	32	63
8050	2-1/2"	60.50	55	115.50
8100	Combination pressure restricting, 1-1/2"	48.50	32	80.50
8150	2-1/2"	109	55	164
8200	Pressure restricting, adjustable, satin brass, 1-1/2"	63	32	95
8250	2-1/2"	105	55	160
8300	Hydrolator, vent and drain, rough brass, 1-1/2"	33.50	32	65.50
8350	2-1/2"	93.50	55	148.50
8400	Cabinet assy, incls. 2-1/2" valve, adapter, rack, hose, nozzle & hydrolator	600	260	860

Figure 18.14

- System balancing
- Quality/complexity adjustments

Some of these factors can be allowed for by using the table in Figure 18.15.

Quality and Complexity

The quality of the systems and prices in books such as *Means Assemblies Cost Data* are in accordance with recognized national building codes and represent good, sound construction. Average complexity is also allowed for. When using the assemblies format for estimating, it becomes necessary at times to make some allowances for situations that differ from the published or assumed quality and complexity. The user of any published cost guides must read any descriptive or qualifying information that may give a clue as to the quality/complexity allowed for in the prices quoted.

Figure 18.15 provides percentage factors that can be added to an estimate for varying quality and complexity. These same percentages can be used for any portion of the estimate, not just mechanical work alone. Figure 18.17 is a typical Fire Protection Estimate Worksheet. Figures 18.18 and 18.19 are completed Estimate Worksheets for Plumbing and Heating and Air Conditioning.

Figure 18.20 summarizes the costs for this division.

Heating Approximations for Quick Estimating

Oil Piping & Boiler Room Piping:
Small System .... 20 to 30% of Boiler
Complex System
with Pumps, Headers, Etc. .... 80 to 110% of Boiler
Breeching With Insulation:
Small .... 10 to 15% of Boiler
Large .... 15 to 25% of Boiler
Coils: .... 15 to 30% of Containing Unit
Balancing (Independent) .... 1/2% of H.V.A.C. Estimating

Quality/Complexity Adjustment: For all heating installations add these adjustments to the estimate to more closely allow for the equipment and conditions of the particular job under consideration.
Economy installation, add .... 0 to 5% of System
Good quality, medium complexity, add .... 5 to 15% of System
Above average quality and complexity, add .... 15 to 25% of System

Figure 18.15

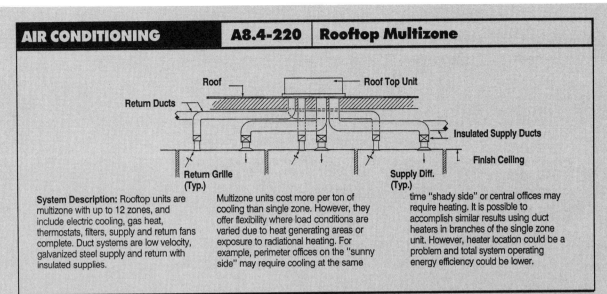

AIR CONDITIONING — A8.4-220 — Rooftop Multizone

System Description: Rooftop units are multizone with up to 12 zones, and include electric cooling, gas heat, thermostats, filters, supply and return fans complete. Duct systems are low velocity, galvanized steel supply and return with insulated supplies.

Multizone units cost more per ton of cooling than single zone. However, they offer flexibility where load conditions are varied due to heat generating areas or exposure to radiational heating. For example, perimeter offices on the "sunny side" may require cooling at the same

time "shady side" or central offices may require heating. It is possible to accomplish similar results using duct heaters in branches of the single zone unit. However, heater location could be a problem and total system operating energy efficiency could be lower.

System Components	QUANTITY	UNIT	COST EACH MAT.	INST.	TOTAL
SYSTEM 8.4-220-1280					
ROOFTOP, MULTIZONE, AIR CONDITIONER					
APARTMENT CORRIDORS, 3,000 S.F., 5.50 TON					
Rooftop multizone unit, standard controls, curb	1.000	Ea.	17,512	2,827	20,339
Ductwork package for rooftop multizone units	1.000	System	4,097.50	7,150	11,247.50
TOTAL			21,609.50	9,977	31,586.50
COST PER S.F.			7.20	3.33	10.53

Note A: Small single zone unit recommended
Note B: A combination of multizone units recommended

8.4-220	Rooftop Multizone Unit Systems		COST PER S.F. MAT.	INST.	TOTAL
1240	Rooftop, multizone, air conditioner				
1260	Apartment corridors, 1,500 S.F., 2.75 ton. See Note A.				
1280	3,000 S.F., 5.50 ton	R8.3 -600	7.20	3.33	10.53
1440	25,000 S.F., 45.80 ton		4.66	2.96	7.62
1520	Banks or libraries, 1,500 S.F., 6.25 ton	R8.4 -010	16.35	7.55	23.90
1640	15,000 S.F., 62.50 ton		7.55	6.35	13.90
1720	25,000 S.F., 104.00 ton		7.55	6.35	13.90
1800	Bars and taverns, 1,500 S.F., 16.62 ton		33.50	11.95	45.45
1840	3,000 S.F., 33.24 ton		25	10.70	35.70
1880	10,000 S.F., 110.83 ton		16	9.70	25.70
2080	Bowling alleys, 1,500 S.F., 8.50 ton		22.50	10.30	32.80
2160	10,000 S.F., 56.70 ton		14.40	9.20	23.60
2240	20,000 S.F., 113.00 ton		10.25	8.60	18.85
2640	Drug stores, 1,500 S.F., 10.00 ton		26	12.10	38.10
2680	3,000 S.F., 20.00 ton		19.55	10.90	30.45
2760	15,000 S.F., 100.00 ton		12.10	10.15	22.25
3760	Offices, 1,500 S.F., 4.75 ton, See Note A				
3880	15,000 S.F., 47.50 ton		8.05	5.15	13.20
3960	25,000 S.F., 79.16 ton		5.75	4.82	10.57
4000	Restaurants, 1,500 S.F., 7.50 ton		19.65	9.05	28.70
4080	10,000 S.F., 50.00 ton		12.70	8.10	20.80
4160	20,000 S.F., 100.00 ton		9.10	7.60	16.70
4240	Schools and colleges, 1,500 S.F., 5.75 ton		15.05	6.95	22
4360	15,000 S.F., 57.50 ton		6.95	5.85	12.80
4441	25,000 S.F., 95.83 ton		6.95	5.85	12.80

Figure 18.16

FIRE PROTECTION ESTIMATE WORKSHEET

TYPE OF COMPONENT	HAZARD OR CLASS	TABLE NUMBER	QUANTITY	UNIT COST	TOTAL COST	COMMENTS
FROM TABLE	Light					
SPRINKLER		8.2-110-0620	10000 SF	1.58	15800	
(Wet Pipe)		8.2-110-0740	20000 SF	1.27	25400	
STANDPIPE		8.2-310-1540	12/10	3900	4680	
(Wet Pipe)		8.2-310-1560	12/10	965	1158	
FIRE SUPPRESS						
CABINETS & COMPONENTS						
Cabinet		8.2-390-0750	3 Ea.	257	771	
Siamese		8.2-390-7200	1 Ea.	468	468	
Alarm		8.2-390-1650	1 Ea.	131.75	132	
Adaptor		8.2-390-0100	3 Ea.	32	96	
Hose,Nozzle,Rack,						
Hydrolator,Valve(Set)			3 Ea.	487	1461	
OTHER						
SUBTOTAL					$49,966	
QUALITY COMPLEXITY						
TOTAL					$49,966	

Figure 18.17

PLUMBING ESTIMATE WORKSHEET

FIXTURE	TABLE NUMBER	QUANTITY	UNIT COST	TOTAL COST	CALCULATIONS
BATHTUBS					
DRINKING FOUNTAIN					
KITCHEN SINK					
LAUNDRY SINK					
LAVATORY	8.1-433-1600	18 Ea.	661	11898	
SERVICE SINK	8.1-434-4380	3 Ea.	1415	4245	
SHOWER					
URINAL	8.1-450-2000	3 Ea.	1005	3015	
WATER COOLER	8.1-460-1880	3 Ea.	1085	3255	
WATER CLOSET	8.1-470-1840	15 Ea.	1370	20550	
W/C GROUP					
WASH FOUNT. GROUP					
LAVATORY GROUP					
URINALS GROUP					
BATHROOMS					
WATER HEATER		3 Ea.	3235	9705	
ROOF DRAINS	8.1-310-4200	4 Ea.	775	3100	
ROOF DRAINS	8.1-310-4320	60 LF	24.20	1452	
(ADDITIONAL LENGTH)	8.1-310-4400	200 LF	27.10	5420	
SUBTOTAL				$62,640	COMMENTS
WATER CONTROL		10%		4296	% of Subtotal
PIPE & FITTINGS		30%		12889	% of Subtotal
QUALITY COMPLEXITY		10%		4296	% of Subtotal
OTHER					
TOTAL				$84,122	

Figure 18.18

HEATING & AIR CONDITIONING ESTIMATE WORKSHEET

EQUIPMENT	TYPE	TABLE NUMBER	QUANTITY	UNIT COST	TOTAL COST	COMMENTS
HEAT SOURCE	Roof Top, Multi-zone	8.4-220-3960	30000 SF	10.57	317100	
PIPE	Incl'd.					
DUCT	Incl'd.					
TERMINALS						
OTHER						
COLD SOURCE	Incl'd.					
PIPE						
DUCT						
TERMINALS						
OTHER Balancing			0.5%		1586	
SUBTOTAL						
QUALITY COMPLEXITY			7%		22197	
TOTAL					$340,883	

Figure 18.19

Assembly Number	Description	Qty	Unit	Total Cost		Cost per S. F.
				Unit	Total	
8.0	**Mechanical**					
	Plumbing					
8.1-433-1600	Lavatory, Vanity Top	18	EA	661	11898	
8.1-434-4380	Service Sink	3	EA	1415	4245	
8.1-450-2000	Urinal	3	EA	1005	3015	
8.1-460-1880	Water Cooler	3	EA	1085	3255	
8.1-470-1840	Water Closet	15	EA	1370	20550	
	Control(10%), Fittings(30%),Qual/Comp(10%)				21481	
8.1-160-1820	Water Heater: 50 gal./37 GPH	3	EA	3235	9705	
8.1-310-4200	Roof Drains 4" CI	4	EA	775	3100	
8.1-310-4320	5" Pipe CI	60	LF	24.20	1452	
8.1-310-4400	6" Pipe CI	200	LF	27.10	5420	
8.2-110-0620	Fire Protection Light Hazard, Wet Pipe	10000	SF	1.58	15800	
8.2-110-0740	Same	20000	SF	1.27	25400	
8.2-310-1540	Standpipe	12/10	FL	3900	4680	
8.2-310-1560	Same	12/10	FL	965	1158	
8.2-390-0750	Cabinet	3	EA	257	771	
	Hose.Nozzle,Rack,Hydrolator,Valve(Set)	3	Sets	487	1461	
8.2-390-7200	Siamese	1	EA	468	468	
8.2-390-1650	Alarm	1	EA	131.75	132	
8.2-390-0100	Adaptor	3	EA	32	96	
	Heating and Air Conditioning					
8.4-220-3960	Roof Top Multi-zone	30000	SF	10.57	317100	
	Balancing (0.5%)				1586	
	Quality/Complexity (7%)				22197	
					$474,969	$15.83
9.0	**Electrical**					

Figure 18.20

Chapter 19

Electrical

The electrical assemblies of a cost estimate are comprised of several subassemblies that are priced out from limited information during the conceptual or preliminary design stage. Before any assemblies-type cost information can be used, the estimator must first do a few simple calculations to determine certain parameters:

- Total building load in watts
- Total amperes

Once these simple calculations are done, the estimate proceeds rather simply. Follow these basic steps:

1. Determine building size and use.
2. Develop total load, in watts, for:
   - Lighting
   - Receptacles
   - Air conditioning
   - Elevators
   - Other power requirements
3. Determine voltage available from the utility company.
4. Determine size of the building service from formulas.
5. Determine costs for service, panels and feeders from tables.
6. Determine costs for above subsystems from tables using calculated loads.

There are several reasons why assemblies estimating can be beneficial, easy to use, and efficient for estimating electrical costs. Very often the assemblies method can be used for comparison purposes or for the selection of a particular system. After spending a little time on familiarization, the estimator can present accurate cost information for the various components in the electrical portion of the building cost estimate. The estimator does not have to be an electrical engineer. The only requirement for an assemblies estimate is a general knowledge of electrical components.

Budget Square Foot Costs

It is possible to obtain a rough approximation of the electrical costs by using the table in Figure 19.1. The prices found in this table are from actual projects that have been bid from 1986 to 1996 in the Northeastern part of the United States. Bid prices have been adjusted to January 1, 1996 price levels. The list of projects is by no means all-inclusive, but by carefully

Cost per S.F. for Electric Systems for Various Building Types

Type Construction	1. Service & Distrib.	2. Lighting	3. Devices	4. Equipment Connections	5. Basic Materials	6. Special Systems		
						Fire Alarm & Detection	Lightning Protection	Master TV Antenna
Apartment, luxury high rise	$1.19	$ .82	$ .60	$ .75	$1.99	$ .39		$.25
Apartment, low rise	.69	.69	.54	.64	1.16	.31		
Auditorium	1.51	4.18	.47	1.11	2.38	.51		
Bank, branch office	1.80	4.60	.79	1.12	2.26	1.44		
Bank, main office	1.36	2.52	.25	.48	2.44	.74		
Church	.92	2.56	.31	.26	1.16	.74		
*College, science building	1.73	3.29	1.03	.87	2.66	.63		
*College library	1.28	1.87	.22	.52	1.47	.74		
*College, physical education center	2.01	2.63	.33	.42	1.14	.42		
Department store	.67	1.80	.22	.74	1.97	.31		
*Dormitory, college	.89	2.32	.23	.48	1.93	.54		.32
Drive-in donut shop	2.53	6.97	1.15	1.13	3.13	–		
Garage, commercial	.34	.86	.15	.34	.66	–		
*Hospital, general	4.88	3.59	1.32	.88	3.90	.46	$.11	
*Hospital, pediatric	4.27	5.35	1.13	3.23	7.23	.53		.40
*Hotel, airport	1.93	2.93	.23	.44	2.85	.42	.22	.36
Housing for the elderly	.54	.69	.32	.85	2.44	.53		.31
Manufacturing, food processing	1.22	3.67	.20	1.63	2.66	.31		
Manufacturing, apparel	.80	1.92	.26	.64	1.43	.27		
Manufacturing, tools	1.83	4.48	.24	.74	2.38	.32		
Medical clinic	.73	1.54	.40	1.16	1.87	.52		
Nursing home	1.26	2.91	.40	.33	2.38	.71		.25
Office Building	1.71	3.92	.20	.64	2.49	.37	.19	
Radio-TV studio	1.22	4.05	.60	1.17	2.94	.49		
Restaurant	4.54	3.84	.73	1.80	3.55	.27		
Retail Store	.95	2.04	.23	.44	1.11	–		
School, elementary	1.61	3.62	.47	.44	2.98	.45		.18
School, junior high	.96	3.02	.23	.79	2.36	.53		
*School, senior high	1.06	2.37	.42	1.04	2.62	.46		
Supermarket	1.09	2.05	.29	1.72	2.28	.21		
*Telephone Exchange	2.66	.86	.15	.74	1.56	.86		
Theater	2.10	2.80	.49	1.51	2.31	.63		
Town Hall	1.27	2.22	.49	.54	3.07	.42		
*U.S.Post Office	3.75	2.85	.50	.82	2.18	.42		
Warehouse, grocery	.70	1.23	.15	.44	1.64	.25		

*Includes cost of primary feeder and transformer. Cont'd. on next page.

Figure 19.1

250

Cost per S.F. for Electric Systems for Various Building Types (cont.)

Type Construction	6. Special Systems, (cont.)						
	Intercom Systems	Sound Systems	Closed Circuit TV	Snow Melting	Emergency Generator	Security	Master Clock Sys.
Apartment, luxury high rise	$ .52						
Apartment, low rise	.36						
Auditorium		$1.36	$ .63		$ .98		
Bank, branch office	.71		1.46			$1.23	
Bank, main office	.40		.30		.80	.67	$.27
Church	.51						
*College, science building	.51				.98		.31
*College, library					.52		
*College, physical education center		.70					
Department store					.21		
*Dormitory, college	.69						
Drive-in donut shop							.11
Garage, commercial							.08
Hospital, general	.53		.21		1.37		
*Hospital, pediatric	3.62	.36	.40		.87		
*Hotel, airport	.53				.51		
Housing for the elderly	.63						
Manufacturing, food processing		.23			1.80		
Manufacturing apparel		.31					
Manufacturing, tools		.40		$.25			
Medical clinic							
Nursing home	1.21				.44		
Office Building		.19			.44	.21	.08
Radio-TV studio	.70				1.12		.50
Restaurant		.31					
Retail Store							
School, elementary		.21					.21
School, junior high		.57			.37		.40
*School, senior high	.49		.31		.52	.27	.28
Supermarket		.24			.46	.31	
*Telephone exchange					4.53	.16	
Theater		.48					
Town Hall							.21
*U.S. Post Office	.47			.08	.51		
Warehouse, grocery	.29						

*Includes cost of primary feeder and transformer. Cont'd. on next page.

Figure 19.1 (cont.)

Cost per S.F. for Total Electric Systems for Various Building Types (cont.)

Type Construction	Basic Description	Total Floor Area in Square Feet	Total Cost per Square Foot for Total Electric Systems
Apartment building, luxury high rise	All electric, 18 floors, 86 1 B.R., 34 2 B.R.	115,000	$ 6.51
Apartment building, low rise	All electric, 2 floors, 44 units, 1 & 2 B.R.	40,200	4.39
Auditorium	All electric, 1200 person capacity	28,000	13.13
Bank, branch office	All electric, 1 floor	2,700	15.41
Bank, main office	All electric, 8 floors	54,900	10.23
Church	All electric, incl. Sunday school	17,700	6.46
*College, science building	All electric, 3-1/2 floors, 47 rooms	27,500	12.01
*College, library	All electric	33,500	6.62
*College, physical education center	All electric	22,000	7.65
Department store	Gas heat, 1 floor	85,800	5.92
*Dormitory, college	All electric, 125 rooms	63,000	7.40
Drive-in donut shop	Gas heat, incl. parking area lighting	1,500	15.02
Garage, commercial	All electric	52,300	2.43
*Hospital, general	Steam heat, 4 story garage, 300 beds	540,000	17.25
*Hospital, pediatric	Steam heat, 6 stories	278,000	27.39
Hotel, airport	All electric, 625 guest rooms	536,000	10.42
Housing for the elderly	All electric, 7 floors, 100 1 B.R. units	67,000	6.31
Manufacturing, food processing	Electric heat, 1 floor	9,600	11.72
Manufacturing, apparel	Electric heat, 1 floor	28,000	5.63
Manufacturing, tools	Electric heat, 2 floors	42,000	10.64
Medical clinic	Electric heat, 2 floors	22,700	6.22
Nursing home	Gas heat, 3 floors, 60 beds	21,000	9.89
Office building	All electric, 15 floors	311,200	10.44
Radio-TV studio	Electric heat, 3 floors	54,000	12.79
Restaurant	All electric	2,900	15.04
Retail store	All electric	3,000	4.77
School, elementary	All electric, 1 floor	39,500	10.17
School, junior high	All electric, 1 floor	49,500	9.23
*School, senior high	All electric, 1 floor	158,300	9.84
Supermarket	Gas heat	30,600	8.65
*Telephone exchange	Gas heat, 300 KW emergency generator	24,800	11.52
Theater	Electric heat, twin cinema	14,000	10.32
Town Hall	All electric	20,000	8.22
*U.S. Post Office	All electric	495,000	11.58
Warehouse, grocery	All electric	96,400	4.70

*Includes cost of primary feeder and transformer.

Figure 19.1 (cont.)

examining the various systems for a particular building type, certain cost relationships emerge. This table can be used to produce a budget square foot cost for the electrical portion of the job that is consistent with the amount of design information normally available at the conceptual estimate stage.

The following explanations apply when using the table in Figure 19.1:

Service & Distribution: The costs shown include the incoming primary feeder from the power company, the main building transformer, metering arrangement, switchboards, distribution panel boards, step-down transformers, and power and lighting panels. Those items marked (*) include the cost of the primary feeder and transformer. In all other projects the primary feeder and transformer are provided by the power company.

Lighting: Costs include all interior fixtures for decor, illumination, exit and emergency lighting. General exterior lighting fixtures are included, but those for area parking illumination are not unless specifically noted.

Devices: Costs include all outlet boxes, receptacles, switches for lighting control, dimmers, and cover plates.

Equipment Connections: Costs include all materials and equipment for making connections for the heating, ventilating, and air conditioning. Food service and other motorized items or systems requiring connections are also included in the costs shown.

Basic Materials: This category includes all disconnect power switches that are not a part of the service equipment. Included are raceways for wires; pull boxes; junction boxes; supports; fittings; grounding materials; wireways; busways; and cable systems.

Special Systems: Costs for installed equipment are included for the particular assemblies shown.

Budget Estimate: Before going into a more detailed analysis of the assemblies costs, use the table in Figure 19.1 to put together a budget estimate for the Three-Story Office Building.

Service & Distribution:	$1.71 × 30,000 S.F.	=	$ 51,300
Lighting:	$3.92 × 30,000 S.F.	=	117,600
Devices:	$0.20 × 30,000 S.F.	=	6,000
Equipment Connections:	$0.64 × 30,000 S.F.	=	19,200
Materials:	$2.49 × 30,000 S.F.	=	74,700
Fire Alarm:	$0.37 × 30,000 S.F.	=	11,100
Emergency Generator:	$0.44 × 30,000 S.F.	=	13,200
Total Budget Cost			$293,100
Budget Cost per Square Foot	$293,100 / 30,000 S.F.	=	$9.77/S.F.

Example: Three-Story Office Building

In Chapter 9, the following loads were developed:

Power Requirements:

Lighting	90,000 watts
Receptacles	60,000 watts
HVAC	141,000 watts
Misc. Motors & Power	36,000 watts
Elevators	51,600 watts
Total Power Required	378,600 watts

Voltage Available: 120V/208V, 3 phase, 4 wire
277V/480V, 3 phase, 4 wire (allows smaller feeders)
Size of Service: 570 amps; use 600 amps (277V/480V)

Using the above calculated loads, do an assemblies square foot estimate for the Three-Story Office Building example:

Lighting:	(Figure 19.2 #0280)		
	30,000 S.F. × $4.89	=	$146,700
Receptacles:	(Figure 19.3 #0640)		
	30,000 S.F. × $2.57	=	77,100
HVAC:	(Figure 19.4 #0280)		
	30,000 S.F. × $0.35	=	10,500
	(Interpolate)		
Misc. Motors &	(Figure 19.5 #0320)		
Power:	30,000 S.F. × $0.18	=	5,400
Elevators:			
2 @ 30 hp	(Figure 19.6 #2200)		
	2 ea. × $2,125	=	4,250
Wall Switches:			
2/1000 S.F.	(Figure 19.7 #0280)		
	30,000 S.F. × $0.26	=	7,800
Service 600 amp:	(Figure 19.8 #0360)		
	1.25 × $6,275	=	7,844
	(Figure 19.8 #0570)		
Panels 600 amp:	(Figure 19.9 #0240)		
	1.20 × $11,900	=	14,280
	(Figure 19.9 #0410)		
Feeders 600 amp:	(Figure 19.10 #0360)		
	100 L.F. × $91.00	=	9,100
Fire Detection:	(Figure 19.1)		
	30,000 S.F. × $0.37	=	11,100
Emergency	(Figure 19.1)		
Generator:	30,000 S.F. × $0.44	=	13,200
Total Cost			$307,274
Cost per Square Foot:	$307,274/30,000 S.F. =		$10.24/S.F.

Cost Analysis

Compare the two cost estimates for the Three-Story Office Building:

Budget Cost per Square Foot: $9.77/S.F.
System Square Foot Cost: $10.24/S.F.

The cost difference is $0.47/S.F. or almost 5%. At the conceptual stage, is it worth spending the additional time to get a figure that is 5% more accurate? Only the estimator can make that decision. For the quick numbers, the budget cost is certainly attractive; however, the more detailed process based on more specific information is more accurate. As long as the building project being estimated is of a general nature—like an office building—the budget cost per square foot method is reasonably accurate. Once the project becomes unique with unusual electrical components, the estimator should rely on the assemblies square foot method or a unit price estimate.

LIGHTING & POWER	A9.2-213	Fluorescent Fixt. (by Wattage)

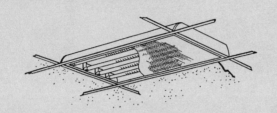

Type C. Recessed, mounted on grid ceiling suspension system, 2' x 4', four 40 watt lamps, acrylic prismatic diffusers.

5.3 watts per S.F. for 100 footcandles.

3 watts per S.F. for 57 footcandles.

System Components	QUANTITY	UNIT	COST PER S.F.		
			MAT.	INST.	TOTAL
SYSTEM 9.2-213-0200					
FLUORESCENT FIXTURES RECESS MOUNTED IN CEILING					
1 WATT PER S.F., 20 FC, 5 FIXTURES PER 1000 S.F.					
Steel intermediate conduit, (IMC) 1/2" diam	.128	L.F.	.15	.45	.60
Wire, 600 volt, type THW, copper, solid, #12	.003	C.L.F.	.02	.10	.12
Fluorescent fixture, recessed, 2'x4', four 40W, w/lens, for grid ceiling	.005	Ea.	.39	.38	.77
Steel outlet box 4" square	.005	Ea.	.03	.09	.12
Fixture whip, Greenfield w/#12 THHN wire	.005	Ea.	.01	.02	.03
TOTAL			.60	1.04	1.64

9.2-213	Fluorescent Fixtures (by Wattage)	COST PER S.F.		
		MAT.	INST.	TOTAL
0190	Fluorescent fixtures recess mounted in ceiling			
0200	1 watt per S.F., 20 FC, 5 fixtures per 1000 S.F.	.60	1.04	1.64
0240	2 watts per S.F., 40 FC, 10 fixtures per 1000 S.F.	1.19	2.04	3.23
0280	3 watts per S.F., 60 FC, 15 fixtures per 1000 S.F	1.80	3.09	4.89
0320	4 watts per S.F., 80 FC, 20 fixtures per 1000 S.F.	2.37	4.07	6.44
0400	5 watts per S.F., 100 FC, 25 fixtures per 1000 S.F.	2.97	5.10	8.07

R9.2 -200

Figure 19.2

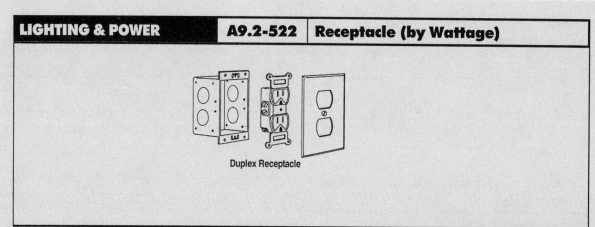

Duplex Receptacle

System Components			COST PER S.F.		
	QUANTITY	UNIT	MAT.	INST.	TOTAL
SYSTEM 9.2-522-0200					
RECEPTACLES INCL. PLATE, BOX, CONDUIT, WIRE & TRANS. WHEN REQUIRED					
2.5 PER 1000 S.F., .3 WATTS PER S.F.					
Steel intermediate conduit, (IMC) 1/2″ diam	167.000	L.F.	.19	.59	.78
Wire 600V type THWN-THHN, copper solid #12	3.382	C.L.F.	.02	.11	.13
Wiring device, receptacle, duplex, 120V grounded, 15 amp	2.500	Ea.	.01	.02	.03
Wall plate, 1 gang, brown plastic	2.500	Ea.		.01	.01
Steel outlet box 4″ square	2.500	Ea.		.04	.04
Steel outlet box 4″ plaster rings	2.500	Ea.		.01	.01
TOTAL			.22	.78	1

9.2-522	Receptacle (by Wattage)	COST PER S.F.		
		MAT.	INST.	TOTAL
0190	Receptacles include plate, box, conduit, wire & transformer when required			
0200	2.5 per 1000 S.F., .3 watts per S.F.	.22	.78	1
0240	With transformer — R9.0 -110	.25	.82	1.07
0280	4 per 1000 S.F., .5 watts per S.F.	.25	.92	1.17
0320	With transformer	.29	.98	1.27
0360	5 per 1000 S.F., .6 watts per S.F.	.31	1.08	1.39
0400	With transformer	.37	1.16	1.53
0440	8 per 1000 S.F., .9 watts per S.F.	.33	1.20	1.53
0480	With transformer	.41	1.31	1.72
0520	10 per 1000 S.F., 1.2 watts per S.F.	.35	1.32	1.67
0560	With transformer	.48	1.50	1.98
0600	16.5 per 1000 S.F., 2.0 watts per S.F.	.41	1.63	2.04
0640	With transformer	.64	1.93	2.57
0680	20 per 1000 S.F.,2.4 watts per S.F.	.44	1.78	2.22
0720	With transformer	.71	2.13	2.84

Figure 19.3

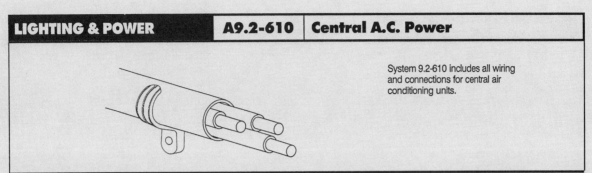

System 9.2-610 includes all wiring and connections for central air conditioning units.

System Components	QUANTITY	UNIT	COST PER S.F.		
			MAT.	INST.	TOTAL
SYSTEM 9.2-610-0200					
CENTRAL AIR CONDITIONING POWER, 1 WATT					
Steel intermediate conduit, 1/2" diam.	.030	L.F.	.03	.11	.14
Wire 600V type THWN-THHN, copper solid #12	.001	C.L.F.	.01	.03	.04
TOTAL			.04	.14	.18

9.2-610	Central A. C. Power (by Wattage)	COST PER S.F.		
		MAT.	INST.	TOTAL
0200	Central air conditioning power, 1 watt	.04	.14	.18
0220	2 watts	.05	.16	.21
0240	3 watts	.06	.17	.23
0280	4 watts	.09	.23	.32
0320	6 watts	.16	.33	.49
0360	8 watts	.20	.35	.55
0400	10 watts	.27	.40	.67

Figure 19.4

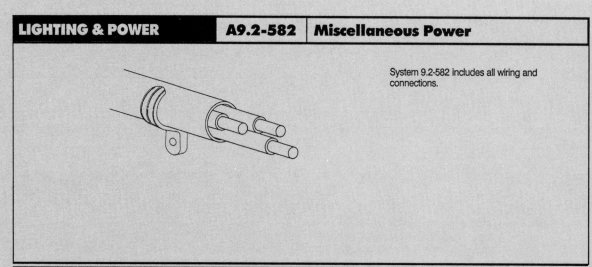

System 9.2-582 includes all wiring and connections.

System Components	QUANTITY	UNIT	COST PER S.F.		
			MAT.	INST.	TOTAL
SYSTEM 9.2-582-0200					
MISCELLANEOUS POWER, TO .5 WATTS					
Steel intermediate conduit, (IMC) 1/2" diam	15.000	L.F.	.02	.05	.07
Wire 600V type THWN-THHN, copper solid #12	.325	C.L.F.		.01	.01
TOTAL			.02	.06	.08

9.2-582	Miscellaneous Power	COST PER S.F.		
		MAT.	INST.	TOTAL
0200	Miscellaneous power, to .5 watts	.02	.06	.08
0240	.8 watts	.02	.08	.10
0280	1 watt	.03	.11	.14
0320	1.2 watts	.04	.14	.18
0360	1.5 watts	.05	.16	.21
0400	1.8 watts	.06	.19	.25
0440	2 watts	.07	.22	.29
0480	2.5 watts	.09	.28	.37
0520	3 watts	.10	.33	.43

Figure 19.5

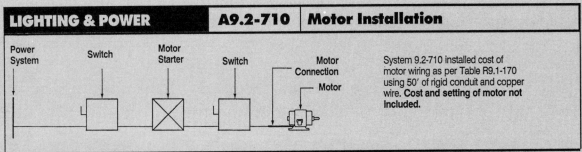

System 9.2-710 installed cost of motor wiring as per Table R9.1-170 using 50' of rigid conduit and copper wire. **Cost and setting of motor not included.**

System Components	QUANTITY	UNIT	COST EACH		
			MAT.	INST.	TOTAL
SYSTEM 9.2-710-0200					
MOTOR INSTALLATION, SINGLE PHASE, 115V, TO AND INCLUDING 1/3 HP MOTOR SIZE					
Wire 600V type THWN-THHN, copper solid #12	1.250	C.L.F.	8.13	40	48.13
Steel intermediate conduit, (IMC) 1/2" diam	50.000	L.F.	57	177	234
Magnetic FVNR, 115V, 1/3 HP, size 00 starter	1.000	Ea.	140	88.50	228.50
Safety switch, fused, heavy duty, 240V 2P 30 amp	1.000	Ea.	72.50	101	173.50
Safety switch, non fused, heavy duty, 600V, 3 phase, 30 A	1.000	Ea.	84.50	111	195.50
Flexible metallic conduit, Greenfield 1/2" diam	1.500	L.F.	.53	2.66	3.19
Connectors for flexible metallic conduit Greenfield 1/2" diam	1.000	Ea.	1.17	4.42	5.59
Coupling for Greenfield to conduit 1/2" diam flexible metalic conduit	1.000	Ea.	.76	7.05	7.81
Fuse cartridge nonrenewable, 250V 30 amp	1.000	Ea.	.90	7.05	7.95
TOTAL			365.49	538.68	904.17

9.2-710	Motor Installation		COST EACH		
			MAT.	INST.	TOTAL
0200	Motor installation, single phase, 115V, to and including 1/3 HP motor size		365	540	905
0240	To and incl. 1 HP motor size		385	540	925
0280	To and incl. 2 HP motor size	R9.1	415	575	990
0320	To and incl. 3 HP motor size	-170	480	585	1,065
0360	230V, to and including 1 HP motor size		375	545	920
0400	To and incl. 2 HP motor size		395	545	940
0440	To and incl. 3 HP motor size		440	585	1,025
0520	Three phase, 200V, to and including 1-1/2 HP motor size		415	600	1,015
0560	To and incl. 3 HP motor size		450	655	1,105
0600	To and incl. 5 HP motor size		490	725	1,215
0640	To and incl. 7-1/2 HP motor size		505	740	1,245
0680	To and incl. 10 HP motor size		845	925	1,770
0720	To and incl. 15 HP motor size		1,125	1,025	2,150
0760	To and incl. 20 HP motor size		1,375	1,200	2,575
0800	To and incl. 25 HP motor size		1,400	1,200	2,600
0840	To and incl. 30 HP motor size		2,225	1,400	3,625
0880	To and incl. 40 HP motor size		2,725	1,675	4,400
0920	To and incl. 50 HP motor size		5,000	1,925	6,925
0960	To and incl. 60 HP motor size		5,125	2,050	7,175
1000	To and incl. 75 HP motor size		6,475	2,350	8,825
1040	To and incl. 100 HP motor size		13,600	2,775	16,375
1080	To and incl. 125 HP motor size		13,900	3,025	16,925
1120	To and incl. 150 HP motor size		16,500	3,550	20,050
1160	To and incl. 200 HP motor size		18,600	4,325	22,925
1240	230V, to and including 1-1/2 HP motor size		400	595	995
1280	To and incl. 3 HP motor size		435	645	1,080
1320	To and incl. 5 HP motor size		475	720	1,195
1360	To and incl. 7-1/2 HP motor size		475	720	1,195
1400	To and incl. 10 HP motor size		775	875	1,650
1440	To and incl. 15 HP motor size		865	960	1,825
1480	To and incl. 20 HP motor size		1,300	1,150	2,450
1520	To and incl. 25 HP motor size		1,375	1,200	2,575

Figure 19.6

9.2-710	Motor Installation	COST EACH		
		MAT.	INST.	TOTAL
1560	To and incl. 30 HP motor size	1,400	1,200	2,600
1600	To and incl. 40 HP motor size	2,675	1,625	4,300
1640	To and incl. 50 HP motor size	2,775	1,725	4,500
1680	To and incl. 60 HP motor size	5,000	1,950	6,950
1720	To and incl. 75 HP motor size	5,875	2,200	8,075
1760	To and incl. 100 HP motor size	6,625	2,450	9,075
1800	To and incl. 125 HP motor size	13,900	2,850	16,750
1840	To and incl. 150 HP motor size	14,800	3,250	18,050
1880	To and incl. 200 HP motor size	15,900	3,600	19,500
1960	460V, to and including 2 HP motor size	485	600	1,085
2000	To and incl. 5 HP motor size	520	650	1,170
2040	To and incl. 10 HP motor size	545	720	1,265
2080	To and incl. 15 HP motor size	760	825	1,585
2120	To and incl. 20 HP motor size	790	880	1,670
2160	To and incl. 25 HP motor size	855	925	1,780
2200	To and incl. 30 HP motor size	1,125	1,000	2,125
2240	To and incl. 40 HP motor size	1,400	1,075	2,475
2280	To and incl. 50 HP motor size	1,525	1,200	2,725
2320	To and incl. 60 HP motor size	2,350	1,400	3,750
2360	To and incl. 75 HP motor size	2,675	1,550	4,225
2400	To and incl. 100 HP motor size	2,900	1,725	4,625
2440	To and incl. 125 HP motor size	5,150	1,950	7,100
2480	To and incl. 150 HP motor size	6,250	2,175	8,425
2520	To and incl. 200 HP motor size	7,100	2,475	9,575
2600	575V, to and including 2 HP motor size	485	600	1,085
2640	To and incl. 5 HP motor size	520	650	1,170
2680	To and incl. 10 HP motor size	545	720	1,265
2720	To and incl. 20 HP motor size	760	825	1,585
2760	To and incl. 25 HP motor size	790	880	1,670
2800	To and incl. 30 HP motor size	1,125	1,000	2,125
2840	To and incl. 50 HP motor size	1,175	1,050	2,225
2880	To and incl. 60 HP motor size	2,325	1,400	3,725
2920	To and incl. 75 HP motor size	2,350	1,400	3,750
2960	To and incl. 100 HP motor size	2,675	1,550	4,225
3000	To and incl. 125 HP motor size	5,050	1,925	6,975
3040	To and incl. 150 HP motor size	5,150	1,950	7,100
3080	To and incl. 200 HP motor size	6,350	2,225	8,575

Figure 19.6 (cont.)

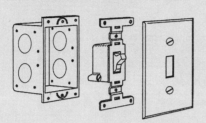

Description: Table 9.2-542 includes the cost for switch, plate, box, conduit in slab or EMT exposed and copper wire. Add 20% for exposed conduit.

No power required for switches.

Federal energy guidelines recommend the maximum lighting area controlled per switch shall not exceed 1000 S.F. and that areas over 500 S.F. shall be so controlled that total illumination can be reduced by at least 50%.

System Components	QUANTITY	UNIT	COST PER S.F.		
			MAT.	INST.	TOTAL
SYSTEM 9.2-542-0200					
WALL SWITCHES, 1.0 PER 1000 S.F.					
Steel intermediate conduit, (IMC) 1/2" diam	22.000	L.F.	25.08	77.88	102.96
Wire 600V type THWN-THHN, copper solid #12	.420	C.L.F.	2.73	13.44	16.17
Toggle switch, single pole, 15 amp	1.000	Ea.	4.40	8.85	13.25
Wall plate, 1 gang, brown plastic	1.000	Ea.	.63	4.42	5.05
Steel outlet box 4" square	1.000	Ea.	1.82	17.70	19.52
Steel outlet box 4" plaster rings	1.000	Ea.	1.10	5.55	6.65
TOTAL			35.76	127.84	163.60
COST PER S.F.			.04	.13	.17

9.2-542	Wall Switch by Sq. Ft.	COST PER S.F.		
		MAT.	INST.	TOTAL
0200	Wall switches, 1.0 per 1000 S.F.	.04	.13	.17
0240	1.2 per 1000 S.F.	.04	.14	.18
0280	2.0 per 1000 S.F.	.05	.21	.26
0320	2.5 per 1000 S.F.	.06	.25	.31
0360	5.0 per 1000 S.F.	.15	.54	.69
0400	10.0 per 1000 S.F.	.30	1.12	1.42

Figure 19.7

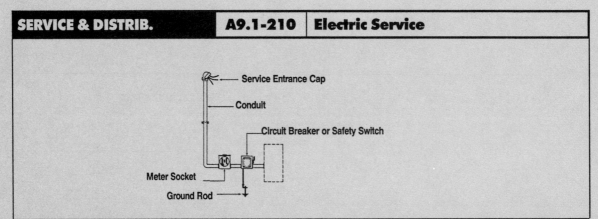

System Components	QUANTITY	UNIT	COST EACH		
			MAT.	INST.	TOTAL
SYSTEM 9.1-210-0200					
SERVICE INSTALLATION, INCLUDES BREAKERS, METERING, 20' CONDUIT & WIRE					
3 PHASE, 4 WIRE, 60 AMPS					
Circuit breaker, enclosed (NEMA 1), 600 volt, 3 pole, 60 amp	1.000	Ea.	300	126	426
Meter socket, single position, 4 terminal, 100 amp	1.000	Ea.	36.50	111	147.50
Rigid galvanized steel conduit, 3/4", including fittings	20.000	L.F.	34.60	88.40	123
Wire, 600V type XHHW, copper stranded #6	.900	C.L.F.	28.35	49.05	77.40
Service entrance cap 3/4" diameter	1.000	Ea.	6.60	27	33.60
Conduit LB fitting with cover, 3/4" diameter	1.000	Ea.	8.55	27	35.55
Ground rod, copper clad, 8' long, 3/4" diameter	1.000	Ea.	26.50	66.50	93
Ground rod clamp, bronze, 3/4" diameter	1.000	Ea.	5.15	11.05	16.20
Ground wire, bare armored, #6-1 conductor	.200	C.L.F.	15.80	39.20	55
TOTAL			462.05	545.20	1,007.25

9.1-210	Electric Service, 3 Phase - 4 Wire		COST EACH		
			MAT.	INST.	TOTAL
0200	Service installation, includes breakers, metering, 20' conduit & wire				
0220	3 phase, 4 wire, 120/208 volts, 60 amp		460	545	1,005
0240	100 amps		590	655	1,245
0280	200 amps		890	1,000	1,890
0320	400 amps	R9.0 -010	1,950	1,850	3,800
0360	600 amps		3,750	2,525	6,275
0400	800 amps		4,500	3,025	7,525
0440	1000 amps		5,650	3,475	9,125
0480	1200 amps		7,375	3,550	10,925
0520	1600 amps		14,200	5,100	19,300
0560	2000 amps		15,600	5,800	21,400
0570	Add 25% for 277/480 volt				
0580					
0610	1 phase, 3 wire, 120/240 volts, 100 amps		310	595	905
0620	200 amps		655	960	1,615

Figure 19.8

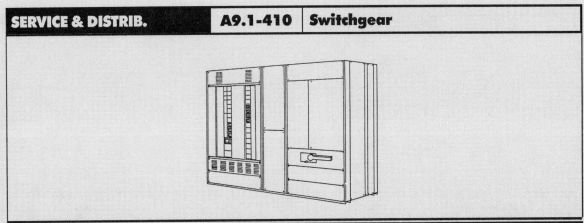

System Components	QUANTITY	UNIT	COST EACH		
			MAT.	INST.	TOTAL
SYSTEM 9.1-410-0240					
SWITCHGEAR INSTALLATION, INCL SWBD, PANELS & CIRC BREAKERS, 600 AMPS					
Panelboard, NQOD 225A 4W 120/208V main CB, w/20A bkrs 42 circ	1.000	Ea.	1,725	1,275	3,000
Switchboard, alum. bus bars, 120/208V, 4 wire, 600V	1.000	Ea.	2,800	705	3,505
Distribution sect., alum. bus bar, 120/208 or 277/480 V, 4 wire, 600A	1.000	Ea.	1,925	705	2,630
Feeder section circuit breakers, KA frame, 70 to 225 amp	3.000	Ea.	2,430	333	2,763
TOTAL			8,880	3,018	11,898

9.1-410	Switchgear		COST EACH		
			MAT.	INST.	TOTAL
0200	Switchgear inst., incl. swbd., panels & circ bkr, 400 amps, 120/208volt		3,950	2,225	6,175
0240	600 amperes		8,875	3,025	11,900
0280	800 amperes	R9.0 -010	11,100	4,300	15,400
0320	1200 amperes		13,800	6,600	20,400
0360	1600 amperes		18,700	9,275	27,975
0400	2000 amperes		23,400	11,800	35,200
0410	Add 20% for 277/480 volt				

Figure 19.9

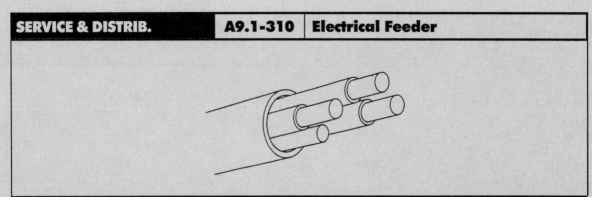

SERVICE & DISTRIB.	A9.1-310	Electrical Feeder

System Components	QUANTITY	UNIT	COST PER L.F.		
			MAT.	INST.	TOTAL
SYSTEM 9.1-310-0200					
FEEDERS, INCLUDING STEEL CONDUIT & WIRE, 60 AMPERES					
Rigid galvanized steel conduit, 3/4", including fittings	1.000	L.F.	1.73	4.42	6.15
Wire 600 volt, type XHHW copper stranded #6	.040	C.L.F.	1.26	2.18	3.44
TOTAL			2.99	6.60	9.59

9.1-310	Feeder Installation		COST PER L.F.		
			MAT.	INST.	TOTAL
0200	Feeder installation 600 volt, including conduit and wire, 60 amperes		2.99	6.60	9.59
0240	100 amperes		5.15	8.70	13.85
0280	200 amperes	R9.1	11.30	13.50	24.80
0320	400 amperes	-140	22.50	27	49.50
0360	600 amperes		47	44	91
0400	800 amperes		63.50	52.50	116
0440	1000 amperes		78.50	67.50	146
0480	1200 amperes		91.50	69	160.50
0520	1600 amperes		127	105	232
0560	2000 amperes		157	135	292

Figure 19.10

Assembly Number	Description	Qty	Unit	Total Cost		Cost per S. F.
				Unit	Total	
8.0	**Mechanical**					
	Plumbing					
8.1-433-1600	Lavatory, Vanity Top	18	EA	661	11898	
8.1-434-4380	Service Sink	3	EA	1415	4245	
8.1-450-2000	Urinal	3	EA	1005	3015	
8.1-460-1880	Water Cooler	3	EA	1085	3255	
8.1-470-1840	Water Closet	15	EA	1370	20550	
	Control(10%), Fittings(30%),Qual/Comp(10%)				21481	
8.1-160-1820	Water Heater: 50 gal./37 GPH	3	EA	3235	9705	
8.1-310-4200	Roof Drains 4" CI	4	EA	775	3100	
8.1-310-4320	5" Pipe CI	60	LF	24.20	1452	
8.1-310-4400	6" Pipe CI	200	LF	27.10	5420	
8.2-110-0620	Fire Protection Light Hazard, Wet Pipe	10000	SF	1.58	15800	
8.2-110-0740	Same	20000	SF	1.27	25400	
8.2-310-1540	Standpipe	12/10	FL	3900	4680	
8.2-310-1560	Same	12/10	FL	965	1158	
8.2-390-0750	Cabinet	3	EA	257	771	
	Hose.Nozzle,Rack,Hydrolator,Valve(Set)	3	Sets	487	1461	
8.2-390-7200	Siamese	1	EA	468	468	
8.2-390-1650	Alarm	1	EA	131.75	132	
8.2-390-0100	Adaptor	3	EA	32	96	
	Heating and Air Conditioning					
8.4-220-3960	Roof Top Multi-zone	30000	SF	10.57	317100	
	Balancing (0.5%)				1586	
	Quality/Complexity (7%)				22197	
					$474,969	$15.83
9.0	**Electrical**					
9.2-213-0280	Lighting	30000	SF	4.89	146700	
9.2-522-0640	Receptacles	30000	SF	2.57	77100	
9.2-610-0280	HVAC	30000	SF	0.32	9600	
9.2-582-0320	Misc. Motors & Power	30000	SF	0.18	5400	
9.2-710-2200	Elevators 2 @ 30 HP	2	EA	2125	4250	
9.2-542-0280	Wall Switches 2/1000 SF	30000	SF	0.26	7800	
9.1-210-0360	Service 600 Amp	1.25	EA	6275	7844	
9.1-210-0570	277v/480v					
9.1-410-0240	Panels 600 Amp	1.2	EA	11900	14280	
9.1-410-0410	277v/480v					
9.1-310-0360	Feeders 600 Amp	100	LF	91	9100	
	Fire Detection	30000	SF	0.37	11100	
	Emergency Lighting	30000	SF	0.44	13200	
					$306,374	$10.21

Figure 19.11

Chapter 20

General Conditions & Profit

In the assemblies format, the General Conditions & Profit division is placed tenth. This part of an estimate, however, is typically performed after all other portions of the building project have been priced out.

If the project estimate is to be complete and comparable to the actual price charged by a building contractor to perform the work, then overhead and profit costs must be added to the estimate. These costs can be added to each division, if different factors are used, or added at the end of the estimate after all items have been totalled.

Overhead & Profit

Means Assemblies Cost Data includes the installing contractor's overhead and profit for each item. The column "Subs Total Overhead & Profit" in Figure 20.1 shows the total percentage and cost added to each trade for the installing contractor's overhead and profit. These costs are developed as explained in the following discussion.

Labor Base Rate
The Base Rate is the figure on which the actual payroll is calculated. The following deductions are made each week with the net balance going to the employee: standard deductions such as withholding taxes; FICA; savings plans; hospitalization; and other voluntary employee-paid deductions.

Base Rate Including Fringe Benefits
Base Rate Including Fringe Benefits is the sum of the Base Rate plus all employer-paid fringe benefits, which might include vacation pay, health and welfare insurance, pension, apprentice training, and industry advancement funds. Figure 20.1 lists the 40 construction trades plus the skilled workers' average and helpers' average for 30 major United States cities. The table is based on union labor rates for January 1, 1996 and is a national average.

Billing Rate
The Base Rate Including Fringe Benefits, when adjusted to include all direct and indirect expenses of the installing contractor, is called the Billing Rate. This rate includes the following direct and indirect expenses.

Workers' Compensation & Employer's Liability Insurance
Workers' Compensation and Employer's Liability Insurance rates vary from state to state and from one trade to another based on the relative safety record for that trade in that particular area during the previous year. Rates also vary from company to company based on experience rating.

Installing Contractor's Overhead & Profit

Below are the **average** installing contractor's percentage mark-ups applied to base labor rates to arrive at typical billing rates.

Column A: Labor rates are based on union wages averaged for 30 major U.S. cities. Base rates including fringe benefits are listed hourly and daily. These figures are the sum of the wage rate and employer-paid fringe benefits such as vacation pay, employer-paid health and welfare costs, pension costs, plus appropriate training and industry advancement funds costs.

Column B: Workers' Compensation rates are the national average of state rates established for each trade.

Column C: Column C lists average fixed overhead figures for all trades. Included are Federal and State Unemployment costs set at 7.3%; Social Security Taxes (FICA) set at 7.65%; Builder's Risk Insurance costs set at 0.34%; and Public Liability costs set at 1.55%. All the percentages except those for Social Security Taxes vary from state to state as well as from company to company.

Columns D and E: Percentages in Columns D and E are based on the presumption that the installing contractor has annual billing of $500,000 and up. Overhead percentages may increase with smaller annual billing. The overhead percentages for any given contractor may vary greatly and depend on a number of factors, such as the contractor's annual volume, engineering and logistical support costs, and staff requirements. The figures for overhead and profit will also vary depending on the type of job, the job location, and the prevailing economic conditions. All factors should be examined very carefully for each job.

Column F: Column F lists the total of Columns B, C, D, and E.

Column G: Column G is Column A (hourly base labor rate) multiplied by the percentage in Column F (O&P percentage).

Column H: Column H is the total of Column A (hourly base labor rate) plus Column G (Total O&P).

Column I: Column I is Column H multiplied by eight hours.

		A		B	C	D	E	F	G	H	I
		Base Rate Incl. Fringes		Workers' Comp. Ins.	Average Fixed Overhead	Overhead	Profit	Total Overhead & Profit		Rate with O & P	
Abbr.	Trade	Hourly	Daily					%	Amount	Hourly	Daily
Skwk	Skilled Workers Average (35 trades)	$25.95	$207.60	20.2%	16.8%	13.0%	10%	60.0%	$15.55	$41.50	$332.00
	Helpers Average (5 trades)	19.25	154.00	21.4		11.0		59.2	11.40	30.65	245.20
	Foreman Average, Inside ($.50 over trade)	26.45	211.60	20.2		13.0		60.0	15.85	42.30	338.40
	Foreman Average, Outside ($2.00 over trade)	27.95	223.60	20.2		13.0		60.0	16.75	44.70	357.60
Clab	Common Building Laborers	19.80	158.40	21.9		11.0		59.7	11.80	31.60	252.80
Asbe	Asbestos Workers	28.55	228.40	19.7		16.0		62.5	17.85	46.40	371.20
Boil	Boilermakers	30.05	240.40	17.7		16.0		60.5	18.20	48.25	386.00
Bric	Bricklayers	25.90	207.20	19.4		11.0		57.2	14.80	40.70	325.60
Brhe	Bricklayer Helpers	20.00	160.00	19.4		11.0		57.2	11.45	31.45	251.60
Carp	Carpenters	25.20	201.60	21.9		11.0		59.7	15.05	40.25	322.00
Cefi	Cement Finishers	24.35	194.80	12.8		11.0		50.6	12.30	36.65	293.20
Elec	Electricians	29.30	234.40	8.0		16.0		50.8	14.90	44.20	353.60
Elev	Elevator Constructors	30.05	240.40	9.6		16.0		52.4	15.75	45.80	366.40
Eqhv	Equipment Operators, Crane or Shovel	26.75	214.00	12.9		14.0		53.7	14.35	41.10	328.80
Eqmd	Equipment Operators, Medium Equipment	25.70	205.60	12.9		14.0		53.7	13.80	39.50	316.00
Eqlt	Equipment Operators, Light Equipment	24.70	197.60	12.9		14.0		53.7	13.25	37.95	303.60
Eqol	Equipment Operators, Oilers	21.90	175.20	12.9		14.0		53.7	11.75	33.65	269.20
Eqmm	Equipment Operators, Master Mechanics	27.55	220.40	12.9		14.0		53.7	14.80	42.35	338.80
Glaz	Glaziers	24.90	199.20	16.0		11.0		53.8	13.40	38.30	306.40
Lath	Lathers	24.95	199.60	13.5		11.0		51.3	12.80	37.75	302.00
Marb	Marble Setters	25.65	205.20	19.4		11.0		57.2	14.65	40.30	322.40
Mill	Millwrights	26.55	212.40	13.2		11.0		51.0	13.55	40.10	320.80
Mstz	Mosaic & Terrazzo Workers	25.25	202.00	11.0		11.0		48.8	12.30	37.55	300.40
Pord	Painters, Ordinary	22.95	183.60	16.8		11.0		54.6	12.55	35.50	284.00
Psst	Painters, Structural Steel	23.95	191.60	62.5		11.0		100.3	24.00	47.95	383.60
Pape	Paper Hangers	23.30	186.40	16.8		11.0		54.6	12.70	36.00	288.00
Pile	Pile Drivers	25.35	202.80	33.6		16.0		76.4	19.35	44.70	357.60
Plas	Plasterers	24.20	193.60	17.4		11.0		55.2	13.35	37.55	300.40
Plah	Plasterer Helpers	20.15	161.20	17.4		11.0		55.2	11.10	31.25	250.00
Plum	Plumbers	30.05	240.40	10.2		16.0		53.0	15.95	46.00	368.00
Rodm	Rodmen (Reinforcing)	27.75	222.00	36.3		14.0		77.1	21.40	49.15	393.20
Rofc	Roofers, Composition	22.55	180.40	37.4		11.0		75.2	16.95	39.50	316.00
Rots	Roofers, Tile & Slate	22.60	180.80	37.4		11.0		75.2	17.00	39.60	316.80
Rohe	Roofers, Helpers (Composition)	15.95	127.60	37.4		11.0		75.2	12.00	27.95	223.60
Shee	Sheet Metal Workers	28.95	231.60	13.8		16.0		56.6	16.40	45.35	362.80
Spri	Sprinkler Installers	31.30	250.40	10.4		16.0		53.2	16.65	47.95	383.60
Stpi	Steamfitters or Pipefitters	30.30	242.40	10.2		16.0		53.0	16.05	46.35	370.80
Ston	Stone Masons	25.90	207.20	19.4		11.0		57.2	14.80	40.70	325.60
Sswk	Structural Steel Workers	27.85	222.80	46.4		14.0		87.2	24.30	52.15	417.20
Tilf	Tile Layers	25.05	200.40	11.0		11.0		48.8	12.20	37.25	298.00
Tilh	Tile Layers Helpers	20.30	162.40	11.0		11.0		48.8	9.90	30.20	241.60
Trlt	Truck Drivers, Light	20.35	162.80	17.0		11.0		54.8	11.15	31.50	252.00
Trhv	Truck Drivers, Heavy	20.70	165.60	17.0		11.0		54.8	11.35	32.05	256.40
Sswl	Welders, Structural Steel	27.85	222.80	46.4		14.0		87.2	24.30	52.15	417.20
Wrck	*Wrecking	19.80	158.40	44.8	▼	11.0	▼	82.6	16.35	36.15	289.20

*Not included in Averages.

Figure 20.1

Average Fixed Overhead

Average Fixed Overhead includes the costs paid by the employer for U.S. and State Unemployment, Social Security (FICA), Builder's Risk, and Public Liability.

Overhead

Every person who operates a business has basic costs just to stay in business. These are costs that represent the inescapable cost of doing business. Building contractors have expenses that must be paid whether or not they are working on a job. These costs do not contribute directly to the actual construction activities of the project but are, nevertheless, necessary to get the job done. The cost of overhead can represent a considerable portion of a contractor's annual business volume. The range used in *Means Assemblies Cost Data* is representative for the type of contractors that work on projects costing $400,000 and more. The overhead range chosen is from 11% to 16% and varies with the type of contractor and annual business volume.

Profit and Contingency

The Profit and Contingency percentage represents the fee added by the installing contractor that offers a return on investment plus an allowance to cover the risk involved in the construction job being bid. The profit percentage varies for each contractor for each job depending on several factors, such as the economic outlook for future work; the contractor's need for work; the time of year (spring, summer, fall, winter); the expected amount of work to be subcontracted; the expected number of bidders; and the estimated risk involved in the job. For estimating purposes, *Means Assemblies Cost Data* has an allowance of 10% built into the prices as a reasonable profit and contingency factor. Realize, however, that contractors set their own profit and contingency based on their own unique situation.

General Conditions

Prices given in *Means Assemblies Cost Data* are those quoted to the general or prime contractor. Consequently, an allowance for the general or prime contractor's general conditions, project management, and markup should be added to the prices in the book. This markup ranges from 5% to 15%. A figure of 10% would be a reasonable allowance for this item in the assemblies square foot estimate.

Project Overhead Costs

To complete an assemblies square foot estimate, it is necessary to identify project overhead items that have not been included in the estimate. Costs of items such as field supervision, tools and minor equipment, field office, sheds, and photos must be accounted for. Figure 20.2 provides factors that may be used for these project overhead items.

Office Overhead

There are certain indirect expense items that are incurred by the requirement to keep the shop doors open, attract business, and bid work. The percentage of main office overhead expense declines with the increased annual volume of the contractor. Overhead is not appreciably increased when there is an increase in the volume of work. Typical main office expenses range from 2% to 20%, with the median about 7.2% of the total volume. This equals 7.7% of the job direct costs.

Figure 20.3 shows approximate percentages for some of the items usually included in a general/prime contractor's main office overhead. These percentages may vary with different accounting procedures.

General Contractor's Overhead

The table below shows a contractor's overhead as a percentage of direct cost in two ways. The figures on the right are for the overhead, markup based on both material and labor. The figures on the left are based on the entire overhead applied only to the labor. This figure would be used if the owner supplied the materials or if a contract is for labor only.

Items of General Contractor's Indirect Costs	% of Direct Costs	
	As a Markup of Labor Only	As a Markup of Both Material and Labor
Field Supervision	6.0%	3.2%
Main Office Expense (see details below)	14.7	7.7
Tools and Minor Equipment	1.0	0.5
Workers' Compensation & Employers' Liability.	20.2	10.6
Field Office, Sheds, Photos, Etc.	1.5	0.8
Performance and Payment Bond, 0.5% to 0.9%.	0.7	0.4
Unemployment Tax. See R10.2-300 (Combined Federal and State)	7.3	3.8
Social Security and Medicare (7.65% of first $61,200)	7.7	4.0
Sales Tax — add if applicable 38/80 x % as markup of total direct costs including both material and labor.		
Sub Total	59.1%	31.0%
*Builder's Risk Insurance ranges from .141% to .586%.	0.3	0.3
*Public Liability Insurance	1.5	1.5
Grand Total	60.9%	32.8%

*Paid by Owner or Contractor

Figure 20.2

Main Office Expense

A General Contractor's main office expense consists of many items not detailed in the front portion of the book. The percentage of main office expense declines with increased annual volume of the contractor. Typical main office expense ranges from 2% to 20% with the median about 7.2% of total volume. This equals about 7.7% of direct costs. The following are approximate percentages of total overhead for different items usually included in a General Contractor's main office overhead. With different accounting procedures, these percentages may vary.

Item	Typical Range			Average
Managers', clerical and estimators' salaries	40%	to	55%	48%
Profit sharing, pension and bonus plans	2	to	20	12
Insurance	5	to	8	6
Estimating and project management (not including salaries)	5	to	9	7
Legal, accounting and data processing	0.5	to	5	3
Automobile and light truck expense	2	to	8	5
Depreciation of overhead capital expenditures	2	to	6	4
Maintenance of office equipment	0.1	to	1.5	1
Office rental	3	to	5	4
Utilities including phone and light	1	to	3	2
Miscellaneous	5	to	15	8
Total				100%

Figure 20.3

Profit Margin

The profit assumed in *Means Assemblies Cost Data* is 10% on material, labor, and equipment. Since this is the profit margin for the installing contractor, an additional percentage must be included to cover the profit of the general or prime contractor.

Contingencies

An allowance for contingencies provides for unforeseen construction difficulties or for oversights during the estimating process. Different factors should be used for the various stages of design completion. The following percentages are guidelines to point the estimator in the right direction.

Conceptual Stage, add:	15% to 20%
Preliminary Drawings, add:	10% to 15%
Working Drawings, 60% Design Complete, add:	7% to 10%
Final Working Drawings, 100% Checked Finals, add:	2% to 7%
Field Contingencies, add:	0% to 3%

As far as the construction contract is concerned, changes in the project can and often will be covered by extras or change orders. The contractor should consider inflationary price trends and possible material shortages that may occur during the course of the job. Escalation factors depend on both economic conditions and the anticipated time between the estimate and actual construction. In the final summary, contingencies are a matter of estimating judgment.

Design and Engineering

Architectural-engineering fees are typical percentage fees for design services. Some fees may vary significantly from those listed in Figure 20.4 because of economic conditions and the scope of work. The architectural fees include the engineering fees in Figures 20.5 and 20.6.

Fees may be interpolated horizontally and vertically. Various portions of the same project requiring different fees should be adjusted

Architectural Fees

Tabulated below are typical percentage fees by project size, for good professional architectural service. Fees may vary from those listed depending upon degree of design difficulty and economic conditions in any particular area.

Rates can be interpolated horizontally and vertically. Various portions of the same project requiring different rates should be adjusted proportionately. For alterations, add 50% to the fee for the first $500,000 of project cost and add 25% to the fee for project cost over $500,000.

Architectural fees tabulated below include Engineering Fees.

Building Types	Total Project Size in Thousands of Dollars						
	100	250	500	1,000	5,000	10,000	50,000
Factories, garages, warehouses, repetitive housing	9.0%	8.0%	7.0%	6.2%	5.3%	4.9%	4.5%
Apartments, banks, schools, libraries, offices, municipal buildings	11.7	10.8	8.5	7.3	6.4	6.0	5.6
Churches, hospitals, homes, laboratories, museums, research	14.0	12.8	11.9	10.9	8.5	7.8	7.2
Memorials, monumental work, decorative furnishings	—	16.0	14.5	13.1	10.0	9.0	8.3

Figure 20.4

proportionately. For alterations, add 50% to the fee for the first $500,000 of the estimated project cost, and 25% for over $500,000.

Taxes

Once the estimate is complete, consideration must be given to sales tax on materials and rental equipment as well as Wheel and Use taxes for the city, county, and/or state where the project will be constructed. Figure 20.7 lists state sales taxes, but not those levied by cities and counties.

Example: Three-Story Office Building

The three-story office building example has the following percentages added for General Conditions:

Overhead	7%
Profit	7%
AE fee	8.5%
General Conditions	5%
Contingency	2%

Engineering Fees

Typical **Structural Engineering Fees** based on type of construction and total project size. These fees are included in Architectural Fees.

Type of Construction	Total Project Size (in thousands of dollars)			
	$500	$500-$1,000	$1,000-$5,000	Over $5000
Industrial buildings, factories & warehouses	Technical payroll times 2.0 to 2.5	1.60%	1.25%	1.00%
Hotels, apartments, offices, dormitories, hospitals, public buildings, food stores		2.00%	1.70%	1.20%
Museums, banks, churches and cathedrals		2.00%	1.75%	1.25%
Thin shells, prestressed concrete, earthquake resistive		2.00%	1.75%	1.50%
Parking ramps, auditoriums, stadiums, convention halls, hangars & boiler houses		2.50%	2.00%	1.75%
Special buildings, major alterations, underpinning & future expansion	↓	Add to above 0.5%	Add to above 0.5%	Add to above 0.5%

For complex reinforced concrete or unusually complicated structures, add 20% to 50%.

Figure 20.5

Mechanical and Electrical Fees

Typical **Mechanical and Electrical Engineering Fees** based on the size of the subcontract. These fees are included in Architectural Fees.

Type of Construction	Subcontract Size							
	$25,000	$50,000	$100,000	$225,000	$350,000	$500,000	$750,000	$1,000,000
Simple structures	6.4%	5.7%	4.8%	4.5%	4.4%	4.3%	4.2%	4.1%
Intermediate structures	8.0	7.3	6.5	5.6	5.1	5.0	4.9	4.8
Complex structures	12.0	9.0	9.0	8.0	7.5	7.5	7.0	7.0

For renovations, add 15% to 25% to applicable fee.

Figure 20.6

The percentages will not be applied to the estimate until the final summary in Chapter 24. For this reason, no totals are shown on the form in Figure 20.8.

Sales Tax by State

State sales tax on materials is tabulated below (5 states have no sales tax). Many states allow local jurisdictions, such as a county or city, to levy additional sales tax.

Some projects may be sales tax exempt, particularly those constructed with public funds.

State	Tax (%)	State	Tax (%)	State	Tax (%)	State	Tax (%)
Alabama	4	Illinois	6.25	Montana	0	Rhode Island	7
Alaska	0	Indiana	5	Nebraska	5	South Carolina	5
Arizona	5.75	Iowa	5	Nevada	6.75	South Dakota	4
Arkansas	4.5	Kansas	4.9	New Hampshire	0	Tennessee	6
California	7.25	Kentucky	6	New Jersey	6	Texas	6.25
Colorado	3	Louisiana	4	New Mexico	5	Utah	5.875
Connecticut	6	Maine	6	New York	4	Vermont	5
Delaware	0	Maryland	5	North Carolina	4	Virginia	4.5
District of Columbia	5.75	Massachusetts	5	North Dakota	5	Washington	6.5
Florida	6	Michigan	6	Ohio	5	West Virginia	6
Georgia	4	Minnesota	6.5	Oklahoma	4.5	Wisconsin	5
Hawaii	4	Mississippi	7	Oregon	0	Wyoming	4
Idaho	5	Missouri	4.225	Pennsylvania	6	Average	4.75%

Figure 20.7

Assembly Number	Description		Qty	Unit	Total Cost		Cost per S. F.
					Unit	Total	
10.0	**General Conditions**						
	Overhead	7%					
	Profit	7%					
	AE Fee	8.5%					
	General Conditions	5%					
11.0	**Special Construction**						
12.0	**Site Work**						
13.0	**Miscellaneous**						

Figure 20.8

Chapter 21

Specialties

The Specialties division includes unique items that either are not included elsewhere as part of a system or have no other place in the estimate. These items are priced per each item, by the square foot, or by the lineal foot. Estimators should know the total installed cost of items such as:

- Appliances
- Bank equipment
- Bathroom accessories
- Chalkboards
- Dental equipment
- Display cases
- Domes
- Flagpoles
- Folding partitions
- Furnishings
- Greenhouses
- Kitchen equipment
- Lockers
- Radio towers
- Signs
- Solar, passive heating
- Swimming pools
- Tanks
- Toilet partitions

Some items may be components of systems that have already been priced and included in the estimate. For example, shower stalls are included in the assembly prices for bathrooms in Division 8, Mechanical. If the estimator needs to account for some additional items, then the prices in the Specialties division can be used.

Some of the specialty items in *Means Assemblies Cost Data* are general in nature and should be used for initial budgets. Once more detailed information is known about the project, more specific costs from suppliers and manufacturers can be used to fine tune the estimate.

Example: Three-Story Office Building

The following specialty items are required to complete the estimate for the three-story office building.

Bathroom Accessories:

Towel Dispenser	(Figure 21.1 #1100)		
	12 ea. @ $ 60.50	=	$ 726
Grab Bar	(Figure 21.1 #1240)		
	12 ea. @ $ 63.60	=	763
Mirror	(Figure 21.1 #1300)		
	18 ea. @ $115.10	=	2,072
Tissue Dispenser	(Figure 21.1 #1440)		
	15 ea. @ $ 30.40	=	456
Vanity	(Allowance)		
	48 L.F. @ $ 75	=	3,600

Other Specialties:

Directory Board (Figure 21.1 #2200) 2 ea. @ $434	=	868
Floor Mats (Figure 21.2 #4740) 48 S.F. @ $ 32.13	=	1,542
Total Cost		$10,027
Cost per Square Foot: $10,027/30,000 S.F.	=	$0.33/S.F.

Figure 21.3 summarizes the costs for this division.

11.1-100	Architectural Specialties/Each	COST EACH		
		MAT.	INST.	TOTAL
1000	Specialties, bathroom accessories, st. steel, curtain rod, 5' long, 1" diam	27.50	25	52.50
1040	1-1/2" diam.	29.50	25	54.50
1100	Dispenser, towel, surface mounted	40.50	20	60.50
1140	Flush mounted with waste receptacle	273	32	305
1200	Grab bar, 1-1/4" diam., 12" long	35	13.40	48.40
1240	1-1/2" diam. 36" long	47.50	16.10	63.60
1300	Mirror, framed with shelf, 18" x 24"	99	16.10	115.10
1340	72" x 24"	620	53.50	673.50
1400	Toilet tissue dispenser, surface mounted, single roll	10.55	10.75	21.30
1440	Double roll	17	13.40	30.40
1500	Towel bar, 18" long	24.50	14	38.50
1540	30" long	30	15.35	45.35
1600	Canopies, wall hung, prefinished aluminum, 8' x 10'	1,150	1,000	2,150
1640	12' x 40'	4,875	2,175	7,050
1700	Chutes, linen or refuse incl. sprinklers, galv. steel, 18" diam., per floor	700	207	907
1740	Aluminized steel, 36" diam., per floor	1,150	259	1,409
1800	Mail, 8-3/4" x 3-1/2", aluminum & glass, per floor	520	145	665
1840	Bronze or stainless, per floor	800	161	961
1900	Control boards, magnetic, porcelain finish, framed, 24" x 18"	100	80.50	180.50
1940	96" x 48"	430	129	559
2000	Directory boards, outdoor, black plastic, 36" x 24"	575	320	895
2040	36" x 36"	660	430	1,090
2100	Indoor, economy, open faced, 18" x 24"	82.50	92	174.50
2140	36" x 48"	150	107	257
2200	Building, aluminum, black felt panel, 24" x 18"	273	161	434
2240	48" x 72"	700	645	1,345
2300	Disappearing stairways, folding, pine, 8'-6" ceiling	82.50	80.50	163
2340	9'-6" ceiling	93.50	80.50	174
2400	Fire escape, galvanized steel, 8'-0" to 10'-6" ceiling	1,125	645	1,770
2440	10'-6" to 13'-6" ceiling	1,425	645	2,070
2500	Automatic electric, wood, 8' to 9' ceiling	5,500	645	6,145
2540	Aluminum, 14' to 15' ceiling	6,300	920	7,220
2600	Display cases, freestanding, glass and aluminum, 3'-6" x 3' x 1'-0" deep	630	80.50	710.50
2640	5'-10" x 4' x 1'-6" deep	1,175	107	1,282
2700	Wall mounted, glass and aluminum, 3' x 4' x 1'-4" deep	1,175	129	1,304
2740	16' x 4' x 1'-4" deep	3,275	430	3,705
2800	Table exhibit, flat, 3' x 4' x 2'-0" wide	755	129	884
2840	Sloping, 3' x 4' x 3'-0" wide	1,125	215	1,340
2900	Fireplace prefabricated, freestanding or wall hung, economy	940	256	1,196
2940	Deluxe	2,675	370	3,045
3000	Woodburning stoves, cast iron, economy	700	510	1,210
3040	Deluxe	2,500	830	3,330
3100	Flagpoles, on grade, aluminum, tapered, 20' high	1,025	380	1,405
3140	70' high	9,350	955	10,305
3200	Fiberglass, tapered, 23' high	1,025	380	1,405
3240	59' high	4,500	850	5,350
3300	Concrete, internal halyard, 20' high	2,650	305	2,955
3340	100' high	24,900	765	25,665
3400	Lockers, steel, single tier, 5' to 6' high, per opening, min.	107	26	133
3440	Maximum	182	30	212
3500	Two tier, minimum	68.50	13.95	82.45
3600	Maximum	85	18.15	103.15
3700	Mail boxes, horizontal, rear loaded, aluminum, 5" x 6" x 15" deep	35	9.45	44.45
3740	Front loaded, aluminum, 10" x 12" x 15" deep	132	16.10	148.10
3800	Vertical, front loaded, aluminum, 15" x 5" x 6" deep	26.50	9.45	35.95
3840	Bronze, duranodic finish	49.50	9.45	58.95
3900	Letter slot, post office	193	40.50	233.50
4000	Mail counter, window, post office, with grille	515	161	676

Figure 21.1

SPECIAL CONSTR.	A11.1-500	Furnishings			

11.1-500	Furnishings/Each	COST EACH		
		MAT.	INST.	TOTAL
1000	Furnishings, blinds, exterior, aluminum, louvered, 1'-4" wide x 3'-0" long	32	32	64
1040	1'-4" wide x 6'-8" long	51.50	36	87.50
1100	Hemlock, solid raised, 1'-4" wide x 3'-0" long	53	32	85
1140	1'-4" wide x 6'-9" long	99	36	135
1200	Polystyrene, louvered, 1'-3" wide x 3'-3" long	40.50	32	72.50
1240	1'-3" wide x 6'-8" long	88	36	124
1300	Interior, wood folding panels, louvered, 7" x 20" (per pair)	24	18.95	42.95
1340	18" x 40" (per pair)	75.50	18.95	94.45
1800	Hospital furniture, beds, manual, economy	785		785
1840	Deluxe	1,375		1,375
1900	All electric, economy	1,300		1,300
1940	Deluxe	3,250		3,250
2000	Patient wall systems, no utilities, economy, per room	785		785
2040	Deluxe, per room	1,450		1,450
2200	Hotel furnishings, standard room set, economy, per room	1,450		1,450
2240	Deluxe, per room	7,650		7,650
2400	Office furniture, standard employee set, economy, per person	455		455
2440	Deluxe, per person	3,975		3,975
2800	Posts, portable, pedestrian traffic control, economy	82.50		82.50
2840	Deluxe	315		315
3000	Restaurant furniture, booth, molded plastic, stub wall and 2 seats, economy	400	161	561
3040	Deluxe	860	215	1,075
3100	Upholstered seats, foursome, single-economy	575	64.50	639.50
3140	Foursome, double-deluxe	3,100	107	3,207
3300	Seating, lecture hall, pedestal type, economy	103	18.40	121.40
3340	Deluxe	320	32	352
3400	Auditorium chair, veneer construction	103	18.40	121.40
3440	Fully upholstered, spring seat	151	18.40	169.40

Figure 21.2

SPECIAL CONSTR.	A11.1-500	Furnishings

11.1-500	Furnishings/S.F.	COST PER S.F.		
		MAT.	INST.	TOTAL
4010	Furnishings, blinds-interior, venetian-aluminum, stock, 2" slats, economy	1.29	.55	1.84
4040	Custom, 1" slats, deluxe	7.60	.73	8.33
4100	Vertical, PVC or cloth, T&B track, economy	7.15	.70	7.85
4140	Deluxe	11.95	.80	12.75
4300	Draperies, unlined, economy	2.01		2.01
4440	Lightproof, deluxe	6		6
4700	Floor mats, recessed, inlaid black rubber, 3/8" thick, solid	18.25	1.63	19.88
4740	Colors, 1/2" thick, perforated	30.50	1.63	32.13
4800	Link-including nosings, steel-galvanized, 3/8" thick	5.70	1.63	7.33
4840	Vinyl, in colors	15.95	1.63	17.58
5000	Shades, mylar, wood roller, single layer, non-reflective	5.60	.47	6.07
5040	Metal roller, triple layer, heat reflective	9.85	.47	10.32
5100	Vinyl, light weight, 4 ga.	.40	.47	.87
5140	Heavyweight, 6 ga.	1.24	.47	1.71
5200	Vinyl coated cotton, lightproof decorator shades	1.24	.47	1.71
5300	Woven aluminum, 3/8" thick, light and fireproof	3.97	.92	4.89

11.1-500	Furnishings/L.F.	COST PER L.F.		
		MAT.	INST.	TOTAL
5510	Furnishings, cabinets, hospital, base, laminated plastic	171	64.50	235.50
5540	Stainless steel	216	64.50	280.50
5600	Counter top, laminated plastic, no backsplash	30	16.10	46.10
5640	Stainless steel	82	16.10	98.10
5700	Nurses station, door type, laminated plastic	198	64.50	262.50
5740	Stainless steel	229	64.50	293.50
5900	Household, base, hardwood, one top drawer & one door below x 12" wide	110	26	136
5940	Four drawer x 24" wide	165	29	194
6000	Wall, hardwood, 30" high with one door x 12" wide	99	29.50	128.50
6040	Two doors x 48" wide	176	35	211
6100	Counter top-laminated plastic, stock, economy	5.20	10.75	15.95
6140	Custom-square edge, 7/8" thick	11	24	35
6300	School, counter, wood, 32" high	136	32	168
6340	Metal, 84" high	210	43	253
6500	Dormitory furniture, desk top (built-in),laminated plastc, 24"deep, economy	22.50	12.90	35.40
6540	30" deep, deluxe	100	16.10	116.10
6600	Dressing unit, built-in, economy	168	53.50	221.50
6640	Deluxe	500	80.50	580.50
7000	Restaurant furniture, bars, built-in, back bar	154	64.50	218.50
7040	Front bar	212	64.50	276.50
7200	Wardrobes & coatracks, standing, steel, single pedestal, 30" x 18" x 63"	71.50		71.50
7300	Double face rack, 39" x 26" x 70"	122		122
7340	Wall mounted rack, steel frame & shelves, 12" x 15" x 26"	51	4.57	55.57
7400	12" x 15" x 50"	40.50	2.39	42.89

Figure 21.2 (cont.)

Assembly Number	Description		Qty	Unit	Total Cost		Cost per S. F.
					Unit	Total	
10.0	**General Conditions**						
	Overhead	7%					
	Profit	7%					
	AE Fee	8.5%					
	General Conditions	5%					
11.0	**Special Construction**						
11.1-100-1100	Towel Dispenser		12	EA	60.50	726	
11.1-100-1240	Grab Bar		12	EA	63.60	763	
11.1-100-1300	Mirror		18	EA	115.10	2072	
11.1-100-1440	Tissue Dispenser		15	EA	30.40	456	
11.1-100-2200	Directory Board		2	EA	434	868	
11.1-500-4740	Floor Mats (2'x4'x6')		48	SF	32.13	1542	
Allowance	Vanity		48	LF	75	3600	
						$10,027	$0.33
12.0	**Site Work**						
13.0	**Miscellaneous**						

Figure 21.3

Chapter 22

Site Work

Site Work is a combination of several different categories: earthwork, paving, drainage, and piping, as well as lawns and plantings. The assemblies square foot approach to site work varies from one project to another because no two sites are precisely alike, and very few are completely predictable.

For assemblies square foot estimates, site work is divided into three separate divisions: earthwork, utilities, and pavements. A fourth, called Site Improvements, could be added, but assemblies costs are not appropriate for this category since most of the items are priced by the unit. At the conceptual stage, most designers and estimators know only that some site improvements are needed. It is unnecessary to develop specific figures at this point. An estimate of the area involved is sufficient to produce a budget or allowance for site improvements. For cost information, any unit price cost data source, such as *Means Building Construction Cost Data,* can provide the necessary information for the majority of site improvements.

Earthwork

Earthwork includes bulk excavation, trench excavation, and backfill. Bulk excavation and backfill, as they relate to the building itself, have been discussed in Chapter 11 and will not be addressed here. Also included in earthwork are site clearing, grubbing, and demolition. As with site improvement, these items are better priced using a unit price estimate if the estimator believes that the work is of significant value in the project to warrant the time required for a detailed estimate. Sheeting, dewatering, drilling and blasting, traffic control, equipment mobilization and demobilization, and any unusual factors pertaining to the site must also be accounted for in the earthwork estimate.

Trenching

The table in Figure 22.1 is representative of the type of pricing information available. The trenching systems are shown on a cost per linear foot basis. The prices include excavation and backfill that is placed and compacted for various depths and trench bottom widths. The trench side slopes represent sound construction practice for most soils.

281

| **A12.3-110** | **Trenching**

Trenching Systems are shown on a cost per linear foot basis. The systems include: excavation; backfill and removal of spoil; and compaction for various depths and trench bottom widths. The backfill has been reduced to accommodate a pipe of suitable diameter and bedding. See System 12.3-310

for bedding costs. The slope for trench sides varies from 0:1 to 2:1.

The Expanded System Listing shows Trenching Systems that range from 2' to 12' in width. Depths range from 2' to 25'.

System Components	QUANTITY	UNIT	COST PER L.F.		
			EQUIP.	LABOR	TOTAL
SYSTEM 12.3-110-1310					
TRENCHING, BACKHOE, 0 TO 1 SLOPE, 2' WIDE, 2' DP, 3/8 C.Y. BUCKET					
Excavation, trench, hyd. backhoe, track mtd., 3/8 C.Y. bucket	.174	C.Y.	.26	.66	.92
Backfill and load spoil, from stockpile	.174	C.Y.	.11	.19	.30
Compaction by rammer tamper, 8" lifts, 4 passes	.014	C.Y.	.01	.03	.04
Remove excess spoil, 6 C.Y. dump truck, 2 mile roundtrip	.160	C.Y.	.49	.33	.82
Total			.87	1.21	2.08

12.3-110	Trenching	COST PER L.F.		
		EQUIP.	LABOR	TOTAL
1310	Trenching, backhoe, 0 to 1 slope, 2' wide, 2' deep, 3/8 C.Y. bucket	.87	1.21	2.08
1320	3' deep, 3/8 C.Y. bucket	1.11	1.80	2.91
1330	4' deep, 3/8 C.Y. bucket	1.34	2.41	3.75
1340	6' deep, 3/8 C.Y. bucket	1.82	3.09	4.91
1350	8' deep, 1/2 C.Y. bucket	2.18	4	6.18
1360	10' deep, 1 C.Y. bucket	3.17	4.82	7.99
1400	4' wide, 2' deep, 3/8 C.Y. bucket	1.82	2.42	4.24
1410	3' deep, 3/8 C.Y. bucket	2.60	3.62	6.22
1420	4' deep, 1/2 C.Y. bucket	2.72	3.77	6.49
1430	6' deep, 1/2 C.Y. bucket	3.66	5.80	9.46
1440	8' deep, 1/2 C.Y. bucket	5.50	7.60	13.10
1450	10' deep, 1 C.Y. bucket	6.55	9.45	16
1460	12' deep, 1 C.Y. bucket	8.35	12.15	20.50
1470	15' deep, 1-1/2 C.Y. bucket	7.55	10.40	17.95
1480	18' deep, 2-1/2 C.Y. bucket	10.15	14.90	25.05
1520	6' wide, 6' deep, 5/8 C.Y. bucket	6.70	7.80	14.50
1530	8' deep, 3/4 C.Y. bucket	8.25	9.80	18.05
1540	10' deep, 1 C.Y. bucket	9.65	11.80	21.45
1550	12' deep, 1-1/4 C.Y. bucket	8.90	9.25	18.15
1560	16' deep, 2 C.Y. bucket	14.60	10.55	25.15
1570	20' deep, 3-1/2 C.Y. bucket	15.70	19.10	34.80
1580	24' deep, 3-1/2 C.Y. bucket	25.50	32	57.50
1640	8' wide, 12' deep, 1-1/4 C.Y. bucket	13.10	12.55	25.65
1650	15' deep, 1-1/2 C.Y. bucket	15.45	15.55	31
1660	18' deep, 2-1/2 C.Y. bucket	22.50	15.70	38.20
1680	24' deep, 3-1/2 C.Y. bucket	35	42.50	77.50
1730	10' wide, 20' deep, 3-1/2 C.Y. bucket	28	32	60
1740	24' deep, 3-1/2 C.Y. bucket	44	53.50	97.50
1780	12' wide, 20' deep, 3-1/2 C.Y. bucket	34	38.50	72.50
1790	25' deep, bucket	59	71.50	130.50
1800	1/2 to 1 slope, 2' wide, 2' deep, 3/8 C.Y. bucket	1.16	1.55	2.71
1810	3' deep, 3/8 C.Y. bucket	1.66	2.59	4.25
1820	4' deep, 3/8 C.Y. bucket	2.17	3.87	6.04
1840	6' deep, 3/8 C.Y. bucket	3.52	6.05	9.57
1860	8' deep, 1/2 C.Y. bucket	5	9.30	14.30
1880	10' deep, 1 C.Y. bucket	8.70	13.05	21.75

Figure 22.1

Utilities

The Site Utilities and Drainage categories account for the costs of pipe bedding, manholes, and catch basins. Electrical cables and ducts, though installed in the site work area, are more properly priced in the Electrical division.

Pipe Bedding Systems

The assemblies shown in Figure 22.2 are for various pipe diameters. Compacted sand is used for the bedding material and fill to 12" over the pipe. No backfill is included in the price. Various side slopes are shown to accommodate different soil conditions. Prices for the pipe are not included and must be added.

Manhole and Catch Basin Systems

Included in Figure 22.3 are: excavation with a backhoe, formed concrete footing, concrete or masonry walls, aluminum steps, and compacted backfill. Inside diameters range from 4' to 6', with depths ranging from 4' to 14'. The construction materials used are concrete, concrete block, precast concrete, or brick.

Pavements

The Roadway and Parking sections include quantities and prices for typical designs with gravel base and asphaltic concrete pavement. Different climate, soil conditions, material availability, and owner's requirements can be adapted to the parameters shown.

Roadway Systems

The roadway prices in Figure 22.4 include: 8" of compacted bank-run gravel, fine grading with a grader and roller, and a bituminous wearing course. The roadway systems are shown on a cost-per-linear-foot basis for the thickness and width indicated.

Parking Lot Systems

The table in Figure 22.5 includes prices for various parking lot assemblies and layouts. The prices include: compacted bank-run gravel, fine grading with a grader and roller, bituminous concrete wearing course, final stall design, and layout with white striping paint. All of the assemblies shown are on a cost-per-car basis and have enough allowance included for the travel areas, ingress, and egress. The prices shown vary according to the thickness of the different materials and the parking angle. The three parking arrangements have different efficiencies and thus have different installed costs, with the 90-degree arrangement being the most efficient and least expensive.

Example: Three-Story Office Building

The following site work is included in the estimate for our example building.

Site Preparation, Clear & Grub (Figure 22.6)	1 Acre × $6,550	=	$ 6,550
Trenching for Strip Footings (Figure 22.1 #1420)	400 L.F. × $6.49	=	2,596
Parking Lot (Figure 22.5 #1520) (23,200 S.F./390 S.F./car = 60 cars)	60 cars × $428	=	25,680
Site Improvements (grass seeding, shrubs & trees) (Allowance)	6,800 S.F. × $1.50	=	13,600
Total Cost			$48,426
Cost per Square Foot: $35,027/30,000		=	$1.61/S.F.

Figure 22.7 summarizes the costs for this division.

A12.3-310 **Pipe Bedding**

The Pipe Bedding System is shown for various pipe diameters. Compacted bank sand is used for pipe bedding and to fill 12" over the pipe. No backfill is included. Various side slopes are shown to accommodate different soil conditions. Pipe sizes vary from 6" to 84" diameter.

System Components	QUANTITY	UNIT	COST PER L.F.		
			MAT.	INST.	TOTAL
SYSTEM 12.3-310-1440 PIPE BEDDING, SIDE SLOPE 0 TO 1, 1' WIDE, PIPE SIZE 6" DIAMETER					
Borrow, bank sand, 2 mile haul, machine spread	.067	C.Y.	.27	.32	.59
Compaction, vibrating plate	.067	C.Y.		.08	.08
TOTAL			.27	.40	.67

12.3-310	Pipe Bedding	COST PER L.F.		
		MAT.	INST.	TOTAL
1440	Pipe bedding, side slope 0 to 1, 1' wide, pipe size 6" diameter	.27	.40	.67
1460	2' wide, pipe size 8" diameter	.59	.87	1.46
1480	Pipe size 10" diameter	.61	.90	1.51
1500	Pipe size 12" diameter	.63	.92	1.55
1520	3' wide, pipe size 14" diameter	1.03	1.51	2.54
1540	Pipe size 15" diameter	1.05	1.52	2.57
1560	Pipe size 16" diameter	1.06	1.55	2.61
1580	Pipe size 18" diameter	1.08	1.57	2.65
1600	4' wide, pipe size 20" diameter	1.56	2.28	3.84
1620	Pipe size 21" diameter	1.58	2.29	3.87
1640	Pipe size 24" diameter	1.62	2.36	3.98
1660	Pipe size 30" diameter	1.65	2.41	4.06
1680	6' wide, pipe size 32" diameter	2.89	4.22	7.11
1700	Pipe size 36" diameter	2.97	4.33	7.30
1720	7' wide, pipe size 48" diameter	3.86	5.65	9.51
1740	8' wide, pipe size 60" diameter	4.82	7.05	11.87
1760	10' wide, pipe size 72" diameter	6.95	10.15	17.10
1780	12' wide, pipe size 84" diameter	9.45	13.80	23.25
2140	Side slope 1/2 to 1, 1' wide, pipe size 6" diameter	.57	.82	1.39
2160	2' wide, pipe size 8" diameter	.94	1.37	2.31
2180	Pipe size 10" diameter	1.02	1.49	2.51
2200	Pipe size 12" diameter	1.09	1.59	2.68
2220	3' wide, pipe size 14" diameter	1.56	2.28	3.84
2240	Pipe size 15" diameter	1.60	2.34	3.94
2260	Pipe size 16" diameter	1.65	2.41	4.06
2280	Pipe size 18" diameter	1.75	2.55	4.30
2300	4' wide, pipe size 20" diameter	2.31	3.37	5.68
2320	Pipe size 21" diameter	2.36	3.45	5.81
2340	Pipe size 24" diameter	2.53	3.70	6.23
2360	Pipe size 30" diameter	2.84	4.15	6.99
2380	6' wide, pipe size 32" diameter	4.18	6.10	10.28
2400	Pipe size 36" diameter	4.48	6.55	11.03
2420	7' wide, pipe size 48" diameter	6.10	8.95	15.05
2440	8' wide, pipe size 60" diameter	7.95	11.65	19.60
2460	10' wide, pipe size 72" diameter	11.15	16.25	27.40
2480	12' wide, pipe size 84" diameter	14.85	21.50	36.35
2620	Side slope 1 to 1, 1' wide, pipe size 6" diameter	.86	1.26	2.12
2640	2' wide, pipe size 8" diameter	1.30	1.89	3.19
2660	Pipe size 10" diameter	1.42	2.07	3.49
2680	Pipe size 12" diameter	1.56	2.28	3.84

Figure 22.2

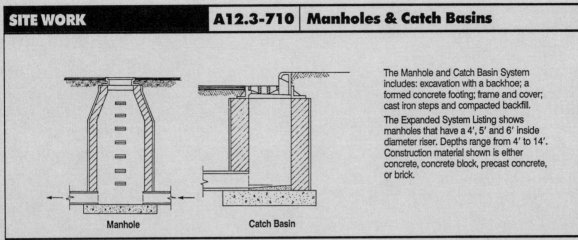

The Manhole and Catch Basin System includes: excavation with a backhoe; a formed concrete footing; frame and cover; cast iron steps and compacted backfill.

The Expanded System Listing shows manholes that have a 4', 5' and 6' inside diameter riser. Depths range from 4' to 14'. Construction material shown is either concrete, concrete block, precast concrete, or brick.

Manhole   Catch Basin

System Components	QUANTITY	UNIT	COST PER EACH		
			MAT.	INST.	TOTAL
SYSTEM 12.3-710-1920					
MANHOLE/CATCH BASIN, BRICK, 4' I.D. RISER, 4' DEEP					
Excavation, hydraulic backhoe, 3/8 C.Y. bucket	14.815	C.Y.		72	72
Trim sides and bottom of excavation	64.000	S.F.		33.92	33.92
Forms in place, manhole base, 4 uses	20.000	SFCA	12	60.20	72.20
Reinforcing in place footings, #4 to #7	.019	Tons	10.93	14.25	25.18
Concrete, 3000 psi, incl. place and vibrate under 1 CY, direct chute	.925	C.Y.	53.19	28.25	81.44
Catch basin or MH, brick, 4' ID, 4' deep	1.000	Ea.	475	575	1,050
Catch basin or MH steps; heavy galvanized cast iron	1.000	Ea.	8.15	8.15	16.30
Catch basin or MH frame and cover	1.000	Ea.	200	133.50	333.50
Backfill with wheeled front end loader	12.954	C.Y.	75.13	19.82	94.95
Backfill compaction, 12" lifts, air tamp	12.954	C.Y.		66.19	66.19
TOTAL			834.40	1,011.28	1,845.68

12.3-710	Manholes & Catch Basins	COST PER EACH		
		MAT.	INST.	TOTAL
1920	Manhole/catch basin, brick, 4' I.D. riser, 4' deep	835	1,025	1,860
1940	6' deep	1,100	1,425	2,525
1960	8' deep	1,375	1,925	3,300
1980	10' deep	1,650	2,375	4,025
3000	12' deep	2,000	2,575	4,575
3020	14' deep	2,375	3,650	6,025
3200	Block, 4' I.D. riser, 4' deep	660	825	1,485
3220	6' deep	830	1,175	2,005
3240	8' deep	1,050	1,600	2,650
3260	10' deep	1,275	1,975	3,250
3280	12' deep	1,575	2,525	4,100
3300	14' deep	1,875	3,075	4,950
4620	Concrete, cast-in-place, 4' I.D. riser, 4' deep	740	955	1,695
4640	6' deep	950	1,275	2,225
4660	8' deep	1,200	1,825	3,025
4680	10' deep	1,450	2,250	3,700
4700	12' deep	1,775	2,825	4,600
4720	14' deep	2,125	3,450	5,575
5820	Concrete, precast, 4' I.D. riser, 4' deep	665	690	1,355
5840	6' deep	865	930	1,795
5860	8' deep	1,075	1,300	2,375
5880	10' deep	1,300	1,600	2,900
5900	12' deep	1,575	1,950	3,525
5920	14' deep	1,900	2,525	4,425

Figure 22.3

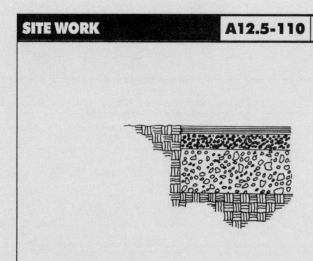

The Roadway System includes surveying; 8″ of compacted bank-run gravel; fine grading with a grader and roller; and bituminous wearing course. Roadway Systems are shown on a cost per linear foot basis.

The Expanded System Listing shows Roadway Systems with widths that vary from 20′ to 34′; for various pavement thicknesses.

System Components			COST PER L.F.		
	QUANTITY	UNIT	MAT.	INST.	TOTAL
SYSTEM 12.5-110-1500					
ROADWAYS, BITUMINOUS CONC. PAVING, 2-1/2″ THICK, 20′ WIDE					
4 man surveying crew	.010	Ea.		10.75	10.75
Bank gravel, 2 mi. haul, place and spread by dozer	.494	C.Y.	2.23	2.34	4.57
Compaction, granular material to 98%	.494	C.Y.		2.09	2.09
Grading, fine grade, 3 passes with grader plus rolling	2.220	S.Y.		3.92	3.92
Bituminous, paving, wearing course 2″ - 2-1/2″	2.220	S.Y.	8.97	2.17	11.14
Curbs, granite, split face, straight, 5″ x 16″	2.000	L.F.	28.80	10.34	39.14
Painting lines, reflectorized, 4″ wide	1.000	L.F.	.07	.11	.18
TOTAL			40.07	31.72	71.79

12.5-110	Roadway Pavement	COST PER L.F.		
		MAT.	INST.	TOTAL
1500	Roadways, bituminous conc. paving, 2-1/2″ thick, 20′ wide	40	31.50	71.50
1520	24′ wide	42.50	33	75.50
1540	26′ wide	43.50	34	77.50
1560	28′ wide	44.50	35	79.50
1580	30′ wide	45.50	36	81.50
1600	32′ wide	47	37	84
1620	34′ wide	48	38	86
1800	3″ thick, 20′ wide	42	31.50	73.50
1820	24′ wide	44.50	33.50	78
1840	26′ wide	45.50	34.50	80
1860	28′ wide	47	35.50	82.50
1880	30′ wide	48.50	36	84.50
1900	32′ wide	49.50	37.50	87
1920	34′ wide	51	38.50	89.50
2100	4″ thick, 20′ wide	46	31.50	77.50
2120	24′ wide	49	33.50	82.50
2140	26′ wide	51	34	85
2160	28′ wide	52.50	35	87.50
2180	30′ wide	54	36	90
2200	32′ wide	56	37	93
2220	34′ wide	57.50	38	95.50
2390				

Figure 22.4

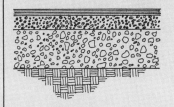

The Parking Lot System includes: compacted bank-run gravel; fine grading with a grader and roller; and bituminous concrete wearing course. All Parking Lot systems are on a cost per car basis. There are three basic types of systems: 90° angle, 60° angle, and 45° angle. The gravel base is compacted to 98%. Final stall design and lay-out of the parking lot with precast bumpers, sealcoating and white paint is also included.

The Expanded System Listing shows the three basic parking lot types with various depths of both gravel base and wearing course. The gravel base depths range from 6″ to 10″. The bituminous paving wearing course varies from a depth of 3″ to 6″.

System Components	QUANTITY	UNIT	COST PER CAR		
			MAT.	INST.	TOTAL
SYSTEM 12.5-510-1500					
PARKING LOT, 90° ANGLE PARKING, 3″ BITUMINOUS PAVING, 6″ GRAVEL BASE					
Surveying crew for layout, 4 man crew	.020	Day		21.50	21.50
Borrow, bank run gravel, haul 2 mi., spread w/dozer, no compaction	5.556	C.Y.	25.06	26.28	51.34
Grading, fine grade 3 passes with motor grader	33.333	S.Y.		24.65	24.65
Compact w/ vibrating plate, 8″ lifts, granular mat'l. to 98%	5.556	C.Y.		23.61	23.61
Bituminous paving, 3″ thick	33.333	S.Y.	160.67	37	197.67
Seal coating, petroleum resistant under 1,000 S.Y.	33.333	S.Y.	20.33	25	45.33
Painting lines on pavement, parking stall, white	1.000	Ea.	.88	2.42	3.30
Precast concrete parking bar, 6″ x 10″ x 6′-0″	1.000	Ea.	25	10.75	35.75
TOTAL			231.94	171.21	403.15

12.5-510	Parking Lots	COST PER CAR		
		MAT.	INST.	TOTAL
1500	Parking lot, 90° angle parking, 3″ bituminous paving, 6″ gravel base	232	171	403
1520	8″ gravel base	240	188	428
1540	10″ gravel base	249	205	454
1560	4″ bituminous paving, 6″ gravel base	291	170	461
1580	8″ gravel base	300	187	487
1600	10″ gravel base	310	203	513
1620	6″ bituminous paving, 6″ gravel base	395	178	573
1640	8″ gravel base	405	195	600
1661	10″ gravel base	410	211	621
1800	60° angle parking, 3″ bituminous paving, 6″ gravel base	266	190	456
1820	8″ gravel base	276	210	486
1840	10″ gravel base	286	229	515
1860	4″ bituminous paving, 6″ gravel base	335	188	523
1880	8″ gravel base	345	208	553
1900	10″ gravel base	355	228	583
1920	6″ bituminous paving, 6″ gravel base	455	198	653
1940	8″ gravel base	465	218	683
1961	10″ gravel base	475	236	711
2200	45° angle parking, 3″ bituminous paving, 6″ gravel base	300	208	508
2220	8″ gravel base	310	231	541
2240	10″ gravel base	325	253	578
2260	4″ bituminous paving, 6″ gravel base	380	207	587
2280	8″ gravel base	390	229	619
2300	10″ gravel base	400	251	651
2320	6″ bituminous paving, 6″ gravel base	520	217	737
2340	8″ gravel base	530	239	769
2361	10″ gravel base	540	400	940

Figure 22.5

The Site Work Division is divided into four sections:

12.1 Earthwork
12.3 Utilities
12.5 Pavement
12.7 Site Improvements

Clear and Grub, Demolition and Landscaping must be added to these costs as they apply to individual sites.

The Earthwork section includes typical bulk excavation, trench excavation, and backfill. To this must be added sheeting, dewatering, drilling and blasting, traffic control, and any unusual items relating to a particular site.

The Parking Lot and Roadway sections have quantities and prices for typical designs with gravel base and asphaltic concrete pavement. Different climate, soil conditions, material availability, and owner's requirements can easily be adapted to these parameters.

The Site Utilities and Drainage section tabulates the installed price for pipe. Excavation and backfill costs for the pipe are derived from the Earthwork secton.

By following the examples and illustrations of typical site work conditions in this division and adapting them to your particular site, an accurate price including O&P can be determined.

Site Preparation

Description	Unit	Total Costs
Clear and Grub Average	Acre	$6,550
Bulk Excavation with Front End Loader	C.Y.	1.25
Spread and Compact Dumped Material	C.Y.	1.70
Fine Grade and Seed	S.Y.	1.80

Figure 22.6

Assembly Number	Description		Qty	Unit	Total Cost		Cost per S. F.
					Unit	Total	
10.0	**General Conditions**						
	Overhead	7%					
	Profit	7%					
	AE Fee	8.5%					
	General Conditions	5%					
11.0	**Special Construction**						
11.1-100-1100	Towel Dispenser		12	EA	60.50	726	
11.1-100-1240	Grab Bar		12	EA	63.60	763	
11.1-100-1300	Mirror		18	EA	115.10	2072	
11.1-100-1440	Tissue Dispenser		15	EA	30.40	456	
11.1-100-2200	Directory Board		2	EA	434	868	
11.1-500-4740	Floor Mats (2'x4'x6')		48	SF	32.13	1542	
Allowance	Vanity		48	LF	75	3600	
						$10,027	$0.33
12.0	**Site Work**						
R12.0-010	Site Preparation, Clear and Grub		1	Acre	6550	6550	
12.3-110-1420	Trenching for Strip Footing		400	LF	6.49	2596	
12.5-510-1520	Parking Lot		60	Cars	428	25680	
	(23000 SF/390 SF/Car = 60 Cars)						
Allowance	Site Improvements (Grass, Trees, etc.)		6800	SF	2.00	13600	
						$48,426	$1.61
13.0	**Miscellaneous**						

Figure 22.7

Chapter 23

City Indexing

Publications such as *Means Assemblies Cost Data* contain building cost information based on a national average of major cities for the year published. Cost indexing to the actual project site is necessary. Estimates created from national average publications must be "localized" so that the costs reflect, as closely as possible, those in the area where the project is to be constructed.

The table in Figure 23.1 is an excerpt from the City Cost Indexes published in the 1996 Edition of *Means Assemblies Cost Data*. The indexes are calculated for 305 cities in the United States and Canada. Both material and installation factors are shown. A reference point of January 1, 1996 for the 30 major city average was used when compiling the cost information published in *Means Assemblies Cost Data*.

\multicolumn{5}{c}{City: Boston, MA}				
Div. No.	**System**	**Material**	**Inst.**	**Total**
1 & 2	Foundation & Substructure	113.3	135.2	127.3
3	Superstructure	110.1	133.1	120.4
4	Exterior Closure	122.1	145.3	133.2
5	Roofing	100.4	140.9	117.8
6	Interior Construction	98.5	143.8	117.0
7	Conveying	100.0	131.2	108.8
8	Mechanical	100.0	126.2	111.6
9	Electrical	101.6	132.4	122.2
11	Special Construction	100.0	139.2	102.6
12	Site Work	105.0	109.7	108.5
1-12	Weighted Average	105.2	133.3	118.8

Figure 23.1

Note: The index figures in Figure 23.1 are based on prices shown in *Means Assemblies Cost Data* being equal to a reference of 100.0. There are certain factors that the indexes cannot take into account. These factors include variations such as the following:

- Productivity
- Labor availability
- Contractor management efficiency
- Competitive conditions
- Automation
- Restrictive union practices
- Client's unique requirements

Example: Three-Story Office Building

Using the index figures in Figure 23.1, adjust the developed cost for the three-story office building for a construction site in Boston, Massachusetts. Since the costs used when developing the estimate were from the "Total" Columns, use the Total index factor. See Figure 23.2.

If the "Weighted Average" for Divisions 1 – 12 had been used, the results would be:

$2,026,835 × 118.8   = $2,407,880, or a difference of $291,011

The new square foot cost   = $80.26

Still another way to use the indexes is to accumulate separate Material and Installation prices, then apply the appropriate factors for each item.

Use City Cost Indexes similar to the one in Figure 23.1 to adjust the cost estimate for the specific location of a building construction project.

Division Number	System	Estimated Cost	Boston Index	Adjusted Cost
1 & 2	Foundation & Substructures	$ 66,521	127.3	$ 84,681
3	Superstructures	281,082	120.4	338,423
4	Exterior Closure	329,708	133.2	439,171
5	Roofing	28,541	117.8	33,621
6	Interior Construction	348,392	117.0	407,619
7	Conveying	138,000	108.8	150,144
8	Mechanical	469,764	111.6	524,257
9	Electrical	306,374	122.2	374,389
11	Special Construction	10,027	102.6	10,288
12	Site Work	48,426	108.5	52,542
	Subtotal	$2,026,835		$2,415,135
	Cost per Square Foot	$67.56/S.F.		$80.50/S.F.

Figure 23.2

Chapter 24

Summary of the Sample Estimate

Some publications list annually-updated square foot costs for various building types. Others list individual breakdown costs for foundations, substructures, superstructures, and so on. Either type of cost manual has a useful application in an assemblies square foot estimate.

Breakdown Square Foot Costs

Means Square Foot Costs, updated annually, includes breakdown costs for many building types and classifications. Figure 24.1 shows typical square foot costs for a three-story office building with a 10′ story height and 58,000 S.F. of floor area. Figure 24.2 shows total costs for similar office buildings with a range of square foot areas and different framing and exterior wall systems. At the bottom of Figure 24.2, a section titled "Common Additives" lists various building elements that provide the user greater latitude when tailoring the estimate.

Breakdown square foot cost tables such as Figures 24.1 and 24.2 allow users to choose specific categories or components from the listings and substitute others. One may wish to customize the exterior wall and interior finishes and use the costs from the tables for the remaining components. For example, the foundation cost may have to be revised if piles or caissons are required because of poor soil conditions.

The models shown in *Means Square Foot Costs* do not show structural bay spacing or superimposed floor and roof loads. Loads used in the models will closely parallel those used in a typical assemblies square foot estimate unless an unusual loading condition exists, but bay sizes may vary widely, thus changing superstructure costs as much as 20%.

Cost Comparisons

Compare the assemblies square foot estimate costs developed in the preceding chapters for the three-story office building with the similar components in a three-story office building as analyzed in *Means Square Foot Costs* (Figure 24.2). Components are factored for a 30,000 square foot building in lieu of the 58,000 square foot building shown in the model (Figure 24.1). The factor was developed as follows using Figure 24.2:

Model 3-Story Office with 10′ story
height:

58,000 S.F.	=	$65.30/S.F.
34,000 S.F.	=	$70.10/S.F.
22,000 S.F.	=	$76.80/S.F.

Model costs calculated for a 3 story building with 12' story height and 58,000 square feet of floor area.

NO.	SYSTEM/COMPONENT	SPECIFICATIONS	UNIT	UNIT COST	COST PER S.F.	% OF SUB-TOT
1.0 FOUNDATIONS						
.1	Footings & Foundations	Poured concrete; strip and spread footings and 4' foundation wall	S.F. Ground	3.18	1.06	
.4	Piles & Caissons	N/A	–	–	–	2.6%
.9	Excavation & Backfill	Site preparation for slab and trench for foundation wall and footing	S.F. Ground	.89	.30	
2.0 SUBSTRUCTURE						
.1	Slab on Grade	4" reinforced concrete with vapor barrier and granular base	S.F. Slab	2.86	.95	1.8%
.2	Special Substructures	N/A	–	–	–	
3.0 SUPERSTRUCTURE						
.1	Columns & Beams	Fireproofing; interior columns included in 3.5 and 3.7	L.F. Columns	21	.57	
.4	Structural Walls	N/A	–	–	–	
.5	Elevated Floors	Open web steel joists, slab form, concrete, columns	S.F. Floor	6.91	4.61	12.8%
.7	Roof	Metal deck, open web steel joists, columns	S.F. Roof	3.21	1.07	
.9	Stairs	Concrete filled metal pan	Flight	4650	.56	
4.0 EXTERIOR CLOSURE						
.1	Walls	Face brick with concrete block backup 80% of wall	S.F. Wall	18.85	5.26	
.5	Exterior Wall Finishes	N/A	–	–	–	
.6	Doors	Aluminum and glass, hollow metal	Each	2212	.23	12.9%
.7	Windows & Glazed Walls	Steel outward projecting 20% of wall	Each	455	1.38	
5.0 ROOFING						
.1	Roof Coverings	Built-up tar and gravel with flashing	S.F. Roof	1.92	.64	
.7	Insulation	Perlite/EPS composite	S.F. Roof	1.15	.38	1.9%
.8	Openings & Specialties	N/A	–	–	–	
6.0 INTERIOR CONSTRUCTION						
.1	Partitions	Gypsum board on metal studs, toilet partitions 20 S.F. Floor/L.F. Partition	S.F. Partition	2.93	1.46	
.4	Interior Doors	Single leaf hollow metal 200 S.F. Floor/Door	Each	524	2.62	
.5	Wall Finishes	60% vinyl wall covering, 40% paint	S.F. Surface	1.06	.85	
.6	Floor Finishes	60% carpet, 30% vinyl composition tile, 10% ceramic tile	S.F. Floor	4.45	4.45	25.2%
.7	Ceiling Finishes	Mineral fiber tile on concealed zee bars	S.F. Ceiling	3.23	3.23	
.9	Interior Surface/Exterior Wall	Painted gypsum board on furring 80% of wall	S.F. Wall	2.82	.79	
7.0 CONVEYING						
.1	Elevators	Two hydraulic passenger elevators	Each	74,530	2.57	4.8%
.2	Special Conveyors	N/A	–	–	–	
8.0 MECHANICAL						
.1	Plumbing	Toilet and service fixtures, supply and drainage 1 Fixture/1320 S.F. Floor	Each	1900	1.44	
.2	Fire Protection	Standpipes and hose systems	S.F. Floor	.20	.20	
.3	Heating	Included in 8.4	–	–	–	23.1%
.4	Cooling	Multizone unit gas heating, electric cooling	S.F. Floor	10.57	10.57	
.5	Special Systems	N/A	–	–	–	
9.0 ELECTRICAL						
.1	Service & Distribution	1000 ampere service, panel board and feeders	S.F. Floor	.87	.87	
.2	Lighting & Power	Fluorescent fixtures, receptacles, switches, A.C. and misc. power	S.F. Floor	6.83	6.83	14.9%
.4	Special Electrical	Alarm systems and emergency lighting	S.F. Floor	.19	.19	
11.0 SPECIAL CONSTRUCTION						
.1	Specialties	N/A	–	–	–	0.0%
12.0 SITE WORK						
.1	Earthwork	N/A	–	–	–	
.3	Utilities	N/A	–	–	–	
.5	Roads & Parking	N/A	–	–	–	0.0%
.7	Site Improvements	N/A	–	–	–	
		SUB-TOTAL			53.08	100%

GENERAL CONDITIONS (Overhead & Profit)				15%	7.96
ARCHITECT FEES				7%	4.26
		TOTAL BUILDING COST			65.30

Figure 24.1

M.460 | **Office, 2-4 Story**

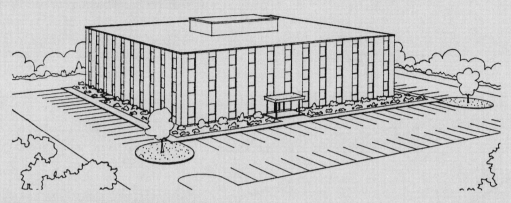

Costs per square foot of floor area

EXTERIOR WALL	S.F. Area	10000	22000	34000	46000	58000	63000	68000	73000	78000
	L.F. Perimeter	246	393	443	543	562	590	603	624	645
Face Brick with Concrete Block Back-up	Wood Joists	88.25	75.55	68.80	66.75	64.05	63.55	62.90	62.50	62.10
	Steel Joists	89.55	76.80	70.10	68.00	**65.30**	64.80	64.20	63.75	63.40
Glass and Metal Curtain Wall	Steel Frame	86.90	75.35	69.50	67.65	65.30	64.90	64.35	64.00	63.65
	R/Conc. Frame	89.75	78.25	72.35	70.50	68.20	67.75	67.20	66.85	66.50
Wood Siding	Wood Frame	76.55	66.30	61.30	59.70	57.75	57.35	56.95	56.60	56.30
Brick Veneer	Wood Frame	82.05	70.25	64.20	62.30	59.90	59.45	58.90	58.50	58.15
Perimeter Adj., Add or Deduct	Per 100 L.F.	10.50	4.80	3.10	2.30	1.85	1.70	1.50	1.40	1.35
Story Hgt. Adj., Add or Deduct	Per 1 Ft.	1.95	1.40	1.00	.90	.75	.75	.70	.65	.65

For Basement, add $19.05 per square foot of basement area

The above costs were calculated using the basic specifications shown on the facing page. These costs should be adjusted where necessary for design alternatives and owner's requirements. Reported completed project costs, for this type of structure, range from $ 37.30 to $ 105.05 per S.F.

Common additives

Description	Unit	$ Cost
CLOCK SYSTEM		
20 room	Each	11,600
50 room	Each	28,000
CLOSED CIRCUIT SURVEILLANCE, One station		
Camera and monitor	Each	1325
For additional camera stations, add	Each	725
DIRECTORY BOARDS, Plastic, glass covered		
30" x 20"	Each	490
36" x 48"	Each	890
Aluminum, 24" x 18"	Each	435
36" x 24"	Each	520
48" x 32"	Each	615
48" x 60"	Each	1350
ELEVATORS, Hydraulic passenger, 2 stops		
1500# capacity	Each	50,700
2500# capacity	Each	52,000
3500# capacity	Each	55,600
Additional stop, add	Each	7600
EMERGENCY LIGHTING, 25 watt, battery operated		
Lead battery	Each	375
Nickel cadmium	Each	620

Description	Unit	$ Cost
SMOKE DETECTORS		
Ceiling type	Each	128
Duct type	Each	276
SOUND SYSTEM		
Amplifier, 250 watts	Each	1450
Speaker, ceiling or wall	Each	125
Trumpet	Each	236
TV ANTENNA, Master system, 12 outlet	Outlet	226
30 outlet	Outlet	146
100 outlet	Outlet	139

Use **Location Factors** in Reference Section

Figure 24.2

Interpolating for a 30,000 square foot building:

2/3 ($70.10−$76.80) + $76.80 = $72.35/S.F.

$$\frac{30,000/\text{S.F.} @ \$72.35/\text{S.F.}}{58,000/\text{S.F.} @ \$65.30/\text{S.F.}} = 1.11 \text{ Cost Multiplier}$$

The 1.11 multiplier would not, in all probability, be proportioned evenly over all components of the estimate, but for purposes of comparison, mark up each component in Figure 24.2 by the size multiplier 1.11.

This is an example of how the numbers were derived:

Footings and Foundations: $1.06/S.F. × 1.11 = $1.18/S.F.

Note: The cost shown for site preparation in Figure 24.3 for the Three-Story Office Building assemblies square foot estimate does not match the cost calculated in Chapter 22, "Site Work." For the following cost comparison to be correct, it was necessary to deduct those portions of the site work that were not included in the model square foot estimate.

Component Comparison

To complete the analysis, we should study further those components that have a large contribution to the total and those that have appreciable differences.

In addition to the cost comparison made in Figures 24.3 and 24.4 through 24.6 (between a model building from *Means Square Foot Costs* and the Three-Story Office Building example in this book), the following items should be considered:

- Foundation square foot costs vary greatly, with both soil capacities and the use of piles or caissons.
- Site preparation is dependent on the contours, soil, rock, and existing vegetation conditions.

System	Means S.F. Cost $/S.F.	% of S.T.	System S.F. Est. $/S.F.	% of S.T.
Footing & Foundations	$ 1.18	2.0	$ 1.26	1.87
Site Preparation	0.33	0.56	1.61	2.38
Substructure	1.05	1.78	0.95	1.41
Superstructure	7.56	12.8	9.37	13.87
Exterior Closure	7.63	13.0	10.99	16.27
Roofing	1.13	1.92	0.95	1.41
Interior Construction	14.87	25.2	11.61	17.19
Elevators	2.85	4.84	4.60	6.81
Mechanical	13.55	23.0	15.66	23.19
Electrical	8.76	14.9	10.21	15.12
Special Construction	—	—	0.33	0.49
Subtotals	$58.91/S.F.		$67.54/S.F.	

Figure 24.3

- Substructure is a function of the selected floor system and the bay sizes.
- Superstructure is also a function of the selected floor system and the bay sizes. In the comparison mentioned above, it has an average 18% contribution to the building cost and a 5% differential.
- Exterior closure has the greatest influence on the square foot cost. It is the most visible component with the greatest variations possible in both material choice and cost. This portion of the estimate should be carefully evaluated. Exterior closure in the comparison mentioned above averages a 13% contribution to the total, with a 7% differential.
- Roofing square foot costs are proportional to the number of stories in the building, with one-story buildings having the highest square foot contribution.
- Interior construction has a large percentage contribution to the total building cost. The variance of finishes in cost, appearance, and durability should be carefully evaluated. The individual items of interior construction do not represent a large percentage of the total building cost. Rather, it is the accumulation of many items that causes this division of the estimate to be so costly. Interior construction in the comparison mentioned above has a 20% average contribution and a differential of 8%.

Superstructure	Means S.F. Costs $/S.F.	Systems S.F. Est. $/S.F.	Remarks
Columns	$0.63	$0.97	Larger Bay Size
Elevated Floors	5.12	5.74	Larger Bay Size
Roof	1.19	1.34	Larger Bay Size
Stairs	0.62	0.90	10' Story Height Fewer Stairs
	$7.56/S.F.	$8.95/S.F.	
Exterior Closure	Means S.F. Costs $/S.F.	Systems S.F. Est. $/S.F.	Remarks
Walls	$5.84	$4.49	
Doors	0.26	0.20	
Windows	1.53	3.74	Aluminum Window Wall in lieu of individual steel sash.
	$7.63/S.F.	$8.43/S.F.	
Owner or Architect's choice of fascia for aesthetics or code compliance.			

Figure 24.4

- Mechanical work represents a large contribution to the total building cost. A large variance may be realized with sophisticated heating, cooling, and control systems. To some degree, the choice of plumbing fixtures will have an effect on the overall cost, but not to the extent of the HVAC systems.
- Electrical work, like mechanical, is a major contributor to the overall square foot cost of the project. Service requirements and the choice of fixtures often represent variables in the estimate that must be resolved.

In the model building from *Means Square Foot Costs,* the share of each division toward the building cost is the Contribution Percentage. The differential is the difference, in percent, between the model building and the Three-Story Office Building example in this book.

From these brief conclusions it is easy to understand the importance of each component in the structure when developing an assemblies square foot estimate.

Interior Construction	Means S.F. Costs $/S.F.	Systems S.F. Est. $/S.F.	Remarks
Partitions & Toilet Partitions	$ 1.62	$ 2.61	10' Floor to Floor vs. 12'
Interior Doors	2.91	0.92	200 S.F. vs. 500 S.F.
Wall Finishes	0.94	0.70	
Floor Finishes	4.94	6.68	
Ceiling Finish	3.58	1.90	Concealed Min. Fiber Tile (12" x 12") vs. Fibrous Glass (24" x 48")
Int. of Ext. Wall	0.88	0.62	
	$14.87/S.F.	$13.43/S.F.	

Adjust Doors: $\dfrac{58,000 \text{ S.F.}}{500 \text{ S.F./door}} = \dfrac{116 \text{ doors} \times \$524/\text{door}}{58,000 \text{ S.F.}} = \$1.05/\text{S.F.}$

Adjust ceiling price by using lower price: $1.90/S.F.

Adjustment $2.95/S.F.

$14.87/S.F. − $2.95/S.F. = 11.92/S.F. versus $13.43/S.F.

Elevator	Means S.F. Costs $/S.F.	Systems S.F. Est. $/S.F.	Remarks
Elevator	$2.85	$4.60	2 @ 1500/lb. vs. 2 @ 3000/lb. Elevators

Figure 24.5

Estimating Accuracy

The degree of accuracy that can be expected in an assemblies square foot estimate depends on several factors, including:

- Accuracy and source of cost information
- Specific design information
- Time available to do the estimate

The first two factors have been dealt with throughout this book. It is the last factor that will be discussed now.

Sufficient time to accomplish an estimate is always, and forever will be, a problem to be addresssed. Whether for a ballpark budget estimate or a detailed estimate for bidding purposes, there is rarely adequate time available for the estimator to make certain that all reasonable factors have been considered.

When preparing an estimate for bidding, the estimator has a responsibility to think the project through from start to finish. For this type of estimate, unit price estimating is usually the method to use. However, a preliminary budget does not warrant the same time or precision. Therefore, assemblies square foot estimating is a more logical approach. Although this method of estimating saves valuable time during the initial decision-making stages of a project, it still has pitfalls. It may be desirable to spend more time on certain portions of the estimate, particularly the expensive parts. A further analysis of the Three-Story Office Building helps to explain this point.

Estimate Summary

What follows is a summary of the assemblies square foot cost estimate before the General Conditions are added for the Three-Story Office Building.

Note: The costs and percentages in Figure 24.7 do not exactly match those shown in the cost comparison made earlier in this chapter. The full cost developed in Chapter 22, "Site Work," is shown, not just the cost of site preparation used earlier.

Comparison

The table from *Means Assemblies Cost Data* shown in Figure 24.8 provides the percentage of total cost for each building system. The figures shown are the "average" percentage contribution of each building system component to the total cost for commercial buildings costing $400,000 and more. The weighted average index is calculated from over 90 materials

Mechanical	Means S.F. Costs $/S.F.	Systems S.F. Est. $/S.F.	Remarks
Plumbing	$1.00	$ 1.86	Additional fixtures for handicapped accessibility.
Fire Protection	0.15	1.53	Assemblies S.F. includes a wet pipe sprinkler.
HVAC	7.45	7.40	
	$8.60/S.F.	$10.79/S.F.	

Figure 24.6

and equipment types and 24 building trades. The table is useful because it identifies the areas where the estimator should spend more time.

To compare the "average" index figures with those just calculated for the Three-Story Office Building, see Figure 24.9.

At first glance the percentages appear to be in general agreement for the most significant divisions: Superstructures, Exterior Closure, Interior Construction, Mechanical, and Electrical. Differences exist because the buildings being compared probably just don't correspond in some areas of design.

One of the most fundamental "rules" in estimating is that the estimator should dwell on those parts of the project where the investment of time counts most. The areas with the greatest potential cost should receive the greatest amount of time and attention. Unfortunately, the easiest part of the estimate or the estimator's specialty usually receives first and foremost attention and time. Unless there are unusual circumstances involved in a project, the most probable order of cost significance will be as follows:

1. Mechanical
2. Superstructures
3. Exterior Closure
4. Interior Construction
5. Electrical
6. Foundations & Substructures
7. Conveying

3-Story Office Building			
Division Number	System	System S.F. Est. $/S.F.	% of S.T.
1	Foundations	1.26	1.87
2	Substructure	0.95	1.41
3	Superstructure	9.37	13.87
4	Exterior Closure	10.99	16.27
5	Roofing	0.95	1.41
6	Interior Construction	11.61	17.19
7	Conveying	4.60	6.81
8	Mechanical	15.66	23.19
9	Electrical	10.21	15.12
11	Special Construction	0.33	0.49
12	Site Work	1.61	2.38
	Subtotal	$67.54/S.F.	

Figure 24.7

8. Site Work

9. Roofing

10. Special Construction

Use the above guidelines like any other rules of thumb: as a guide or reference. There will always be exceptions.

Variables

Using published historical or model square foot costs for preliminary "guesstimates" is acceptable when the projects are similar. If the projects are not similar, even the most accurate cost data may be useless. Cost data may apply only to projects that are identical in all respects to the project for which the cost data was compiled. Producing realistic building costs in just a few hours that are tailored to the particular project and its scope is by far the most desirable approach that members of the construction industry—developers, architects, engineers, builders, municipalities, and government agencies—can use to prepare preliminary budget estimates.

Given a little time for familiarization, this method can be the greatest aid to speeding up the estimating process. It is during the troublesome and ever-changing preliminary design stage of a building construction project when the single most significant event occurs: preparation of the initial budget. Unfortunately, the first cost figure written down or discussed is

General: The following information on current city cost indexes is calculated for 305 cities in the United States and Canada. Index figures for both material and installation are based upon the 30 major city average of 100 and represent the cost relationship on July 1, 1995.

In addition to index adjustment, the user should consider:
1. productivity
2. management efficiency
3. competitive conditions
4. automation

5. restrictive union practices
6. owner's unique requirements

The weighted average index is calculated from about 72 materials and equipment types and 21 building trades.

If the systems component distribution of a building is unknown, the weighted average index can be used to adjust the cost for any city.

Labor, Material and Equipment Cost Distribution for Weighted Average Listed by System Division

Division No.	Building System	Percentage	Division No.	Building System	Percentage
1 & 2	Foundation & Substructure	6.5 %	7	Conveying	3.5%
3	Superstructure	18.0	8	Mechanical	21.0
4	Exterior Closure	12.1	9	Electric	11.9
5	Roofing	2.8	11	Equipment	1.9
6	Interior Construction	17.3	12	Site Work	5.0
				Total weighted index (Div. 1-12)	100.0%

Figure 24.8

usually the one that someone, usually the decision maker, latches on to. No amount of discussion will cause those involved in the project to forget it, even when the scope or use shifts, or changes are made in design that significantly affect the final cost.

Most non-technical owners or clients fail to see how seemingly small changes can drastically affect cost. Once a number that even remotely looks like a construction cost leaves someone's lips or is written down, that is "the" cost.

Don't be caught in the position of having to live with a construction budget based on a vague or "off-the-cuff" figure. Estimating building construction costs is an unusual business where the estimator is often cast in the role of either prophet or fool. Certain factors are often beyond the estimator's control—factors that do nothing but taint the image of the estimating profession. Take control by making use of all the tools and advantages available. Assemblies square foot estimating is one such tool that, if used wisely and with creativity, provides one with an advantage over competitors. Assemblies square foot estimating may be just the right competitive edge needed in a fast-paced construction industry where accurate preliminary cost estimates have taken on a new significance.

Final Summary: Three-Story Office Building

Figure 24.10 is the final summary of costs for the Three-Story Office Building. Before adding up the divisions for a total cost figure, several factors should be considered.

Where applicable, a sales tax on all materials should be added. This additional cost has not been added to the figures in the example, since there is no specific location for the project. To derive the cost of sales tax for an assemblies square foot estimate, it is reasonable to assume that 50% of the subtotal cost is the material cost, and 50% is the labor cost. Multiply the appropriate cost by the sales tax percentage to derive the sales tax cost.

Division Number	System	Estimate % of S.T.	Table % of S.T.
1 & 2	Foundations & Substructure	3.28	6.5
3	Superstructure	13.87	18
4	Exterior Closure	16.27	12.1
5	Roofing	1.41	2.8
6	Interior Construction	17.19	17.3
7	Conveying	6.81	3.5
8	Mechanical	23.19	21
9	Electrical	15.12	11.9
11	Special Construction	0.49	1.9
12	Site Work	2.38	5

Figure 24.9

Sales tax in most states is applied to material only, although labor and equipment are sometimes taxed.

After the sales tax is accounted for, the General Conditions should be added in. If time and accuracy warrant, it is permissable to list the cost of several items to be included in General Conditions, rather than use a simple percentage. For example, the cost of bonds, building permits, superintendents, etc., can be added to the table of costs in the space just under "General Conditions." In Chapter 20, "General Conditions," a 5% factor was derived for the Three-Story Office Building.

After the General Conditions are added, the General Contractor's Overhead and Profit, Architect's Fee, and Contingency should be included.

Although not included in the Three-Story Office Building example, space is available on the form to apply the City Cost Index. All items are then added up to arrive at Total Cost.

SYSTEMS ESTIMATING FORM | Cost Summary

Project	Office Building	Total Area		Sheet No.	
Location		Total Volume		Estimate No.	
Architect		Cost per SF		Date	
Owner		Cost per CF		No. of Stories	3
Quantities By		Extensions By		Checked By	

NO.	DESCRIPTION	SUB TOTAL COST	COST PER SF	%
1.0	FOUNDATIONS	37921	1.26	1.87
2.0	SUBSTRUCTURE	28600	0.95	1.41
3.0	SUPERSTRUCTURE	281082	9.37	13.83
4.0	EXTERIOR CLOSURE	329708	10.99	16.23
5.0	ROOFING	28541	0.95	1.40
6.0	INTERIOR CONSTRUCTION	348392	11.61	17.14
7.0	CONVEYING	138000	4.60	6.79
8.0	MECHANICAL	474969	15.83	23.37
9.0	ELECTRICAL	306374	10.21	15.08
11.0	SPECIAL CONSTRUCTION	10027	0.33	0.49
12.0	SITE WORK	48426	1.61	2.38
	BUILDING SUB TOTAL	$2,032,040		

		$2,032,040
Sales Tax: % x Sub Total/2	=	

10.0   GENERAL CONDITIONS   (BREAKDOWN)

General Conditions(%) 5%	x Sub Total	=	$101,602
		Sub Total 'A'	$2,133,642
Overhead(%) 7%	x Sub Total 'A'	=	$149,355
		Sub Total 'B'	$2,282,997
Profit(%) 7%	x Sub Total 'B'	=	$159,810
		Sub Total 'C'	$2,442,807
City Cost Index(%)	x Sub Total 'C'	=	
		Adjusted Building Cost	$2,442,807
Architects Fee(%) 8.5%	x Adjusted Building Cost	=	$207,639
Contingency(%) 2%	x Adjusted Building Cost	=	$48,856
		TOTAL COST	$2,699,302

SQUARE FOOT COST = TOTAL COST/BUILDING AREA $89.98 /SF
CUBIC FOOT COST = TOTAL COST/BUILDING VOLUME $7.50 /CF

Figure 24.10

Chapter 25

Using UNIFORMAT II in Preliminary Design & Planning

by Robert P. Charette, P.E., CVS and Anik Shooner, Architect

The UniFormat system, based on assemblies (systems or elements) according to their use or function, is particularly well suited to square foot estimating. The systems format is a concept that is relatively easy to use since it reflects the instinctive way that designers and builders think about a building construction project. This approach makes it possible to define each building system, determine how each fits in the overall scheme, and establish realistic budgets that can effectively be monitored during design.

This chapter provides detailed historical background information on the UniFormat system, its evolution, and efficient use to increase design team performance.

The UniFormat system that is widely used in the construction industry was revised by the American Society of Testing Materials (ASTM) in 1993. UNIFORMAT II defines *building elements* as "major components, common to most buildings, that perform a given function, regardless of the design specification, construction method, or materials used." In practice, an *element* is really any distinct component of a whole—such as the assemblies referred to throughout this book.

To control building project scope and costs, good coordination and communication among all team members is necessary. It can be argued that team performance is greatly enhanced when the quantity and format of the information exchanged is timely, consistent, and reliable. This chapter addresses this objective as it relates to the early phases of project design.

During the program phase, the owner defines the project content, quality, cost, and schedule for the designers. These criteria will be used to measure the design team's performance so that timely corrective action can be taken when necessary to keep the project on track, without any delays in the design schedule, and within the allocated professional fees and the established project budget.

The challenge facing the owner and design team leader is to ensure efficient dissemination of information, which will foster clear communication and efficient coordination among all participants so that any potential problems or misunderstandings surface early on and are dealt with expeditiously. To meet this challenge, the quality of design documents and estimates, and their format, must be appropriate and consistent from project to project. Documentation presented to the owner may be based on guidelines established by the owner if they exist, or in-house practices developed by consultants based on their own experiences. In either case, the documentation may be excellent, but there is a general lack of standardization in the process that contributes to inefficiency and for which the owner and the construction industry inevitably pay.

This lack of standardization has the added effect of further reducing overall performance by constraining the value engineering role as well.

New Tools to Improve Communication, Coordination, and Productivity During Design

Two recent publications promulgated by the ASTM (American Society of Testing Materials) and CSI/CSC (Construction Specifications Institute/Construction Specifications Canada) can contribute to improving communication, coordination, and productivity during design:

- ASTM Standard E1557-93 "UNIFORMAT II—A Classification of Building Elements and Related Sitework".
- CSI/CSC Practice FF/180 "Preliminary Project Descriptions and Outline Specifications" (The CSC equivalent is "Developing an Outline Specification—Volume A").

The ASTM document (see Figure 25.1) provides a standard classification of building elements and related sitework that is universal and has many applications, including program and design estimates, specifications, and CAD layering.

The CSI/CSC document (see Figure 25.2) outlines the application of the classification of building elements for "concept specifications," referred to

---

 Designation: E 1557 – 93

Standard Classification for
Building Elements and Related Sitework—UNIFORMAT II[1]

This standard is issued under the fixed designation E 1557; the number immediately following the designation indicates the year of original adoption or, in the case of revision, the year of last revision. A number in parentheses indicates the year of last reapproval. A superscript epsilon (ε) indicates an editorial change since the last revision or reapproval.

1. Scope

1.1 This standard establishes a classification of building elements and related sitework. Elements, as defined here, are major components common to most buildings. Elements usually perform a given function, regardless of the design specification, construction method, or materials used. The classification serves as a consistent reference for analysis, evaluation, and monitoring during the feasibility, planning, and design stages of buildings. Using UNIFORMAT II ensures consistency in the economic evaluation of buildings projects over time and from project to project. It also enhances reporting at all stages in construction—from feasibility and planning through the preparation of working documents, construction, maintenance, rehabilitation, and disposal.

Figure 25.1

as "Preliminary Project Descriptions (PPD)" in CSI/CSC terminology. The latter expression is more correct, because at the concept phase "specifications" as such are not really produced.

The application of these two documents, which allow elemental design estimates and preliminary project descriptions to be linked in a common reference structure, provides a framework for more efficient, effective, and productive planning and decision making. Time and money are saved for all parties on a project as a result of improved coordination, an increase in programming and design team performance, and a significant reduction in the probability of costly delays related to project scope "creep" and cost overruns.

The History of Elemental Classifications

The development of the first elemental classification is usually attributed to the British Ministry of Education immediately after World War II; at that time, quantity surveyors developed an "elemental classification" to control the cost of the accelerated post-war school expansion program.

From the UK, the methodology was exported to other British Commonwealth countries such as Canada, South Africa, and Singapore, who adapted the classification to their needs. The methodology was exported from Canada to the U.S. in the 1970s, and resulted in the adoption of the UniFormat classification by the American Institute of Architects (AIA) and the U.S. General Services Administration (GSA). In 1993, the original UniFormat classification was revised by ASTM and promulgated as Standard E1557-93—UNIFORMAT II following four years of task group meetings with members from other organizations such as CSI/CSC, the American Association of Cost Engineers (AACE), the Tri Services Committee, and the General Services Administration. At about that time, CSI also

Manual of Practice

Construction Documents Fundamentals and Formats Module
Preliminary Project Descriptions and Outline Specifications

FF/180
Preliminary Project Descriptions and Outline Specifications

Well-prepared documents improve coordination, communication, and productivity during all phases of a project. Two written documents recommended for use during the initial design phases are "Preliminary Project Descriptions" and "Outline Specifications." This chapter describes these documents and how they apply to a conventional project delivered under a lump-sum, single prime contract. This chapter does not discuss multiple prime contracts, construction management, or design-build delivery methods.

Source: Reprinted from the Construction Specifications Institute (CSI) *Manual of Practice*, (FF Module, 1992 edition) with permission from CSI, 1995.

Figure 25.2

recommended that Preliminary Project Descriptions at the schematic phase be structured according to a modified version of the UniFormat elemental classification.

Today, elemental classifications are used all over the world, primarily for design estimates. In Canada, virtually all Federal Departments and most Provincial Departments call for mandatory elemental design estimates. In the U.S., the General Services Administration and the Military Services have been the main proponents of UniFormat, though a few state public works and education departments call for it as mandatory.

Additional applications for UNIFORMAT II are being developed, such as for design-build contractual documents, layering of CAD drawings, and building condition assessments. Since ASTM and CSI support the use of UniFormat in North America, it is anticipated that the classification will gain widespread acceptance as a common standard, thus saving the construction industry significant sums of money.

For additional background information on elemental classifications, refer to the National Institute of Standards and Technology (NIST) Special Publication 841 "UNIFORMAT II—A Recommended Classification for Building Elements and Related Site Work."

Elemental Program and Design Estimates

To illustrate the concept of an elemental unit cost, let us consider the outside wall of a building. In a conventional estimate, each component would be priced separately, without any relation to function (e.g., brick, vapor retarder, insulation, block back-up, furring). With the elemental approach, a single composite rate per square foot would be used that includes all component costs, thus greatly simplifying the estimating task. Such elemental cost data is available from *Means Assemblies Cost Data* or Southam's *Yardsticks for Costing.* These publications are currently based on the original version of UniFormat; however, plans call for further revisions to UNIFORMAT II in the near future.

At the program phase, estimates or cost plans can be prepared using the cost modeling techniques outlined in *Project Budgeting for Buildings,* by Don Parker and Al Dell'Isola. Developing a model based on gross floor area and an appropriate massing configuration provides quantities for most building elements; this allows a realistic program estimate to be prepared, though elements are not defined other than by performance and levels of quality (e.g., the structure could be steel or concrete, the exterior enclosure curtain wall or brick with block back-up).

With such a model, unit costs may be allocated to elements based on cost manuals, historical cost data from similar projects, and the level of quality called for in the facilities program or permissible within the owner's anticipated cost for the building. For example, even without a design, the exterior wall for an insurance company's head office could be set at $42.00/S.F., an anticipated high level of quality, versus $12.00/S.F., which would be appropriate only for an industrial-type building. A detailed approach to developing costs when parameter quantities have been established can be found in this book.

Such realistic program estimates will permit a Design-To-Cost (DTC) approach for each discipline that can readily be monitored element by element; designers must commit to the targeted cost, fully aware that cost issues are of prime importance to the team and the client. It also will facilitate conducting an effective Value Engineering (VE) workshop at this stage, which would be particularly cost effective if the budget exceeds the anticipated project cost. It is at the program phase that major budget problems should be resolved by ensuring that the program and budget are compatible, thus avoiding costly redesigns and delays after design has been initiated.

In a similar fashion, designers should, as a matter of routine, validate any given budget to which they will be contractually committed. If the projected cost is higher than the budget, and the difference is validated, the program scope and quality levels can be reviewed to reach an agreement on the budget before time has been expended on design work—a saving for everyone that fosters a harmonious designer-client relationship.

During the design of a project, schematic design estimates must be prepared to monitor and control costs from the schematic phase through the completion of working drawings to ensure that the program budget is respected. This task is greatly facilitated by the parameters generated within elemental estimates that allow effective cost analysis; e.g., all elements are expressed in quantity per unit GFA (gross floor area), cost per unit of measurement of the element, or a percentage of total cost. Such in-depth analysis would not be possible with costs based on trade or MasterFormat Divisions 1 to 16. For Value Engineering workshops, elemental estimates also serve as a checklist for all the components of a project; provide unit costs and quantities; facilitate the team's comprehension of the project; allow directed brainstorming of alternatives; and allow the rapid calculation of cost differentials for proposed alternatives, whatever their scope.

At the completion of design, schematic design estimates are usually converted to trade estimates or MasterFormat estimates to accelerate bid analysis; the relationship of UniFormat and MasterFormat Divisions 1 – 16 is illustrated in Figure 25.3.

As a general rule, function-oriented schematic design estimates permit effective cost analysis during the program and design phases of a project, and as a result, significantly increase the performance and return on investment for Value Engineering studies.

A typical UNIFORMAT II schematic design cost estimate summary is presented in Figure 25.4.

Elemental Preliminary Project Descriptions (Schematic Design Phase)

The quality of project documentation at the schematic design and design development phases varies greatly. Few organizations have adopted standard guidelines that adequately define what should be presented.

The absence of such documentation is usually reflected in inaccurate design estimates resulting from lack of information, misunderstandings between clients and designers on the nature of the project, delays in decision-making as to baseline systems selection, and difficulty in coordinating design team tasks. In most cases, designers know what they plan to incorporate in their design, but this information is not usually documented for other disciplines because it is not mandatory; in general, no two consultants submit schematic or design development specifications in a similar way, making it more difficult for the project manager and owner to analyze submittals.

The CSI has recognized this situation, being fully aware that MasterFormat Divisions 1 to 16 are not suitable to describe a project at such an early stage—the final selection of products and materials has not been made, or is likely to change during design. As a result, CSI now recommends in their Practice FF/180 that Preliminary Project Descriptions (schematic specifications) be structured according to an elemental classification that will describe a project in terms of building systems or elements. The use of the Preliminary Project Description based on the UNIFORMAT II elemental classification links the specification and the estimate in a common reference structure and provides a framework for more efficient and productive planning and decision-making. Though baseline systems have been identified and their cost estimated at this early phase of design, changes are still possible within the budget allocated as the design progresses.

In practice FF/180, CSI also recommends the use of a condensed MasterFormat specification at the preliminary design phase—referred to

Relationship of UNIFORMAT to MASTERFORMAT

Design Uniformat Level 2	Level 3	Construction UCI	01 General Requirements	02 Sitework	03 Concrete	04 Masonry	05 Metals	06 Wood + Plastic	07 Thermal and Moisture Protect	08 Doors and Windows	09 Finishes	10 Specialties	11 Equipment	12 Furnishings	13 Special Construction	14 Conveying Systems	15 Mechanical	16 Electrical
01 Foundations	011 Standard Foundations			■	■		■											
	012 Spec Foundation Cond			■	■													
02 Substructure	021 Slab On Grade				■													
	022 Basement Excavation			■														
	023 Basement Walls				■	■												
03 Superstructure	031 Floor Construction				■		■	■										
	032 Roof Construction				■		■	■										
	033 Stair Construction				■		■											
04 Ext. Closure	041 Exterior Walls				■	■	■	■	■									
	042 Ext. Doors & Windows									■								
05 Roofing									■									
06 Int. Const.	061 Partitions					■	■	■			■							
	062 Interior Finishes										■							
	063 Specialties											■						
07 Conveying Sys.																■		
08 Mechanical	081 Plumbing																■	
	082 H.V.A.C.																■	
	083 Fire Protection																■	
	084 Spec. Mechanical Systems																■	
09 Electrical	091 Service & Distribution																	■
	092 Lighting And Power																	■
	093 Spec. Electrical System																	■
10 Gen.Cond. OH&P			■															
11 Equipment	111 Fixed & Movable Equip.												■					
	112 Furnishings													■				
	113 Special Construction														■			
12 Sitework	121 Site Preparation			■														
	122 Site Improvements			■														
	123 Site Utilities			■													■	■
	124 Off-Site Work			■														

Note: CSI's MASTERFORMAT has superseded the UCI (Uniform Construction Index).

Source: The American Institute of Architects, "Chapter B5 - - Design and Construction Cost Management," Architect's Handbook of Professional Practice (Washington, DC: American Institute of Architects, 1984).

Figure 25.3

as "Outline Specifications." This notion is gaining acceptance in North America, in that CSI, AIA, and W2 Consultants (in Edmonton) offer such computerized specifications at a modest cost of $250. The use of this type of specification at the design development phase again pushes the decision-making up front, reduces the risk of future misunderstandings, and accelerates the preparation of final working drawings and MasterFormat specifications. Though an Outline Specification is proposed at the design development phase in the FF/180 Practice, an elemental specification could also be considered to complement drawings that are less detailed than normal.

Designers may not be called upon contractually to submit Preliminary Project Descriptions and Outline Specifications; but because of the resulting benefits, owners should consider, if necessary, offering a supplementary fee for this service.

Benefits of Elemental Estimates and Preliminary Project Descriptions:

- At the schematic design phase, the project manager, owner, and end users will be presented with, in layman's language, a clear description of what is proposed for each element of the project, i.e., a clear scope definition. This will allow them to make early comments on any changes that may be required and in effect commit to accepting the proposed design. This approval will result in minimizing changes and delays that are usually at the expense of the consultants.
- The architect is reassured that all engineering disciplines have expended the time necessary to consider alternatives for all systems and selected baseline systems at the schematic design

UNIFORMAT II BUILDING ELEMENTAL COST SUMMARY

ELEMENTS Levels 2 & 3		RATIO QTY/GFA	ELEMENT COST			ELEMENT AMOUNT	COST per UNIT GFA	% Trade Cost
			UM	QUANTITY	UNIT PRICE			
A10	FOUNDATIONS					87 081	1,19	2,54%
A1010	Standard Foundations	0,25	FPA	18 200	1,93	35 211	0,48	
A1020	Special Foundations	0,00		0		0	0,00	
A1030	Slab on Grade	0,25	SF	18 200	2,85	51 870	0,71	
A20	BASEMENT CONSTRUCTION					65 813	0,90	1,92%
A2010	Basement Excavation	0,25	SF	18 200	1,43	26 026	0,36	
A2020	Basement Walls	0,04	SF	3 200	12,43	39 787	0,54	
B10	SUPERSTRUCTURE					669 695	9,15	19,57%
B1010	Floor Construction	0,75	SF	55 000	10,25	563 641	7,70	
B1020	Roof Construction	0,26	SF	18 900	5,61	106 054	1,45	
B20	EXTERIOR CLOSURE					435 496	5,95	12,73%
B2010	Exterior Walls	0,17	SF	12 100	7,47	90 421	1,24	
B2020	Exterior Windows	0,34	SF	25 060	13,50	338 310	4,62	
B2030	Exterior Doors	0,00	LVS	6	1127,50	6 765	0,09	
B30	ROOFING					48 558	0,66	1,42%
B3010	Roof Coverings	0,26	SF	18 900	2,47	46 752	0,64	
B3020	Roof Openings	0,00	EA	3	602,00	1 806	0,02	
C10	INTERIOR CONSTRUCTION					171 549	2,34	5,01%
C1010	Partitions	0,41	SF	29 940	4,23	126 654	1,73	
C1020	Interior Doors	0,00	EA	66	296,00	19 536	0,27	
C1030	Specialities	1,00	SF	73 200	0,35	25 359	0,35	

Figure 25.4

phase that appear most appropriate; this pushes the decision-making up front, and minimizes the possibility of major unexpected changes as the design proceeds. Similarly, the architect also must reflect on each architectural element and document his or her intentions vis-à-vis what he or she is providing the client. The classification, in essence, becomes a checklist for all building and sitework elements to monitor design and facilitate technical reviews.

- The cost manager, having been given descriptions of all building elements, is now in a position to provide a relatively accurate cost plan, element by element, that conforms exactly to the standard classification descriptions provided. The Preliminary Project Description saves the cost manager and design team members a great deal of time in collecting design data in writing instead of verbally from each consultant. As a result, the cost manager can now spend his or her time more effectively, analyzing costs rather than estimating.

- The presentation of the project description and estimated cost in the same format facilitates and accelerates the design review and approval process; it also allows the taking of corrective actions related to scope or cost at the earliest possible time, without consuming a disproportionate amount of design fees, and without significant delays in the design schedule.

- With a common understanding at the schematic design phase between client and designers as to the scope and cost of the project, the design development phase may be initiated with fewer uncertainties. As a result, the schedule may be accelerated. Redesigning should also be minimized, adding to the profitability for designers, thus creating a good environment that encourages teamwork.

- Coordination, communication, and productivity are improved significantly at all phases of a project. Many projects run into trouble because of costly redesigns resulting from scope "creep" and cost overruns. Many of these are really the result of a lack of effective communication among the client, project managers, and designers at the program and schematic design phases, when the problems should have been identified and addressed.

Other Applications of the UNIFORMAT II Classification

The UNIFORMAT II classification has other practical applications when standardized reporting is desirable during the design and operation life of the building. Some of these include:

- Value Engineering function/cost models
- A checklist for technical reviews
- A checklist for brainstorming VE function/system alternatives
- Layering of CAD drawings
- Life cycle costing
- Building condition assessment
- Budgeting for major future cyclical renewal tasks
- Filing of technical literature for assemblies
- Financial evaluation of buildings
- Master schedules for design and construction
- Defining the scope of contracts for each consultant
- Classifying construction details

Conclusion   Improving communication and coordination among project team members
(including the client) will lead to a significant increase in design team
performance. This can be achieved in a consistent manner from project to
project through the application of the ASTM UNIFORMAT II standard
classification of building elements for estimates and early design
specifications prepared as oultined in CSI Practice FF/180. These documents
link the technical program, design estimates, and specifications in a
common reference structure that provides framework for more efficient,
effective, and productive planning and decision making; i.e., they will
increase design team efficiency.

Appendix

Overview of Square Foot/Assemblies Estimating Software

Building Systems Design (BSD)

An effective assemblies cost database not only saves the cost estimator time in the preparation of cost estimates, but also provides cost estimating standards and a consistent methodology; an easy and fast method for editing and adding assemblies; and an organized system for storing and retrieving assembly costs.

BSD's *Composer Gold* Cost Estimating Software Systems can be used in conjunction with *Means Assemblies Cost Data.* The cost data file may be stored within the *Composer Gold* program and will contain the work activity with its associated tasks and ratios. When a task, such as the installation of ceramic tile, is required for a cost estimate, the user can look up the assembly database to find the ceramic tile assembly. The user can then copy the assembly and all its related cost item tasks (floor preparation, mastic, grouting, cleaning, and sealing) to the estimate. The quantity of tile is multiplied by the ratio for each task, saving time for the user, as five cost item tasks are placed with their associated quantities and costs in the estimate. In a simple assembly such as this, the ratios are obvious — for example, for each square foot of tile to be laid, there is also one square foot of preparation, cleaning, and sealing.

To add or edit an assembly in the database, the user must first decide where the assembly should be placed in the database file. In the *Composer Gold* program, each assembly is identified by a nine-digit code. The first two digits are the Level 1 System Code, which refers to the construction activity rather than CSI divisions. For example, System 06–Interior Construction may contain assemblies using cost codes from CSI Division 03 (for concrete flooring); 06 (for wood flooring); or 09 (for finishes including ceramic tile, terrazzo, resilient tile, carpet, and special floor finishes).

To add new ceramic tile assemblies, the user would open the assemblies database to System 06–Interior Construction; select Floor Finishes; select Tile & Covering; and identify the specific assembly for editing. The Software Copy feature allows the user to add a new ceramic tile assembly by editing an existing assembly. To add a new assembly, the user enters a nine-digit assembly code, assembly name, and unit of measure. The user can also add notes to further describe the assembly.

For each assembly the cost item tasks are also listed along with their respective factors. From the assembly library the user can search the Unit Cost Database to find desired items. Only the cost codes are stored within the assembly library; material costs are maintained in the Unit Cost Database. When material costs are updated in the Unit Cost Database, those costs will also be calculated correctly in the assembly database. Thus the user can add new assemblies by editing or creating, and can develop new estimating standards.

Assemblies

The R.S. Means Assemblies Cost Database File may be stored within the *Composer Gold* Cost Estimating Software System. This cost data file will contain the work activity with associated tasks and ratios for use when the task, such as the ceramic tile, is required for the cost estimate. If a project calls for 1750 S.F. of ceramic tile for bathrooms, the computer software allows the user to "lookup" in the assembly database the ceramic tile assembly, and copy the assembly and all its related cost item tasks (floor preparation, mastic, etc.) to the estimate. The quantity of 1750 S.F. is multiplied by the ratio for each task, and the user has saved time because he or she now has five cost item tasks with associated quantities and costs in the estimate.

To add or edit a new assembly in the assembly database, the user must first decide where the assembly should be placed in the database file. The computer software and cost database organization assist the user in this process. Each assembly is identified by a nine-digit code with the first two digits being the Level 1 System Code. The 12 system codes are oriented toward the construction activity rather than CSI division classification. System 06–Interior Construction may contain assemblies using cost codes from: CSI Divisions 03, for concrete flooring; 06, for wood flooring; or 09, finishes, which would include ceramic tile, terrazzo, resilient tile, carpet and special floor finishes.

To add new ceramic tile assemblies the user opens the assemblies databases to Division 6–Interior Construction. The user then steps to the next level, Floor Finishes; to the next level, Tile and Covering; and then identifies the specific assembly for editing. The software copy feature will allow the user to add a new ceramic tile assembly by editing an existing assembly. To add rather than edit, the user is prompted to enter the nine-digit assembly code, an assembly name, and the unit of measure. A "notes" field is also available to further describe the assembly and explain in what situations the

```
Thu 16 Nov 1995              Building Systems Design, Inc. - NOT For RESALE                    TIME 14:20:23
Eff. Date  06/14/95          PROJECT CHAPEL:  Chapel Family Life Center
                                          Bid Estimate                              SUMMARY PAGE    1
                             ** PROJECT DIRECT SUMMARY - System **
```

	QUANTITY	UOM	MANHRS	LABOR	EQUIPMNT	MATERIAL	OTHER	TOTAL CST	UNIT COST
AA Primary Facilities									
AA.01 Chapel Family Life Center									
AA.01.01 Substructure	17175.00	SF	2,400	37,023	1,772	33,793	0	72,587	4.23
AA.01.02 Superstructure	17175.00	SF	3,128	71,054	7,044	142,413	0	220,511	12.84
AA.01.03 Exterior Closure	17175.00	SF	4,601	83,975	1,273	108,294	0	193,541	11.27
AA.01.04 Roofing	17175.00	SF	1,143	25,344	660	50,610	0	76,614	4.46
AA.01.05 Interior Construction	17175.00	SF	5,190	101,228	1,877	126,311	0	229,416	13.36
AA.01.06 Interior Finishes	17175.00	SF	2,068	36,903	670	50,872	0	88,445	5.15
AA.01.08 Plumbing	17150.00	SF	1,475	38,420	831	32,200	0	71,451	4.17
AA.01.09 HVAC	17175.00	SF	2,787	81,332	1,561	187,979	0	270,872	15.77
AA.01.10 Fire Protection	17175.00	SF	766	18,624	1,654	11,672	3,163	35,112	2.04
AA.01.11 Electric Power And Lighting	17175.00	SF	2,989	123,185	525	96,667	0	220,377	12.83
AA.01.12 Electrical Systems	17175.00	SF	1,611	67,148	338	49,673	0	117,159	6.82
AA.01.15 Special Construction	17175.00	SF	661	12,300	423	50,157	0	62,880	3.66
TOTAL Chapel Famimly Life Center	17175.00	SF	28,820	696,538	18,627	940,640	3,163	1,658,967	96.59
TOTAL Primary Facilities	1.00	EA	28,820	696,538	18,627	940,640	3,163	1,658,967	1658967

Sample Report from Building Systems Design

assembly would be most applicable. At the level beneath the assembly title, the cost item tasks are listed along with their respective factors. From the assembly library, the user can "lookup" to search the Unit Cost Database using the computer software to find the desired items. Only the cost codes are stored in the assembly library; material costs are maintained in the Unit Cost Database. When material costs are updated in the Unit Cost Database, those costs will also be calculated correctly in the assembly database. The user has now added new assemblies by editing or creating and has developed new estimating standards.

For more information, contact Building Systems Design at 1-800-875-0047, or 1175 Peachtree St., N.E., Suite 1600, Atlanta, GA 30361-6205.

Dataquire Facility Auditor

Dataquire Facility Auditor (DFA) is a computer software application that can be used by facility managers to manage critical information about their physical plant assets. *DFA* is designed to help integrate physical plant management, financial planning, capital acquisition, resource allocation, and accountability.

By doing a life-cycle analysis of each of the component systems that make up a facility, including both buildings and site features, *DFA* yields information that will help the user manage buildings and finances more effectively. The program comes "pre-loaded" with nationally recognized data about life cycles and component assembly costs. Component costs are adjusted for the user's locality. The user identifies the components that make up each building from a "point and click" list of over 1,600 available component assemblies. Then, using the program-generated data collection forms, the user can survey his or her facility to gather data on the quantity of each component assembly and its date of last replacement. After these data are entered into the program, a variety of reports can be produced that give useful information about the facility, building by building and component by component, as well as summary tables and other useful future projections about the facility.

As projects are done, the user can edit the existing data, adding components or substituting for original materials or equipment, and entering new replacement dates as work is completed. This way the information in *DFA* is always current and updated reports can be easily produced. This feature saves the user from repeatedly doing expensive and time-consuming inspection-based audits. The program provides an accounting system for management of physical plant assets that is always up to date.

DFA calculates the Current Replacement and Current Deferral values for the physical plant (each building and each component). *DFA* calculates a Deferred/Replacement Ratio for the whole facility and for each building, allowing the user to track the effects of capital replacement programs over time. Two kinds of reports—building reports and component reports—provide back-up data for summary reports. Reports are available in tabular and graph formats. The program also provides reports for projections of replacement costs. Using time intervals and cost inflation parameters that the user defines, he or she can produce projections of the capital costs of maintaining the facility, either building by building or component by component. Notes about the components can be kept handy in printed logs.

DFA is a sophisticated facility accounting software program. Its cost is based on the number of sites and buildings the user wishes to "audit" and account for. Special purchase plans for members of associations and distributorships are available.

For more information, contact Dataquire at (401) 253-8969 or 24 Beach Rd., Bristol, RI 02809.

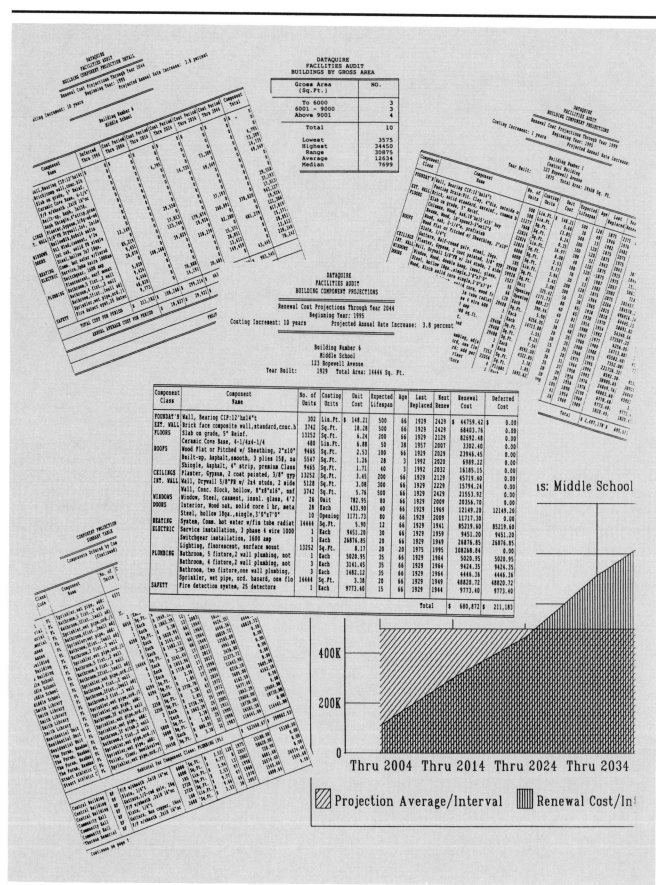

DATAQUIRE
FACILITIES AUDIT
BUILDINGS BY GROSS AREA

Gross Area (Sq.Ft.)	NO.
To 6000	3
6001 - 9000	3
Above 9001	4
Total	10
Lowest	3575
Highest	34450
Range	30875
Average	12634
Median	7699

DATAQUIRE
FACILITIES AUDIT
BUILDING COMPONENT PROJECTIONS

Renewal Cost Projections Through Year 2044
Beginning Year: 1995
Costing Increment: 10 years   Projected Annual Rate Increase: 3.8 percent

Building Number 6
Middle School
123 Hopewell Avenue
Year Built: 1929   Total Area: 14444 Sq. Ft.

Component Class	Component Name	No. of Units	Costing Units	Unit Cost	Expected Lifespan	Age	Last Replaced	Next Renew	Renewal Cost	Deferred Cost
FOUNDAT'N	Wall, Bearing CIP:12'hx14"t	302	Lin.Ft.	$ 148.21	500	66	1929	2429	$ 44759.42	$ 0.00
EXT. WALL	Brick face composite wall,standard,conc.b	3742	Sq.Ft.	18.28	500	66	1929	2429	68403.76	0.00
FLOORS	Slab on grade, 5" Reinf.	13252	Sq.Ft.	6.24	200	66	1929	2129	82692.48	0.00
	Ceramic Cove Base, 4-1/4x4-1/4	480	Lin.Ft.	6.88	50	38	1957	2007	3302.40	0.00
ROOFS	Wood Flat or Pitched w/ Sheathing, 2"x10"	9465	Sq.Ft.	2.53	100	66	1929	2029	23946.45	0.00
	Built-up, Asphalt,smooth, 3 plies 15#, na	5547	Sq.Ft.	1.26	28	3	1992	2020	6989.22	0.00
	Shingle, Asphalt, 4" strip, premium Class	9465	Sq.Ft.	1.71	40	3	1992	2032	16185.15	0.00
CEILINGS	Plaster, Gypsum, 2 coat painted, 3/8" gyp	13252	Sq.Ft.	3.45	200	66	1929	2129	45719.40	0.00
INT. WALL	Wall, Drywall 5/8"FR w/ 2x4 studs, 2 side	5128	Sq.Ft.	3.08	300	66	1929	2229	15794.24	0.00
	Wall, Conc. Block, hollow, 8"x8"x16", unf	3742	Sq.Ft.	5.76	500	66	1929	2429	21553.92	0.00
WINDOWS	Window, Steel, casment, insul. glass, 4'2	26	Unit	782.95	80	66	1929	2009	20356.70	0.00
DOORS	Interior, Wood oak, solid core 1 hr, meta	28	Each	433.90	40	66	1929	1969	12149.20	12149.20
	Steel, hollow 18ga.,single,3'0"x7'0"	10	Opening	1171.73	80	66	1929	2009	11717.30	0.00
HEATING	System, Comm. hot water w/fin tube radiat	14444	Sq.Ft.	5.90	12	66	1929	1941	85219.60	85219.60
ELECTRIC	Service installation, 3 phase 4 wire 1000	1	Each	9451.20	30	66	1929	1959	9451.20	9451.20
	Switchgear installation, 1600 amp	1	Each	26876.85	20	66	1929	1949	26876.85	26876.85
	Lighting, fluorescent, surface mount	13252	Each	8.17	20	20	1975	1995	108268.84	0.00
PLUMBING	Bathroom, 5 fixture,2 wall plumbing, not	1	Each	5020.95	35	66	1929	1964	5020.95	5020.95
	Bathroom, 4 fixture,2 wall plumbing, not	3	Each	3141.45	35	66	1929	1964	9424.35	9424.35
	Bathroom, two fixture,one wall plumbing,	3	Each	1482.12	35	66	1929	1964	4446.36	4446.36
	Sprinkler, wet pipe, ord. hazard, one flo	14444	Sq.Ft.	3.38	20	66	1929	1949	48820.72	48820.72
SAFETY	Fire detection system, 25 detectors	1	Each	9773.40	15	66	1929	1944	9773.40	9773.40
								Total	$ 680,872	$ 211,183

is: Middle School

400K

200K

0

Thru 2004   Thru 2014   Thru 2024   Thru 2034

Projection Average/Interval   Renewal Cost/In

Sample Reports from Dataquire

Advanced Construction Estimating (ACE)

Advanced Construction Estimating can handle projects ranging from light construction to high-rises, accommodating up to 20 estimators at once. It includes tools such as a quick look-up function, a pop-up calculator, and a built-in notepad. The program also creates on-screen blueprints and calculations for a given job using an electronic digitizer that allows 1/1000 of an inch precision.

ACE comes loaded with a catalog of specific items and assemblies. It also enables the user to define his or her own assemblies, including line formulas and variables, and then call on them any time with a single key stroke. The program creates extensions, and automatically projects costs and quantities based on dimensions (e.g., height and length of a frame wall assembly) specified by the user. Items can be sorted according to material type, vendor, rental vs. company-owned, etc.

In addition to the items and assemblies included in *ACE*, users can access and incorporate other trade-specific databases and sources, including those from R.S. Means. *ACE* also interfaces with F.W. Dodge's Dodge/SCAN for plans, and with Primavera Project Planner and SureTrak for project scheduling. In addition, *ACE* can be integrated with The Construction Manager, Software Shop's total job-cost accounting system.

ACE also assists with basic bid preparation—allocating labor costs, adding special material costs, comparing vendor and subcontractor quotes, assigning direct job expenses for items such as equipment rental, and adding mark-ups by job, phase, division, or bid item. The program also converts takeoff quantities to purchasing quantities (e.g., linear feet to board feet).

ACE contains standard formulas for calculating square footage, cubic yards, and so on. The user can also build his or her own formulas, including nested formulas, for more complex takeoff problems.

ACE allows the user to define his or her own divisions, phases, and bid items, and group related takeoff items under them. CSI formats can also be used for cost-coding by division, subdivision, and task or item. The user tells *ACE* when and where to apply overhead and mark-up to specific codes.

ACE also includes standard report forms such as Takeoff Audit Trails, Bills of Material, Labor Distribution, and Unit Costs. The user can also design his or her own forms using this program.

Software Shop service and support includes training, maintenance, program updates, and custom report design.

For more information, contact Software Shop Systems at (800) 554-9865, or 1340 Campus Parkway, P.O. Box 1973, Wall, NJ 07719.

ACE is a trademark of Software Shop Systems, Inc.
Dodge/SCAN, Primavera Project Planner, and SureTrak are trademarks or registered trademarks of their respective companies.

AUDIT TRAIL

The Audit Trail provides a detailed hard copy of the takeoff. Each item taken off is displayed, including: Component description (and size), Quantity taken off, unit of measure, Price per unit and extended amount, Labor hours per unit and total Labor hours. Takeoff Detail gives each input used to take off an Assembly as well as estimator's comments.

```
START OF ASSEMBLY Slab on Grade  3500
DEPTH SAND (IN)=      4.000  DEPTH CONC.(IN)=        6.00
AREA          =  10000.000  VOLUME SAND          123.33
VOLUME CONC.  =    105.185  TOTAL VOLUME    =    308.51
```

DIMENSIONS ARE POSTED TO AUDIT TRAIL

```
JOB     : SECOND COMMERCIAL BLDG.
ADDRESS : 555 STATE ST.
ADDRESS : SAN FRANCISCO, CA. 94040
```

```
                                                                        PAGE # 1
                                                                        DATE: 09-26-1989
                                                                        DUE DATE: 09-28-1989
```

TAKEOFF#	DESCRIPTION	(SIZE)	QUANTITY	UNIT	PRICE	DISC%	TOTAL	UNIT	HOURS	DISC%	TOTAL
					<--MATERIAL-->			<--LABOR-->			
	START OF ASSEMBLY Slab on Grade 3500										
	DEPTH SAND (IN)= 4.000 DEPTH CONC.(IN)=		6.00								
	AREA = 10000.000 VOLUME SAND =		123.33								
	VOLUME CONC. = 105.185 TOTAL VOLUME =		308.51								
0100005	3500 P.S.I.		105.19	CY	43.89	/E	8127.99		0.00	/E	0.00
0100006	Fine Sand		123.33	CY	9.00	/E	1109.97		0.00	/E	0.00
0100007	30# -6X6 W2.1 X W2.1		11000.00	SF	5.00	/C	550.00		0.00	/E	0.00
0100008	VAP BAR 4 MIL POLY		12000.00	SF	15.20	/M	182.40		0.00	/E	0.00
0100009	6 Man Pour Crew-Days		10.00	DAY	0.00	/E	0.00	DAY	8.00	/E	80.00
	END OF ASSEMBLY Slab on Grade 3500										
	START OF ASSEMBLY 2x4x92-5/8 16 OC Wall										
	LENGTH (FT) = 400.000										
0100013	CARPENTER		10.00	DAY	0.00	/E	0.00	DAY	8.00	/E	80.00
0100014	2X4 WOL. PLATE		400.00	LF	330.00	/M	132.00		0.00	/E	0.00
```

*Sample Report from Advanced Construction Estimating*
```

Eagle Point's Quantity Take-Off

Quantity Take-Off is a computer program used to enhance Eagle Point Software's architecture, building, civil engineering, survey and hydrology and hydraulics modules. This software package is fully integrated with all modules, yet modular in style. This allows the user to choose the level of estimating detail from assemblies to sub-assembly components, to detailed unit price line items. Some of *Quantity Take-Off's* features include a database manager, R.S. Means' seamless interface, material assessment, dynamic modification, *Take-Off* extents, back-up utility, detailed results table and context-sensitive on-line help.

The program organizes and estimates the materials. The user needs only assign item numbers to project entities and specify the parameters of the report. The Database Manager compiles and displays the items available for projects, and stores MeansData or provides access to user-defined data. Options on the Database Manager allow the user to further customize this database to fit specific needs by adding, removing or modifying items. The Database Manager tracks the item number, description, unit of measure and two separate unit costs. An unlimited number of entries is allowed. The *Search* feature aids in finding specific entries. The database can have files added to it, modified or removed, all within the Database Manager dialog box.

Every entity in a drawing (for example, curbing, rebar, windows, walls, pipes and site furnishings) is automatically tagged. Once an item number is assigned, it becomes the default until the entity material is changed.

Quantity Take-Off performs two kinds of costing: unit and assembly. Unit Costing counts all items in the project as individual items. The specific items for costing calculation can be selected through the Database Manager. Assembly Costing prices groups of items, such as the components that create stairs, as whole units. *Quantity Take-off* incorporates 6,000 Means assembly line items, which are derived from Means' unit price database. The Sub-Assembly Management feature captures costs and quantities of subcomponents of assembly-priced entities (walls, stairs, etc.).

The *Calculate* option sets up the criteria and extent of the take-off, as well as executing the calculation. The options on the Calculate dialog box allow a customized take-off containing only the information desired. Select from *Quantities Only, Quantities with Unit Cost 1* and *Quantities with Unit Cost 2* as the take-off criteria. *Eagle Point/User Entities* or *Eagle Point Entities Only* are the choices for the Take-Off Extents. There are five different search methods for non-tagged items. The first is *All Entities;* this program searches for all entities found in the current drawing. *By Selection Set* calculates all the entities graphically selected by the user. Costs can be calculated by layer or by linetype with the *By Layer* and *By Linetype* options. The *By Block* option calculates costs by listing blocks in your drawing.

Final take-off tabulation is displayed in the *Bill of Materials* dialog box. The user can add, modify or delete entries to produce customized results. The item number, description, quantity, unit of measure, unit cost and extended cost are displayed on the report. The *Bill of Materials* may be altered by the user by adding entries or modifying existing entries.

For more information on the *Quantity Take-Off* module please contact Ed Graham, Eagle Point Software, 4131 Westmark Drive, Dubuque, IA 52002, 1-800-678-6565.

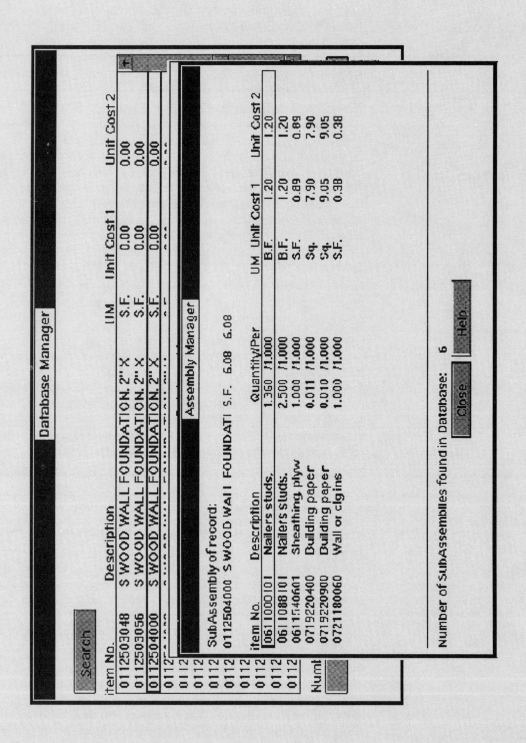

Sample Report from Eagle Point

Timberline's Precision Collection of Estimating Software

Designed for more precise square foot estimates, Timberline's *Model Estimating* is a new conceptual estimating tool available in its Precision Extended Software. It enables users to build a model for an office building, warehouse, retail outlet or any other type of facility. It also provides the option of basing the model on square footage. *Model Estimating* provides more precise square foot estimates because the assumptions made about a project are based on details such as quantities, crews, hours, and waste factors.

For example, *Model Estimating* enables users to determine much more about a building's estimated electrical costs. Rather than figuring a simple cost of $1 per square foot, the model provides the exact number of fixtures and fittings, footage of conduit, and other details associated with electrical work for the building. Users need only enter the building's square footage to obtain this detail.

If the type of fixture needs to be changed, the model will logically "think" through the change and take appropriate action. Change a building's exterior facing and the model selects a new foundation size and backup wall. Stretch the height of a concrete tilt-up, and the model specifies a different type of rebar.

```
┌─────────────────────────────────────────────────────────────────────┐
│                        Holland Warehouse                              │
├─────────────────────────────────────────────────────────────────────┤
│ ┌Model takeoff──────────────────────────────────────────────────────┐│
│ │ 110              Tiltup Warehouse - Total Cost per SQFT            ││
│ │    Reference   SQFT costs for plan revision: 2Ac                   ││
│ ├───────────────────────────────────────────────────────────────────┤│
│ │                                                                    ││
│ │ Assumptions:                                                       ││
│ │ 1.   The tilt up walls rest on top of the footings.                ││
│ │ 2.   The slab is 6" thick with 4" of gravel with a                 ││
│ │       vapor barrier using 4000 psi concrete.                       ││
│ │                                                                    ││
│ │ BUILDING SHELL SPECIFICATIONS:                                     ││
│ │      Enter the area enclosed by the building (SF).    10,000.000   ││
│ │   Enter area of interior space to be finished (SF).    2,500.000   ││
│ │                                                                    ││
│ │ SITEWORK SPECIFICATIONS                                            ││
│ │      Area of other exterior concrete paving (SF).     1,600.000    ││
│ │                 Area of asphalt paving (SF). _            .000     ││
│ └───────────────────────────────────────────────────────────────────┘│
├─────────────────────────────────────────────────────────────────────┤
│ F1 Accept model    F3 Calculator      F5                F7 Select options│
│ F2 Cancel model    F4                 F6                F8             │
└─────────────────────────────────────────────────────────────────────┘
```

Sample Report from Precision Extended's Model Estimating program

Transition to Final Estimate

With *Model Estimating,* the conceptual estimate can be used to prepare the final estimate. As the design progresses, the model can be refined — changing assumptions into specifications — until the estimate is complete.

Model Estimating can also be used for fast track projects. It allows users to complete estimate details on one design phase before finalizing the entire estimate. Estimate changes can also be made on later phases without affecting earlier work.

For more information, contact Timberline Software Corporation at (503) 626-6775, or 9600 SW Nimbus Avenue, Beaverton, Oregon 97008-7163.

G2 Estimator Assemblies Capabilities

G2 Estimator is the only system that treats a template, estimate, historical data, and commercially available databases (like R.S. Means' Assemblies packages) in the same manner. All of the above are defined as a "PROJECT" and can be accessed without leaving the project you are currently working on, using a mouse or a hot key.

G2 defines "Assemblies or Work Packages" as composites. A *G2* composite is an individual resource that may include up to 999 line items. Each of these line items may also contain up to 999 line items (nested composites). The user has the option of showing the composite as a single line item or exploding the composite detail. Additionally, *G2 Estimator* features "Intelligent" composite capabilities ("if then — else" technology). These composites apply user-defined logic to check conditional requirements for a task and to choose the best material resources to accomplish the task.

In summary, combining the Assemblies databases from R.S. Means with *G2 Estimator* provides some unique database management capabilities.

For more information contact Craig Lindquist at (208) 384-1300, or write to G2 Inc., 2399 S. Orchard, Suite 200, Boise, ID 83705.

G2 Inc.
R.S.MEANS GEN (R.S.MEANS GEN)
Worksheet: 001 - MEANS ASSEMBLIES

Worksheet Listing

Number: 001	Description: MEANS ASSEMBLIES
Quantity: 1.00 Unit: LS	Production: 1.00 /
Estimator: G2 Revision:	Start Date End Date

Work Codes

Field	Code	Description
1	1	

Notes

All R.S. Means Square Foot Cost Data in electronic format is available in G2 ESTIMATOR.
For additional information contact:
G2 INC.
2399 South Orchard
Suite 200
Boise, ID 83705
(208) 384-1300 phone
(208) 384-0300 fax

Line	Resource	Description	Quantity Unit	Manhours	Material	Labor	Equipment	Sub Contracts	(Not Used)	(Not Used)	(Not Used)	Total Cost
1.00	0022262600	GRAVEL UNDER FLOOR SLAB, 6" DEEP, COMPACTED	500.00 S.F.	3	85	57	8					150
1.01	0025122010	FINE GRADE AREA TO BE PAVED,SMALL AREA	55.00 S.Y.	1		23	38					61
1.02	0031132200	EXPANSION JOINT, PREMOLDED BITUMINOUS FIBER, 1/2" X 6"	110.00 L.F.	2	54	53						107
1.03	0031170300	EDGE FORMS IN PLACE FOR SLAB ON GRADE TO 6" HIGH, 4 USES	15.00 L.F.	1	4	17	1					21
1.04	0033125035	CONCRETE, REDI MIX REG WT 4500 PSI	10.00 C.Y.	1	564							564
1.05	0033134030	CURING WITH SPRAYED MEMBRANE CURING COMPOUND	5.00 C.S.	1	10	15						25
1.06	0033172430	PLACE AND VIBRATE CONCRETE FOR SLAB ON GRADE, 4" THICK, DIRECT CHUTE	10.00 C.Y.	4		83	6					89
1.07	0033454025	FINISHING FLOOR, MONOLITHIC STEEL TROWEL FINISH FOR FINISH FLOOR	500.00 S.F.	8		171	34					205
1.08	0033454085	FLOOR FINISH GRANOLITHIC TOPPING & FINISH 1" THICK	500.00 S.F.	21	145	444	64					653
1.09	0033454210	FINISH FLOOR, METALLIC HARDENER, 2.0 PSF, HEAVY SERVICE	500.00 S.F.	6	180	136	27					343
1.10	0071922090	POLYETHYLENE VAPOR BARRIER, STANDARD, .006" THICK	5.00 Sq.	1	9	25						33
		Sheet Totals	1.00 LS	48	1050	1023	177					2250

G2 ESTIMATOR (TM)

Date: 95/12/11 Time: 16:25:42

Sample Report from G2 Estimator

US Cost Building Parametric Quantity Models

US Cost offers *SUCCESS*, an estimating and cost management system for Windows®. *SUCCESS* allows you to create, save and retrieve Parametric Quantity Models for facilities of all types. Using the Quantity Factors associated with each Level of the Workspace Tree and each Detail Item, relationships can be established which automatically generate the Unit Quantities for each Level and Detail Item.

Creating Parametric Relationships Manually

Once the relationships among Level and Detail Items have been established using the Quantity Factors, any change to a Level Unit Quantity will automatically modify the Unit Quantities for all Levels and Detail Items related to the changed Level in the Workspace Tree. This method is used when measured quantities are not available for all construction elements and you wish to rely on typical or historical relationships to quickly establish approximate quantities and costs.

Creating Parametric Relationships Automatically

Sometimes, historical data may not be available to produce quantity factors for construction elements. In this case, you can enter and quantify all the detail items for a level, then use the Factor button in the level dialog box to automatically add quantity factors to each of the detail items.

Saving and Retrieving Models

All of the model relationships created using the Quantity Factors are saved when the project is saved to a permanent file. All or part of the model can be retrieved and modified at any time. For example, suppose you had completed and saved the entire model for the Life Sciences Building and wished to apply the model to create a budget estimate for a similar project with an area of 100,000 square feet. You would open the file and change the Unit Quantity for the project level "LIFE SCIENCES BUILDING" from 62,400 to 100,000 square feet. All other quantities in the project will automatically change based on the previously defined quantity factors.

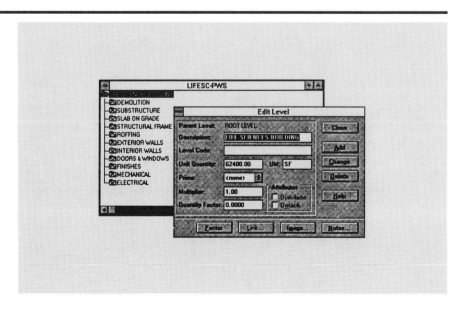

Sample Report showing US Cost's SUCCESS *parametric relationship*

Creating Assemblies

With the user-defined Level Structure and full "cut and paste" capability, *SUCCESS* can also be used to create models at any level of detail. For example, groups of Detail Items can be aggregated under Assembly Levels for specific types of systems or other construction features. You may wish to save a number of different exterior wall types in an Exterior Wall project file. As new and varied projects arise, the appropriate wall type can be selected from the file and pasted in the current project. All of the Quantity Factor relationships will remain in place, thus eliminating the need to measure quantities for each Detail Item.

For more information, contact US Cost at 1-800-955-1385 or 5605 Glen Ridge Drive, Suite 850, Atlanta, GA 30342.

Abbreviations

amp	ampere
apt.	apartment
avg.	average
BF	board foot
bldg.	building
Btu	British Thermal Unit
Btu/hr.	Btu per hour
C	hundred; centigrade
c	conductivity
C.C.F.	hundred cubic feet
C.F.	cubic foot
CFM, cfm	cubic feet per minute
C.I.	cast iron
C.I.P.	cast in place
C.L.F.	hundred linear feet
CMU	concrete masonry unit
col.	column
conc.	concrete
cont'd.	continued
CPE	chlorinated polyethylene
CPVC	chlorinated polyvinyl chloride
C.S.F.	hundred square feet
CSPE	chlorosulfenated polyethylene
C.Y.	cubic yard (27 cubic feet)
C.Y./hr.	cubic yards per hour
demob.	demobilization
dia.	diameter
dp.	deep
ea.	each
elec.	electrician; electrical
elev.	elevator
EMT	thin wall conduit
EPDM	ethylene propylene diene monomer
equip.	equipment
est.	estimate
excav.	excavation
exp.	expansion
ext.	exterior
F	Fahrenheit

fixt.	fixture
fl.	floor
fndtn.	foundation
fpm	feet per minute
ft.	foot; feet
ftg.	footing
ft. lb.	foot pound
ga	gauge
gal.	gallon
galv.	galvanized
gen.	general
gpd	gallons per day
gph	gallons per hour
gpm	gallons per minute
hdr	header
horiz	horizontal
hp	horsepower
hr.	hour
hrs./day	hours per day
HVAC	heating, ventilating and air conditioning
hvy.	heavy
in.	inch
inst.	installation
insul	insulation
K	thousand
kip	thousand pounds
KCF	kips per cubic foot
KLF	kips per linear foot
KSF	kips per square foot
KSI	kips per square inch
kva	kilovolt ampere
kW	kilowatt
lb.	pound
lb./hr.	pounds per hour
lb./L.F.	pounds per linear foot
L.F.	linear foot
LL	live load
lt. ga.	light gauge
lt. wt.	light weight
mat.; mat'l.	material
max.	maximum
med.	medium
mfg.	manufacturing
min.	minimum
mo.	month
NA	not available
no.	number
O.C.	on center
O & P	overhead and profit
opng.	opening
oz.	ounce
pcf	pounds per cubic foot
PIB	polyisobutylene
pkg.	package
pr.	pair
prefab.	prefabricated
prefin.	prefinished
psf	pounds per square foot
psi	pounds per square inch

ptn.	partition
PVC	polyvinyl chloride
reinf.	reinforcing
req'd.	required
S.F.	square foot
S.F.C.A.	square foot contact area
sq. in.	square inch
S.O.G.	slab on grade
sq.	square; 100 square feet
std.	standard
S.T.	subtotal
subs.	subcontractors
S.Y.	square yard
THK	thick
unfin.	unfinished
V	volt
vent.	ventilator; ventilate
vert.	vertical
V.L.F.	vertical linear foot
vol.	volume
W	watt
W.C.	water closet
WWM	welded wire mesh
%	percent
~	approximately
@	at
#	pound; number

Index

Index

Notes

344

Notes

Notes

348

Notes

Notes

Notes

Notes

Notes